Springer Proceedings in Business and Economics

Springer Proceedings in Business and Economics brings the most current research presented at conferences and workshops to a global readership. The series features volumes (in electronic and print formats) of selected contributions from conferences in all areas of economics, business, management, and finance. In addition to an overall evaluation by the publisher of the topical interest, scientific quality, and timeliness of each volume, each contribution is refereed to standards comparable to those of leading journals, resulting in authoritative contributions to the respective fields. Springer's production and distribution infrastructure ensures rapid publication and wide circulation of the latest developments in the most compelling and promising areas of research today.

The editorial development of volumes may be managed using Springer Nature's innovative EquinOCS, a proven online conference proceedings submission, management and review system. This system is designed to ensure an efficient timeline for your publication, making Springer Proceedings in Business and Economics the premier series to publish your workshop or conference volume.

This book series is indexed in SCOPUS.

Adina Letiţia Negruşa · Monica Maria Coroş ·
Mark Anthony Camilleri
Editors

Sustainable and Resilient Businesses in the Global Economy

4th International Conference on Modern Trends in Business, Hospitality and Tourism, Cluj-Napoca, Romania, May 16–18, 2024

Springer

Editors
Adina Letiția Negrușa
Faculty of Business
Babeș-Bolyai University
Cluj-Napoca, Romania

Monica Maria Coroș
Department of Hospitality Services
Faculty of Business
Babeș-Bolyai University
Cluj-Napoca, Romania

Mark Anthony Camilleri
Department of Corporate Communication
University of Malta
Msida, Malta

ISSN 2198-7246 ISSN 2198-7254 (electronic)
Springer Proceedings in Business and Economics
ISBN 978-3-031-75882-9 ISBN 978-3-031-75883-6 (eBook)
https://doi.org/10.1007/978-3-031-75883-6

This Springer imprint is published by the registered company Springer Nature Switzerland AG
The registered company address is: Gewerbestrasse 11, 6330 Cham, Switzerland

Preface

The first edition of the International Conference Modern Trends in Business, Hospitality, and Tourism took place four years ago, at the beginning of May 2021. Centered around the conference theme, *Challenges in Overcoming Current and Future Crises*, it was a fully online event, addressing the challenges faced by businesses in general, in the pandemic context, and by the tourism and hospitality sector, in particular. Three national and international keynote reputable speakers honored the event. A total of 58 papers were presented by scholars and researchers from 6 continents; of these, 14 were prepared by undergraduate and graduate students, under the supervision of academic staff.

The second edition was the first one to be organized in hybrid mode. It focused on *Remodelling Businesses for Sustainable Development*, and it provided the opportunity to share knowledge and research findings through the 7 keynote presentations and 68 papers presented at the conference. The student section included 13 works. A total of 13 selected papers were published in the first Springer Book of Proceedings associated with the conference.

The third edition continued to mark the development of the conference, inviting contributions dedicated to *Sustainable Approaches and Business Challenges in Times of Crisis*. Three highly cited keynote speakers enriched the conference, which provided the environment for conceptual, knowledge, and research findings exchanges and sharing for 99 submissions, of which the students' sections included 15. Academics and researchers from 6 continents joined the event onsite or online. In total, 17 papers were selected and included in the second Springer Book of Proceedings associated with the conference.

The most recent edition of the conference, the fourth one was an anniversary one, being linked to the celebration of 30 years of Business School at Universitatea Babeș-Bolyai. Between the 16th and the 18th of May, it hosted onsite or online attendees from 5 continents, 26 countries, and 71 universities. Bringing together leading researchers, academics, and practitioners from around the world, the conference provided discussion opportunities for 115 papers, authored and co-authored by 211 unique authors and co-authors, of which 80 were international ones, addressing topics related to the central theme, *Sustainable and Resilient Businesses in the Global Economy*. Bachelor and master students contributed 40 research-focused presentations. In total, 20 papers were selected for the current edition of the Springer Book of Proceedings, addressing the indicated four topics from among the 12 conference topics: *Smart and Sustainable Development Issues in the Digital Economy, Challenges and Opportunities in the International Business Context, Sustainable and Resilient Business Models, Business and Wellbeing, Innovative and Sustainable Entrepreneurship, Changing Dimensions of Modern Business and Hospitality, Financing Alternatives for SMEs, Modern Customer Experiences, Human Capital Leadership & Organizational Behavior, Global Capital Markets and their Role in Investment Financing, Corporate Governance, Classic and Alternative Approaches in the fields of Accounting, Banking, Business Law, Economics, and Finance.*

Five reputable keynote speakers honored the event with their valuable contributions, sharing both scholarly and practice-oriented insights. Professor Stroma Cole, Ph.D., from Westminster University, UK shared a presentation on *Gender and Sustainability in the New World of Tourism Work*. Assoc. Professor Nicolás Depetris-Chauvin, Ph.D., from HES-SO/HEG-Genève, Switzerland led the audience into the topic of *Promoting Exports from Emerging Markets to OECD Countries*. Further, Associate Professor Amir Hasnaoui, Ph.D., representing Excelia Business School, La Rochelle, France, discussed *Global Fintech Innovations: Navigating the Future of Finance through Sustainability and Decarbonization*. Associate Professor Mark Anthony Camilleri, Ph.D., from the Department of Corporate Communication, University of Malta, provided an inspirational online keynote presentation on the *Factors Affecting the Acceptance and Usage of ChatGPT*. A practice-oriented perspective was shared by the Honorary Consul of France at Cluj-Napoca, Mr. Pascal Fesneau, who made a presentation entitled *French Investments in Romania: About Shared Values*, addressing shared values and joint business developments.

Consequently, this volume comprised the 20 selected papers that discuss topics related to *Smart and Sustainable Development Issues in the Digital Economy, Sustainable and Resilient Business Models, Innovative and Sustainable Entrepreneurship*, and *Modern Customer Experiences*.

Overall, the conference is a growing event that successfully manages to achieve its objective of bringing together researchers, practitioners, and scholars from economic fields to exchange their research results and experiences and to discuss the most recent trends, challenges, and solutions in the fields of business, hospitality, and tourism. We are all looking forward to the upcoming fifth edition of the conference and to its participants' valuable and inspiring contributions.

Monica Maria Coroș
Department of Hospitality Services,
Faculty of Business
Babeș-Bolyai University
Cluj-Napoca, Romania

Mark Anthony Camilleri
Department of Corporate
Communication
University of Malta
Msida, Malta

Contents

Analysis of Resilience Among European Tourism Businesses

Matúš Marciš, Eva Zabudská, and Kristína Pompurová

Abstract This paper aims to delineate the resilience of tourism businesses across European countries throughout the first year of the Covid-19 pandemic. Using Eurostat data, this study assesses the resilience of the tourism businesses by examining fluctuations in turnover and employment within accommodation and food services, as well as among travel agencies and tour operators, compared to the five-year period preceding the pandemic. The impact of the pandemic is evaluated through various lenses including the rigour of lockdown measures, the tally of virus-attributable deaths, healthcare spending, changes in international travel income, and household expenditure on restaurant and hospitality services. Initially, this paper employs non-hierarchical cluster analysis to facilitate inter-country comparisons and to categorise countries into clusters exhibiting analogous trends. Subsequently, through linear and multilinear regression analyses, this study examines the effect of individual pandemic-related factors on the resilience of European tourism businesses. The findings suggest that, given similar changes in pandemic-induced conditions such as lockdown severity, health care expenditures and virus- identified deaths, it is possible to delineate six distinct groups. The first group covers 5 countries, the second comprises 8 countries, the third group includes 4 countries, the fourth group contains 4 countries, the fifth group contains 4, and the sixth covers 4 countries. The application of statistical hypothesis testing, operationalised through linear regression models, disclosed that particular factors bore statistical significance exclusively impacting the turnover streams of accommodation and hospitality establishments located within specific European countries during the COVID-19 health crisis. However, the analysis did not demonstrate any discernible effect on the turnover figures of the travel agencies, based on the variables examined. Further exploration by multilinear regression analysis revealed a mediating variable (household expenditures on hotel and restaurant services), which forged a link between the independent variables (overnight stays by domestic tourists and fluctuations in international travel income) and the dependent variables (turnover of accommodation and food service sectors).

M. Marciš (✉) · E. Zabudská · K. Pompurová
Faculty of Economics, Matej Bel University, Banská Bystrica, Slovakia
e-mail: matus.marcis@umb.sk

A. L. Negrușa et al. (eds.), *Sustainable and Resilient Businesses in the Global Economy*,
Springer Proceedings in Business and Economics,
https://doi.org/10.1007/978-3-031-75883-6_1

Keywords Tourism businesses · COVID-19 pandemic · Business resilience · European countries

Introduction

In 2020, the tourism sector worldwide encountered an unprecedented and abrupt crisis (Gössling et al., 2020). The onset of the Covid-19 pandemic precipitated profound disturbances in the international economic landscape, precipitating the most severe economic downturn observed in more than a century. The implementation of stringent government policies substantially impacted the operational dynamics of tourism businesses, frequently resulting in diminished turnover, workforce deficiencies, and instances of insolvency (Neise et al., 2021). During the inaugural year of the pandemic, international travel fell by 72%, signifying the year with the most detrimental impact on tourism history. This downturn led to a shortfall of 1.1 billion international tourists worldwide, a figure reminiscent of tourism levels seen three decades earlier (UNWTO, 2023). At the same time, in 2020, the industry was characterised by hasty decisions, uncoordinated actions, and widespread uncertainty, affecting both consumers and businesses alike (Hailemariam & Dzhumashev, 2023).

Continually, with the onset and consequences of the pandemic, the number of articles examining the resilience of tourism businesses has increased exponentially (Hu & Xu, 2022; Badoc-Gonzales et al., 2022; Hall et al., 2023; Aydogan et al., 2024). However, there is a scarcity of studies that address the resilience of tourism businesses from a pan-European perspective (Roman et al., 2022). Therefore, this paper seeks to elucidate the operational dynamics of European tourism businesses during the initial year of the pandemic. Analysis of operational performance, particularly in terms of turnover and employment in the context of Covid-19, may lighten vulnerabilities or, conversely, the resilience of tourism businesses (Ntounis et al., 2022).

This paper aims to delineate the resilience of tourism businesses across European countries during the first year of the Covid-19 pandemic. Using Eurostat data, this study assesses the resilience of tourism businesses by examining fluctuations in turnover and employment within accommodation and food services, as well as between travel agencies and tour operators, compared to the 5-year period preceding the pandemic. The impact of the pandemic is evaluated through various lenses, including the rigour of lockdown measures, the count of virus-attributable fatalities, healthcare spending, changes in international travel income, and household expenditure on restaurant and hospitality services. Initially, this paper employs non-hierarchical cluster analysis to facilitate intercountry comparisons and to categorise countries into clusters exhibiting analogous trends. Subsequently, through linear and multilinear regression analyses, this study examines the effect of individual pandemic-related factors on the resilience of European tourism businesses.

Literature Review

According to MacKinnon and Derickon (2013), the concept of resilience has transitioned from the natural and physical sciences to the social sciences. This change occurred due to the recognition of global threats such as economic crises, climate change, and international terrorism, which have highlighted the importance of assessing the vulnerabilities and adaptive capacities of (social) systems. Following this and given its diverse interpretations across disciplines, resilience has become a subject of interest for many researchers, including tourism. The concept of resilience was first introduced into tourism research by Holling (1973), and since then it has continued to attract increasing attention, predominantly in recent years reflecting the Covid-19 pandemic (Hu & Xu, 2022; Badoc-Gonzales et al., 2022).

There are different approaches to access resilience in tourism research. According to Hall et al. (2023), the main foundations of resilience research include engineering, ecological and socioecological perspectives. The engineering perspective focusses on how a system is impacted by and resists disturbances, as well as the speed at which it returns to its previous equilibrium. The ecological perspective focusses on the magnitude of a shock that could be absorbed by a system before it switches to another equilibrium. Regarding socioecological resilience, it refers to the capacity of the system for self-organisation, learning, adaptation, and innovation (Prayag, 2023).

In tourism research, the focus is predominantly on engineering perspective (Hall et al., 2023). Therefore, for this paper, drawing from the engineering perspective, resilience can be defined as the ability to return to a previous state or condition after an external stressor or shock (Orchistion et al., 2016; Prayag, 2023). This approach is described as a 'bouncing back' with the ability to return to an original form. However, many studies highlight the 'bouncing forward' approach, which, in contrast to the previous, extends resilience beyond mere recovery (Pascariu et al., 2021). This approach views crises or disruptions as opportunities for positive change and growth. The bouncing forward approach to resilience involves learning from challenges and creating a stronger, more adaptable tourism sector for the long term (Mihotić et al., 2023; Prayag, 2023). Instead of returning to business as usual, it envisions a better and more robust system. However, due to the lack of available data sufficiently representing changes and growth in the system, and for the purpose of this paper, we have chosen to adopt the bouncing back approach in this research.

Within the tourism system, Prayag (2023) distinguishes four levels of resilience, specifically individual, organisational (business), community and destination-level resilience. Individual resilience may refer to psychological resilience, for example, of employees, residents, or entrepreneurs (Hall et al., 2018). Organisational or business resilience refers to the capacity to withstand major business disruptions due to unforeseeable, unexpected, or catastrophic events, leading organisational systems beyond planned service limits without serious losses (Neise et al., 2021). At the community level, resilience can be associated with the ability of the community to absorb negative economic, sociocultural, and environmental impacts of tourism (Prayag, 2023). Destination resilience can occur at different spatial scales, from local to national and transnational, with a focus on the adaptation to negative impacts in a way that leads to a more adaptive and resilient tourism (eco)system (WTTC & ICF, 2022). To improve resilience at the individual, organisational, community, and destination levels, it is crucial to diversify research approaches and management tools.

As delineated earlier, and for the purpose of this paper, business resilience is the ability to endure significant disruptions in operations caused by unforeseen, unexpected, or catastrophic events, allowing it to persevere through various shocks and operate effectively in the face of ongoing threats (Wiig & Fahlbruch, 2019). Resilient businesses are able to maintain their operations or quickly resume them in the face of crises. Additionally, business resilience is not only a key factor in ensuring the survival of an organisation, but it also plays a significant role in improving its competitive edge (Lee et al., 2013).

Research of business resilience within the tourism realm has expanded in recent years (Hall et al., 2023) and incorporates diverse approaches, from the qualitative, quantitative and mix-method perspective. For instance, Usher et al. (2020) evaluated the resilience and vulnerability of tourism businesses through face-to-face interviews with business owners and managers. They discovered that, even though not all businesses prioritised employee training in preparation for potential vulnerabilities, they recognised the importance of offering support to their staff during severe events. Roman et al. (2022) assessed the impact of the Covid-19 pandemic in 31 European countries using secondary data from Eurostat and Our World in Data. Their study focused on several key indicators: changes in arrivals by air passenger transport, changes in the occupancy rate at tourist accommodation establishments, changes in the number of nights spent at tourist accommodation establishments by foreign visitors, changes in arrivals at tourist accommodation establishments, and the average stringency index for 2020. Based on these variables, they identified four groups that showed the most similarity in terms of changes in the tourism sector between 2019 and 2020. More recently, Wieczorek-Kosmala (2022) explored the resilience of tourism businesses from four Central European countries based on cash-driven capabilities. The author found that businesses with greater cash-driven resilience capabilities are distinguished by higher profitability and are less financially constrained.

One of the topics in tourism business resilience research is its relationship with vulnerability. Resilience can be thought of as the opposite of vulnerability (Holling, 2001). Therefore, the relationship between business resilience and business vulnerability is like two sides of the same coin. A high-resilience business is likely to have low vulnerability, as it is better equipped to manage risks and recover from disruptions. In contrast, a business with high vulnerability is likely to have low resilience, as it is less capable of managing risks and recovering from disruptions (Imeni & Edalatpanah, 2023). As Bec et al. (2018) point out, a resilient organisation is not immune to change, but rather it is capable of managing impacts and outcomes. Furthermore, Doğantan and Kozak (2019) confirm that increased resilience capacity in tourism businesses reduces vulnerabilities. Therefore, resilience helps businesses cope and recover from vulnerabilities. Ntounis et al. (2022) assessed the vulnerability of tourism businesses during the Covid-19 lockdown period with impact on turnover and indicated that a loss of turnover can result in businesses becoming less resilient and more vulnerable. Using this approach, this paper examines fluctuations in turnover and employment across the tourism businesses in European countries during the Covid-19 pandemic to pinpoint businesses that exhibit vulnerability or resilience.

Building upon the aforementioned context, the research question addressed in this paper is as follows: How did tourism businesses in Europe perform in terms of employment and turnover during the first year of the Covid-19 pandemic?

Research Methodology

Employing data sourced from Eurostat, this research rigorously assesses the resilience of tourism businesses by analysing fluctuations in turnover and employment within the accommodation and food services sectors, as well as among travel agencies and tour operators during the initial year of the pandemic, in comparison with the five-year period prior to the pandemic's onset. The study was limited to European countries for which Eurostat provides comprehensive and detailed data, ensuring comparability across the regions and the period considered.

The impact of the pandemic is evaluated through a multifaceted lens, considering factors such as the rigour of lockdown measures (intensity), the count of virus-attributable fatalities, healthcare spending, changes in international travel income, and household expenditure on restaurant and hospitality services. The variables used for this research, as well as its metrics and applications, are described in Table 1.

The intensity of COVID-19 lockdowns was assessed using data sourced from the Our World in Data platform and analyses performed by the Oxford Coronavirus Government Response Tracker (OxCGRT) project. The stringency index, as calculated by OxCGRT, is a composite metric derived from nine response indicators, which include measures such as school closures, workplace closures, and travel bans. This index is determined daily as the mean score of these indicators, each ranging from 0 to 100, where a higher score represents a stricter governmental response (100 being the strictest possible measure). Due to the variations in lockdown intensity throughout the year, the average stringency index for the entirety of 2020 was computed for each country to facilitate this research.

Secondary data obtained from Eurostat were processed in MS Excel, imported into the statistical software SPSS (version 29), and evaluated using multiple sophisticated statistical tests.

In the initial phase, this study leverages non-hierarchical cluster analysis to enable the identification of underlying patterns within the data by minimising the variance within clusters while maximising the variance between clusters. This results in the grouping of entities that are similar to each other within the same cluster and dissimilar to those in other clusters, thereby elucidating the natural groupings within the dataset. In addition, non-hierarchical cluster analysis contributes to the simplification of complex data, making it more interpretable and actionable for decision-making

Table 1 Description of included variables

Variable	Metric	Source	Application
Turnover of accommodation and food service activities	As % for the 2020 compared to pre-covid 5-year period (average for 2015–2019)	Eurostat	Vulnerability/resilience of businesses in terms of turnover changes
Turnover of travel agencies, tour operators and other reservation services			
Employment in accommodation and food service activities			Vulnerability/resilience of businesses in terms of employment changes
Employment in travel agencies, tour operators and other reservation services			
Final consumption expenditure of households on restaurants and hotel services			Changes in households' expenditures on tourism services
International trade in services—travel credit from balance of payments			Changes in income from international tourism
Overnights of domestic tourists	As a % of all overnights for 2020		Domestic tourism importance
COVID-19, virus identified deaths	Per 100 000 inhabitants		Prioritization of healthcare
Health care expenditure	As a % of GDP for 2020		
Average Stringency index in 2020	Value from 0 to 100 (100 = strictest) for 2020	Our World in Data	Covid lockdown intensity

Source Authors

processes. It aids in the discovery of hidden patterns and relationships, facilitating a deeper understanding of the data structure and dynamics.

This method enabled the categorisation of European countries into groups that exhibit similar trends based on the impacts of the COVID-19 pandemic. The inclusion of countries in specific clusters was determined by a set of key variables that captured the impact of the pandemic: the stringency of lockdown measures, the number of deaths attributed to the virus, healthcare expenditures, changes in international tourism income, and household spending on restaurant and hospitality services.

The clustering process involved analysing data obtained from Eurostat and other reliable sources to ensure uniformity and comparability. The data for each country was standardised to minimise variance within clusters while maximising variance between clusters, thereby elucidating the natural groupings within the dataset. This approach allowed for the objective grouping of countries that responded similarly to the pandemic's challenges, thus facilitating a nuanced analysis of the resilience mechanisms at play within the European tourism industry.

Subsequently, the investigation delves into the examination of the impact exerted by individual pandemic-related variables on the resilience of European tourism businesses through the application of linear and multilinear regression analyses. These methodologies facilitate the quantification of relationships between one dependent variable and one or more independent variables, allowing one to determine the degree and direction of associations. A fundamental advantage of linear regression analysis lies in its simplicity and interpretability. The linear model provides a clear and straightforward representation of how changes in the independent variables correlate with changes in the dependent variable, which makes it particularly valuable for understanding and predicting outcomes within a given data set.

Multilinear regression analysis, which expands the capabilities of linear regression, accommodates multiple independent variables, thereby offering a more comprehensive view of the factors influencing the dependent variable. This multivariate approach allows for the examination of complex relationships and the isolation of individual variable effects while controlling for the influence of other variables. This is instrumental in disentangling the intertwined effects that multiple predictors may exert on the outcome variable.

This methodological approach facilitates a granular understanding of how each factor, in isolation and in conjunction, influences the robustness of the tourism sector in the face of unprecedented global challenges.

Results

Segmentation of European countries based on the impacts of the COVID-19 Pandemic

To improve understanding of the diversity and similarity in the approaches adopted by European countries in response to the COVID-19 pandemic and its impact on tourism businesses, cluster analysis was used to identify relatively homogeneous groups of European countries. This identification was based on data on the stringency of lockdown measures, the number of deaths attributable to the virus, healthcare expenditures, changes in international tourism income, and household spending on restaurant and hospitality services. The statistical software facilitated the proposition of suitable country segments that exhibited a natural tendency to cluster. However, the clusters formed were observed to lack sufficient stability, necessitating the gradual exclusion of several characteristics from the criteria to achieve more robust groupings.

The final clusters were derived from three variables: the COVID-19 virus identified deaths (per 100,000 inhabitants), healthcare expenditures (as a percentage of GDP), and the intensity of COVID-19 lockdowns (as an average stringency index for the whole 2020 year). In the cluster analysis, the European countries were reclassified into six clusters based on their mutual similarity through these variables, as illustrated in Fig. 1.

The analysis reveals differences among clusters, reflecting varied pandemic experiences and policy responses (Table 2). The first cluster stands out with anomalously high average values for COVID-10 identified deaths. On the contrary, clusters 3 and 6 show the highest average health care expenditures. Furthermore, cluster 4 represents the most influential governmental responses on the stringency of lookdown measures.

In the wide-ranging analysis of the changes in tourism business turnover in response to the COVID-19 pandemic in 2020, it was observed that the clusters exhibited diverse degrees of economic impact, particularly in accommodation and food service activities and travel agencies and tour operators. Cluster 2 and 3 manifested the most significant downturn in turnover among accommodation and food service businesses, with an almost 30% decrease (Table 3).

Regarding travel agencies, tour operators and other reservation services, these businesses experienced the highest decrease in turnover, ranging from 63.39% (Cluster 2) to 69.68% (Cluster 5) and 71.41% (Cluster 4 consisting of Croatia, Lithuania, Czechia, Malta). The findings of Cluster 4 align with the highest average stringency of the lockdown measures. Various strict government responses have affected the operational dynamics of travel agencies and tour operators, often leading to reduced turnover. The substantial decrease in turnover indicates the vulnerability of these businesses in the face of the first year of the pandemic.

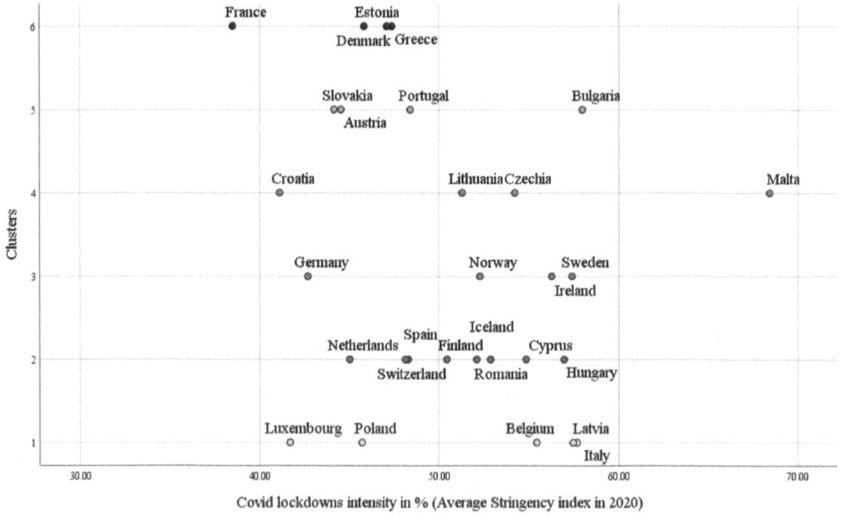

Fig. 1 Segments of European countries based on the impacts of the COVID-19 pandemic. *Source* Authors

Table 2 Average values of evaluated variables related to the COVID-19 pandemic by individual clusters

Clusters	COVID-19: virus-identified deaths (per 100,000 Inhabitants)	Health care expenditures (as a % of GDP)	COVID-19 lockdown intensity (as an average stringency index)
1. Cluster	97.40	8.07	51.57
2. Cluster	76.15	9.34	51.07
3. Cluster	44.98	10.63	52.14
4. Cluster	83.05	8.75	53.75
5. Cluster	80.19	9.39	48.76
6. Cluster	42.16	9.95	44.69

Source Authors

Table 3 Country clusters and the average rate of turnover decline for tourism businesses

Clusters	Countries	Turnover decline in accommodation and food service activities (%)	Turnover decline in travel agencies, tour operators, and other reservation services (%)
1. Cluster	Luxembourg, Poland, Belgium, Latvia, Italy	−21.87	−68.26
2. Cluster	Netherlands, Spain, Finland, Switzerland, Iceland, Romania, Cyprus, Hungary	−29.60	−63.39
3. Cluster	Germany, Norway, Sweden, Ireland,	−29.82	−64.44
4. Cluster	Croatia, Lithuania, Czechia, Malta	−28.84	−71.41
5. Cluster	Slovakia, Austria, Portugal, Bulgaria	−20.36	−69.68
6. Cluster	France, Estonia, Denmark, Greece	−26.54	−67.23

Source Authors

The results of the cluster analysis were also used to elucidate the impact of the pandemic on the economic performance of tourism businesses in various European countries (Table 4).

The results indicate a decrease in household spending on restaurant and hotel services in all clusters, again, with the fourth cluster showing the most significant decline. This notable reduction in consumption expenditure highlights the acute vulnerability of tourism businesses to the economic consequences of the pandemic.

The resilience of tourism businesses can be observed through the lens of employment trends, which offer a nuanced view of the sector's adaptability in the face of adversity. In particular, the first cluster that included Luxembourg, Poland, Belgium,

Table 4 Average values of the characteristic features indicative of the economic resilience in tourism businesses across clusters

Clusters	Decline in household final consumption expenditure by purpose (as % compared to pre-covid period)	Decline in employment within accommodation and food service activities (as % compared to pre-covid period)	Decline in employment within travel agencies, tour operators, and other reservation services (as % compared to pre-covid period)
1. Cluster	−24.87	−1.14	−7.67
2. Cluster	−25.93	−9.37	−11.06
3. Cluster	−24.57	−10.66	−4.63
4. Cluster	−34.83	−4.99	−0.26
5. Cluster	−18.60	−10.27	−7.50
6. Cluster	−26.82	−6.82	−11.45

Source Authors

Latvia, Italy, demonstrated fewer vulnerabilities to employment among hospitality and food businesses. This suggests that businesses in these countries were better able to adapt to the new circumstances. In contrast, the third and fifth clusters faced more significant employment reductions in food and accommodation services. Additionally, significant employment reductions were observed in clusters 2 and 6 within travel agencies and tour operators. Cluster 4, although it is a group with a major decrease in turnover among tour operators and travel agencies, shows relative adaptability with smaller declines in terms of employment. These differences could be the result of various government responses to protect their labour market despite the lower demand for tour operator services, such as the antivirus programme (so-called kurzarbeit) in the Czech Republic and Croatia.

Factors influencing the decline in turnover of tourism businesses

For the purpose of determining which factors influenced the decline in turnover in accommodation and hospitality businesses and travel agencies operating in European countries during the pandemic period, several secondary data sources were relied upon. Through the use of statistical testing through linear regressions, it was found that only certain factors were statistically significant, exclusively affecting the turnover of accommodation and hospitality businesses located in individual countries during the COVID-19 pandemic, while no impact on the turnover of travel agencies was demonstrated through the variables examined.

Specifically, these factors include the rate of growth/decline in overnight stays by domestic visitors (p-value = 0.001), the rate of growth/decline in turnover from international travel income (p-value = 0.004), and the rate of growth/decline in household consumption in tourism for accommodation and hospitality services (p-value = 0,000). It can be seen that these factors individually influenced the turnover of accommodation and hospitality businesses during the pandemic period. The coefficients of determination (R2) showed that the general expenditure of households had a more intensive effect on the turnover of businesses (54%) compared to the development of overnight stays of domestic visitors (30%) and the development of income from international trade in services (26%) (Table 5).

However, multilinear regression analysis revealed the possibility of the occurrence of a mediating variable, which creates a relationship between independent and dependent variables (Table 6).

In linear regression (Table 5), the factors of the number of overnight stays and income from international tourism exhibited a p-value less than 0.05, while in multilinear regression (Table 6), these factors showed a p-value greater than 0.05. However, the expenditures of the households in both regression analyses showed a p-value less than 0.05, indicating that this factor can act as a third variable (mediator) creating a relationship between the independent variables (number of overnight stays and income from international tourism) and the dependent variable (turnover of accommodation and hospitality businesses). Subsequently, through a mediation analysis, it was verified that the expenditures of households actually serve as a mediator through which the factors of the number of overnight stays and income from international

Table 5 Linear regression of individual factors

Overnights—% of domestic tourists	International travel income	Consumption expenditure of households
R2 = 0.301	R2 = 0.256	R2 = 0.541
$\beta = 0.287$	$\beta = 0.300$	$\beta = 0.496$
Sig. = 0,001	Sig. = 0,004	Sig. = 0,000
Sig. < 0,05 (α)	Sig. < 0,05 (α)	Sig. < 0,05 (α)

Source Authors

Table 6 Multilinear regression analysis

Factors	Beta coefficient (unstandardized β)	Sig	Testing dependence at the significance level of 0.05
Overnights—% of domestic tourists	−0.033	0.673	Sig. > 0,05 (α)
International travel income	0.093	0.169	Sig. > 0,05 (α)
Consumption expenditure of households	0.660	0.000	Sig. < 0,05 (α)
$R^2 = 0.767$	$\hat{y} = -1.97 - -0.033 \times 1 + 0,093 \times 2 + 1,660 \times 3 + (1 - R2)$		

Source Authors

tourism indirectly influence the turnover of accommodation and hospitality businesses during the pandemic. Figure 2 illustrates not only the direct effect of household expenditures on the turnover of these businesses during the pandemic but also the relationship between the number of overnight stays by domestic tourists and the turnover of accommodation and hospitality businesses, mediated by a third variable (total households' expenditures of households on accommodation and hospitality services).

Figure 3 also shows the influence of individual variables on the pace of turnover development in accommodation and hospitality establishments located in European countries.

The intensity of the indirect effect between the rate of change in overnight stays and the rate of change in turnover within accommodation and hospitality businesses during the pandemic was subsequently examined using the Sobel test, as well as between the rate of change in income from international tourism and the rate of change in turnover within accommodation and hospitality businesses. The P-values in the Sobel test, in both instances, indicated a value less than 0.001. Based on such results, it can be posited that the indirect influence between the variables (number of overnight stays and income from international tourism) and the rate of change in turnover of accommodation and hospitality businesses across European countries through the mediator (households' expenditures) is statistically significant. The estimate of the first indirect effect (2.71) can be economically interpreted as follows: If, during the pandemic, the growth rate of overnight stays accelerated by 1%, then, through the direct impact on the growth rate of household expenditures, an increase in the growth

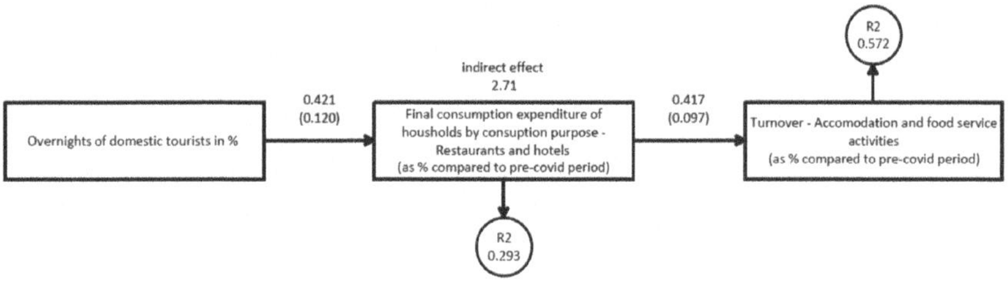

Fig. 2 Mediation model with the influence of overnights of domestic tourists. *Source* Authors

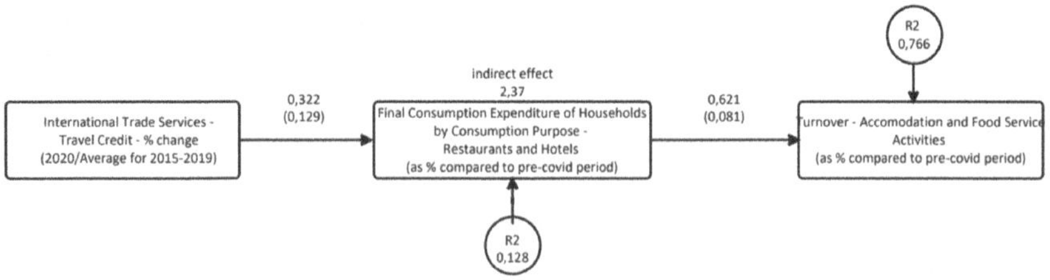

Fig. 3 Mediation model with the influence of the income from international tourism. *Source* Authors

rate of the turnover of the businesses turnover was induced by 2.71%. The second interpretation suggests that if the growth rate of income from international tourism increased by 1% during the pandemic, then, through the direct impact on the growth rate of total households' expenditures, an increase in the growth rate of business turnover was prompted by 2.37%.

Conclusions and Discussion

The pandemic has had a significant impact on the global economy, with tourism being one of the hardest hit sectors. In this context, a comprehensive analysis of the resilience of European tourism businesses becomes crucial. Specifically, this study delves into the resilience of these businesses during the first year of the pandemic. It uses Eurostat data to assess turnover and employment trends of tourism businesses, as well as the impacts of pandemic through various variables, including the rigour of lockdown measures, the total of virus-attributable deaths, and healthcare spending.

Segmentation of European countries based on pandemic impacts shows significant differences between clusters. Specific clusters experience more pronounced declines in tourism-related turnover and employment, reflecting the variable intensity of the crisis, different sectoral compositions, economic structures, and the quality of institutional settings between countries. Similarly to the results of Roman et al. (2022), understanding these variations is essential for policymakers. Tailored strategies can help mitigate the economic consequences in specific clusters and tourism sectors, particularly, as found in this research, for travel agencies and tour operators. Policymakers must recognise that a one-size-fits-all approach will not suffice (Škare et al., 2021). Initiatives such as targeted financial support, flexible regulations, and innovative marketing approaches can support the recovery of tourism businesses.

The study also uses linear and multilinear regression analyses to assess the influence of pandemic-related variables on the resilience of tourism businesses. These analyses underscore the importance of household expenditures as a potential mediating factor that affects the relationship between various independent variables and the turnover of restaurant and accommodation businesses. This finding is consistent with the research by Franco (2022) that indicates that responses of household consumption to epidemics can significantly affect economic outcomes, including tourism.

Household expenditures emerge as a critical factor, and mediation analysis confirms its role in influencing indirect effects between the number of overnight stays, international tourism income and the turnover of accommodation and hospitality businesses. This follows the results of Arbulú et al. (2021) and suggests that household spending patterns during the pandemic have had a significant impact on the tourism sector, particularly in countries where tourism is a major contributor to the economy.

The findings of this study present several critical managerial implications for leaders of the tourism sector in Europe, especially during the initial year of the

COVID-19 pandemic. First, the various impacts observed in different European clusters underscore the need for strategic flexibility and adaptability. Managers of tourism businesses located in clusters with severe lockdown measures and high virus-attributable fatalities should develop comprehensive contingency plans and flexible operational strategies to mitigate these impacts. This strategic flexibility is essential to ensure business continuity in the face of unexpected disruptions. Second, the study highlights the importance of efficient resource allocation and financial management during crises. Managers should prioritise financial buffers, ensure access to emergency funding, and maintain strong cash reserves to improve their resilience to future crises. Next, variations in employment impacts across clusters indicate the need for robust employee support systems. Investing in employee training and development, particularly in areas such as crisis management and customer service, can enhance business resilience and ensure a quicker recovery post-crisis. By supporting their workforce, managers can maintain operational stability and improve overall organisational resilience. Lastly, the differences in government responses and their impacts on tourism businesses suggest that managers should actively engage with policymakers. Collaborating with industry associations to advocate for sector-specific support measures, such as financial aid, tax relief, and regulatory adjustments, can significantly benefit tourism businesses. This collaborative policy engagement is crucial for shaping supportive frameworks that address the unique challenges facing the tourism sector.

Despite the comprehensive analysis provided, this study has several limitations that must be acknowledged. First, the reliance on secondary data from Eurostat and other sources presents inherent limitations in terms of accuracy, completeness, and timeliness. Future research could benefit from incorporating primary data collection to provide more detailed and up-to-date insights into the resilience of tourism businesses. Second, the temporal scope of the study is limited to the first year of the COVID-19 pandemic. Longitudinal studies extending beyond this period are necessary to understand the long-term impacts and resilience dynamics within the tourism sector. Such studies could provide a more comprehensive view of how businesses adapt and recover over time. Third, while the study covers multiple European countries, it does not account for intra-country variations that might affect resilience differently in various regions within the same country. More localised studies focussing on regional differences could offer deeper insights into specific challenges and resilience strategies at the regional level. Fourth, the primary focus on accommodation, food services and travel agencies ignores other important segments of the tourism industry, such as transportation, entertainment, and cultural attractions, which also experienced significant impacts during the pandemic. A broader sectoral analysis that includes these segments would provide a more holistic understanding of the resilience of the industry. Fifth, although household expenditures were identified as a mediator, other potential mediators such as government support measures, digital transformation, and innovation adoption were not explored. Future research should consider these variables to enhance understanding of resilience mechanisms and their influence on tourism businesses. Finally, the use of non-hierarchical cluster analysis and regression models, while robust, may not capture the full complexity of interactions between variables. Advanced econometric techniques and machine learning models could be used in future research to uncover more intricate patterns and causal relationships. Addressing these limitations in future studies can provide more detailed and actionable insights, further improving the resilience and sustainability of tourism businesses in Europe and beyond.

The study contributes valuable information to understand the resilience of European tourism businesses during the first year of the Covid-19 pandemic. Recognising the nuanced challenges faced by different industries and countries, new sustainable and resilient business models could be developed to ensure the long-term sustainability of the tourism sector in Europe.

Acknowledgements The paper was elaborated within the framework of the project VEGA 1/0136/23 from resilience to sustainability. The impact of data on sustainable and competitive development of tourism.

References

Arbulú, I., et al. (2021). Can domestic tourism relieve the COVID-19 tourist industry crisis? The case of Spain. *Journal of Destination Marketing & Management, 20*, 100568. https://doi.org/10.1016/j.jdmm.2021.100568.

Aydogan, M., de Esteban Curiel, J., Antonovica, A., & Cetin, G. (2024). Characteristics of crises resilient organizations in the hospitality and tourism industry. *International Journal of Contemporary Hospitality Management, 36*(4), 1432–1453. https://doi.org/10.1108/IJCHM-12-2022-1630.

Badoc-Gonzales, B. P., Mandigma, M. B. S., & Tan, J. J. (2022). SME resilience as a catalyst for tourism destinations: A literature review. *Journal of Global Entrepreneurship Research, 12*, 23–44. https://doi.org/10.1007/s40497-022-00309-1.

Bec, A., Moyle, B., & Moyle, Ch. I. (2018). Resilient and sustainable communities. *Sustainability, 10*(2), 4810. https://doi.org/10.3390/su10124810..

Doğantan, E., & Kozak, M., A. (2019). Resilience capacity in different types of tourism businesses. *Tourism: An International Interdisciplinary Journal, 67*(2), 126–146.

Eurostat. Retrieved December 10, 2023, from https://ec.europa.eu/eurostat/en/web/main/home. (Online)

Franco, M. (2022). Changes in consumption patterns of tourists after the COVID-19 pandemic. In D. Vrontis, A. Thrassou, Y. Weber, S. M. R. Shams, E. Tsoukatos, L. Efthymiou (Eds.), *Business under crisis volume I. Palgrave studies in cross-disciplinary business research* (pp. 85–100) . Association with EuroMed Academy of Business. Cham: Palgrave Macmillan. https://doi.org/10.1007/978-3-030-76567-5_5.

Gössling, S., Scott, D., & Hall, C. M. (2020). Pandemics, tourism and global change: A rapid assessment of COVID-19. *Journal of Sustainable Tourism, 29*(1). https://doi.org/10.1080/09669582.2020.1758708.

Hailemariam, A., & Dzhumashev, R. (2023). The impact of pandemic-induced uncertainty shock on tourism demand. *Current Issues in Tourism, 26*(16), 2575–2581. https://doi.org/10.1080/13683500.2022.2113044.

Hall, C. M., Safonov, A., & Naderi Koupaei, S. (2023). Resilience in hospitality and tourism: Issues, synthesis and agenda. *International Journal of Contemporary Hospitality Management, 35*(1), 347–368. https://doi.org/10.1108/IJCHM-11-2021-1428.

Hall, C. M., Prayag, G. & Amore, A. (2018). *Tourism and resilience. Individual, organisational and destination perspectives*. Channel View Publications.

Holling, C. (2001). Understanding the complexity of economic, ecological, and social systems. *Ecosystems, 4*(5), 390–405.

Holling, C. (1973). Resilience and stability of ecological systems. *Annual Review of Ecology, Evolution and Systematics, 4*(1). https://doi.org/10.1146/annurev.es.04.110173.000245.

Hu, H., & Xu, K. (2022). Visualizing the development of research on tourism resilience with mixed methods. *SAGE Open, 12*(3). https://doi.org/10.1177/21582440221126684.

Imeni, M., & Edalatpanah, S. A. (2023). Resilience: Business sustainability based on risk management. In H. Garg (Ed.), *Advances in reliability, failure and risk analysis*. Industrial and Applied Mathematics. Singapore: Springer. https://doi.org/10.1007/978-981-19-9909-3_9.

Lee, A. V., Vargo, J., & Seville, E. (2013). Developing a tool to measure and compare organizations resilience. *Nature Hazard Review, 14*(1), 29–41. https://doi.org/10.1061/(ASCE)NH.1527-6996.0000075.

MacKinnon, D., & Derickson, K. D. (2013). From resilience to resourcefulness: A critique of resilience policy and activism. *Progress in Human Geography, 37*(2), 253–270. https://doi.org/10.1177/0309132512454775.

Mihotić, L., Raynard, M., & Sinčić Ćorić, D. (2023). Bouncing forward or bouncing back? How family firms enact resilience in times of crisis. *Journal of Family Business Management, 13*(1), 68–86. https://doi.org/10.1108/JFBM-03-2022-0047.

Neise, T., Verfürth, P., & Franz, M. (2021). Rapid responding to the COVID-19 crisis: Assessing the resilience in the German restaurant and bar industry. *International Journal of Hospitality Management, 96*, 102960. https://doi.org/10.1016/j.ijhm.2021.102960.

Ntounis, N., et al. (2022). Tourism and Hospitality industry resilience during the Covid-19 pandemic: Evidence from England. *Current Issues in Tourism, 25*(1). https://doi.org/10.1080/13683500.2021.1883556.

Orchistion, C., Prayag, G., & Brown, C. (2016). Organizational resilience in the tourism sector. *Annals of Tourism Research, 56*, 128–163.

Our World in Data. *COVID-19: Stringency index*. Retrieved February 10, 2024, from https://ourworldindata.org/covid-stringency-index. (Online)

Pascariu, G. C., Ibănescu, B. C., Nijkamp, P., & Kourtit, K. (2021). Tourism and economic resilience: Implications for regional policies. In S. Suzuki, K. Kourtit, P. Nijkamp (Eds.), *Tourism and regional science* (Vol. 53). New Frontiers in Regional Science: Asian Perspectives. Singapore: Springer. https://doi.org/10.1007/978-981-16-3623-3_8.

Prayag, G. (2023). Tourism resilience in the new normal: Beyond jingle and jangle fallacies? *Journal of Hospitality and Tourism Management, 54*, 513–520. https://doi.org/10.1016/j.jhtm.2023.02.006.

Roman, M., Roman, M., Grzegorzewska, E., Pietrzak, P., & Roman, K. (2022). Influence of the COVID-19 pandemic on tourism in European countries: Cluster analysis findings. *Sustainability, 14*, 1602. https://doi.org/10.3390/su14031602.

Škare, M., Riberio-Soriano, D., & Porada-Rochoń, M. (2021). Impact of COVID-19 on the travel and tourism industry. *Technological Forecasting and Social Change, 164*, 120469. https://doi.org/10.1016/j.techfore.2020.120469.

UNWTO. (2023). *Impact assessment of the COVID-19 outbreak on international tourism*. Retrieved February 15, 2024, from https://www.unwto.org/impact-assessment-of-the-covid-19-outbreak-on-international-tourism. (Online)

Usher, L. E., Yusuf, J. -E. (W)., & Covi, M. (2020). Assessing tourism business resilience in Virginia Beach. *International Journal of Tourism Cities, 6*(2), 397–414. https://doi.org/10.1108/IJTC-02-2019-0019.

Wieczorek-Kosmala, M. (2022). A study of the tourism industry's cash-driven resilience capabilities for responding to the COVID-19 shock. *Tourism Management, 88*, 104396. https://doi.org/10.1016/j.tourman.2021.104396.

Wiig, S., & Fahlbruch, B. (Eds.). (2019). *Exploring resilience: A scientific journey from precise to theory.* Cham: Springer international Publishing. https://doi.org/10.1007/978-3-030-03189-3.

WTTC & ICF. (2022). *Enhancing resilience to drive sustainability in destinations.* Retrieved February 20, 2024, from https://research.wttc.org/enhancing-resilience-to-drive-sustainability-in-destinations. (Online)

Driving Tourism Development: Multicriteria Decision-Making in Rio de Janeiro's Tourism Sector

Fernanda A. Ferreira, José Augusto da Fonseca Pinto Coutinho, Carlos Francisco Simões Gomes, and Marcos dos Santos

Abstract As operations stabilized after the onset of the Covid-19 pandemic, many economic sectors have gradually resumed activities, with some nearing pre-pandemic levels. Tourism is no exception, as it continues to be used as a source of resource acquisition and socioeconomic development. The purpose of this study is to promote the development of the tourism sector in the municipality of Rio de Janeiro by applying the Analytic Hierarchy Process (AHP) and AHP-Gaussian (AHP-G) multicriteria analysis methods to support potential public and/or private investments in tourism. The findings of this study will help a travel agency make a strategic decision about where to open a new branch, taking into account the sector's potential in Rio de Janeiro. The study found that São Conrado and Ipanema are ideal neighbourhoods for establishing a new travel agency branch. Furthermore, the methodology used in this study can be used as a guideline for other organizations in the field looking to undertake similar evaluations, providing a structured approach to decision-making and investment evaluation.

Keywords Tourism · Operational research · AHP · AHP- Gaussian · Location analysis

F. A. Ferreira (✉)
UNIAG, CiTUR, ESHT, Polytechnic of Porto, rua D. Sancho I, n.º 981, 4480-876 Vila do Conde, Portugal
e-mail: faf@esht.ipp.pt

J. A. da Fonseca Pinto Coutinho · C. F. S. Gomes
Universidade Federal Fluminense, rua Passo da Pátria, n.º 156, Niterói, Brazil

M. dos Santos
Instituto Militar de Engenharia, Praça Gen. Tibúrcio, 80 - Urca, Rio de Janeiro, Brazil

A. L. Negrușa et al. (eds.), *Sustainable and Resilient Businesses in the Global Economy*, Springer Proceedings in Business and Economics,
https://doi.org/10.1007/978-3-031-75883-6_2

Introduction

Various cities employ tourism as a method to boost local development. According to Oliveira et al. (2022), tourism is a socioeconomic phenomenon that has attracted attention since its integration into Brazil's national higher education system in 1971. To drive city development, public administrators must adopt effective public policies that support long-term sector growth. Goeldner and Brent Ritchie (2016) defines tourism public policy as "a set of regulations, rules, guidelines, directives, objectives, and development and promotion strategies that provide a framework in which collective and individual decisions are made that directly affect tourism development and daily activities within a destination." Tourism investment decisions are also included. In any organizational segment, the accuracy of investments is of utmost importance for achieving long-term success, and this holds true for public investments in tourism. Any governmental action that is of social and economic interest tends to be questioned by society (Da Silva et al., 2021), which highlights the need for deeper analysis to underpin tourism investment decisions.

The COVID-19 pandemic presented unprecedented obstacles to the tourism industry, including unemployment, bankruptcies, large income reductions, changes in customer behaviour, and new health and safety regulations. As a result, the need for more sustainable and resilient tourist practices was reaffirmed, with long-term solutions prioritizing health and safety while maintaining economic growth through collaboration among sector players (Spadaro et al., 2023).

One option to provide scientific underpinnings for these public policies and private activities is to use widely acknowledged Operational Research (OR) methods from the literature. In this context, this study seeks to support the development of tourism in Rio de Janeiro, Brazil, by helping to identify the most suitable location for opening a new branch of a travel agency through the use of the AHP and AHP-G method multicriteria analysis.

The focus of this study was to apply the AHP and AHP-G methods. AHP is a pairwise comparison-based methodology, while AHP-G does not require pairwise comparisons. Instead, it calculates weight distribution based on the Gaussian factor and its sensitivity analysis (Santos et al., 2021). This paper demonstrates how these methods can aid in deciding where to establish a new travel agency branch in Rio de Janeiro.

Literature Review

Because it is a service sector that combines other service networks, the tourism industry has several features that improve economic production and create new jobs (Ozkaya & Demirhan, 2022).

Tourism's definition is vast and extensive, and it varies greatly depending on who defines it owing to several specifics. Tourism is a complex, open, and dynamic system, continually evolving due to the actions of various social agents. This constant motion makes it challenging for researchers to fully capture and understand the phenomenon, rendering their work perpetually incomplete (Fratucci, 2014).

To grow tourism in a city, a strategic plan is required, which includes research that prepare the city for potential future scenarios. According to Walton et al. (2019), prospective studies do not aim to predict the future. Instead, they explore various plausible future scenarios to prepare organisations to face them or create conditions to alter the likelihood of uncertain situations occurring or minimise their effects.

Business expansion is critical for growth and success in any industry's competitive marketplace. Exploring new markets and growth prospects is an important approach for expanding client base and revenue (Farida & Setiawan, 2022). This also applies to tourism-related businesses. Expanding enables a tourist operator to vary its service offerings and destinations, providing to a greater range of customer demands and preferences.

AHP is a tool for enhancing the decision-making process and is applied across various fields due to its effectiveness particularly in the tourism industry. Božić et al. (2018) used AHP to rank the attractiveness of six cultural heritage sites in Phuket, Thailand. Kumari et al. (2010) employed AHP to identify potential ecotourism sites in West District, Sikkim, using 21 geospatial layers. Crouch and Ritchie (2005) applied AHP to illustrate the competitiveness of a destination region. Raha et al. (2021) used AHP to demarcate the Ecotourism Potential Zones of West Bengal, India.

The AHP-G method can be utilised across various industries and applications, including healthcare, engineering, and finance. For instance, AHP-Gaussian was employed to select a medium-sized warship to build by the Brazilian Navy, considering three possible choices and nine criteria, including economic and operational factors (Santos et al., 2021).

Research Methodology

In this study, the Webibliomining model was employed, and an adapted version of the model proposed by Costa (2016) and Da Silva et al. (2015) was applied. A strategy was tested to locate documents discussing the development of prospective scenarios as a tool for guiding company decision-making, related to the research theme, in the Scopus and Web of Science databases. By applying this technique, relevant information was extracted from extensive databases of scientific articles, thereby identifying patterns and trends related to prospective scenarios and multi-criteria decision support methods. This contribution aids in making strategic decisions in the tourism sector. A total of 1067 articles addressing the analysed topics were identified with TITLE-ABS-KEY (((multi-criteria OR MCDA OR MCDM OR "multiple criteria") AND tourism) OR ((multicritério OR AMD) AND turismo)) (Fig. 1).

In the research conducted in April 2023, no articles were found that encompassed the utilization of the Gaussian AHP (Analytic Hierarchy Procedure) method in

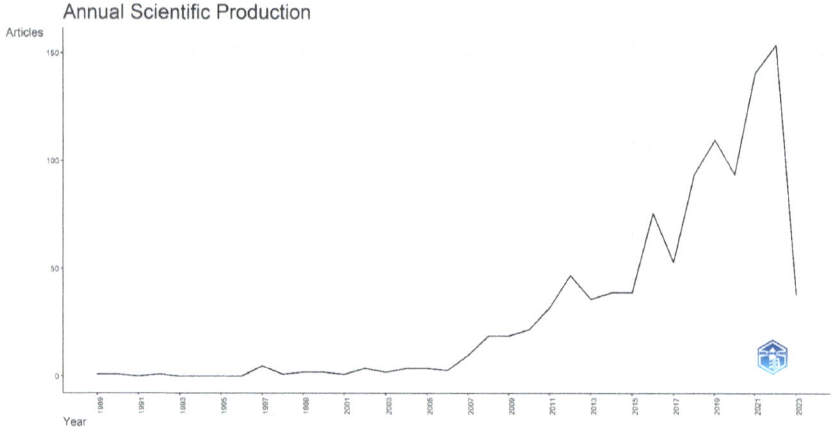

Fig. 1 Distribution of articles by year (Multi-criteria and tourism). *Source* Authors (2023)

conjunction with tourism, TITLE-ABS-KEY (("AHP Gaussiano" or "Analytic Hierarchy Process-Gaussian" or "AHP-Gaussian") and (turismo or tourism)), rendering the study more challenging. The undertaking of research on topics not yet explored in the literature of scientific articles is of paramount importance for advancing knowledge and science. Through such studies, new ideas are developed, and novel solutions are proposed for existing problems. Additionally, exploring previously uncharted topics facilitates the discovery of new research areas and novel approaches to already known subjects.

Figure 1 illustrates the distribution of articles according to the year of publication. Records began in 1989, with only one article.

The publication rate remained low until 2006, with several years devoid of publication and only a peak in 1997, with five articles. After this period, the curve began to rise more markedly, albeit moderately. The number of published articles significantly increased from 2018 to 2022, a period with one of the highest annual publication rates. In 2022, the highest number of publications occurred.

This work presents a practical application of the AHP and AHP-G methods to determine the optimal neighbourhood for opening a new branch of a travel agency in the municipality of Rio de Janeiro, Brazil. The AHP is a decision analysis method that combines quantitative and qualitative factors and is based on a hierarchical structure.

This study falls under applied research, showcasing real-world application.

Results and Discussions

Opening additional branches in various areas can help the company boost its presence in strategic regions, increase visibility, and expand its market reach. Business expansion can also benefit the region where the branch is located by creating jobs, strengthening the local economy, and promoting socioeconomic development.

In this case, multi-criteria analysis tools are used to determine the best location for establishing a new branch of a tourism company, which falls under the National Classification of Economic Activities (CNAE) number 7912-1/00 (Legal Operators). This CNAE covers activities such as designing and constructing tourism packages and organized tours, which are marketed by travel agencies or directly to clients, as well as providing tour guides. These vacations might include a wide range of activities, such as transportation, lodging, meals, visits to museums, historical and cultural sites, theatre presentations, music performances, and sporting events.

AHP Method

The AHP, developed in the 1970s by mathematician Thomas Saaty, is structured as a hierarchy of alternatives based on defined criteria. The top-ranked level is represented by an objective function, such as "best investment" in this case. AHP is based on a hierarchical approach, involving the decomposition of a problem into a structure of criteria and alternatives (Bhushan & Rai, 2004). The hierarchy is organized into levels, starting with the overall goal, followed by specific criteria and sub-criteria, and finally, the alternatives to be evaluated.

To use the AHP, it is necessary to establish a comparison matrix between the elements in the hierarchy. This matrix is filled with the decision maker's relative preferences, comparing each element with others in terms of importance or performance. These comparisons are made using a scale of values, typically ranging from 1 to 9, where 1 indicates equal importance and 9 indicates significantly greater importance (Saaty, 1987).

After constructing the comparison matrix, mathematical calculations are performed to determine the relative weights of each element in the hierarchy. These calculations involve normalizing the matrix values, determining the eigenvectors, and obtaining the corresponding eigenvalues. Based on these results, it is possible to determine the relative weights of criteria and alternatives.

Once the relative weights are obtained, a consistency analysis is performed to verify the reliability of the decision maker's comparisons. This analysis considers the consistency ratio (CR), which indicates the consistency of the comparison matrix values. Acceptable CR values are generally less than 0.1.

The result of the AHP is a ranking of alternatives based on the obtained relative weights. This ranking can support decision-making in various contexts, such as selecting a supplier, project, or resource allocation, where multiple alternatives need to be evaluated according to different criteria.

AHP-G Method

The AHP-G method is a multicriteria analysis approach based on an adaptation of the original AHP method. This new method was introduced by Santos et al. (2021) and incorporates a sensitivity analysis that uses the Gaussian factor as a key element. Therefore, AHP-G seeks to improve the accuracy and robustness of decision-making by considering the influence of criteria and their interrelationships more efficiently.

Derived from the well-established AHP, the multicriteria AHP-G method relies on the data of the criteria themselves to determine the criterion weights among the alternatives. This eliminates the need to calculate Saaty's inconsistency ratio and avoids the pairwise comparisons for alternatives and criteria (Dodevska et al., 2023).

The AHP-G method is structured into seven main steps (Santos et al., 2021):

1. Determination of the Decision Matrix.
2. Calculation of the average of alternatives in each criterion.
3. Calculation of the standard deviation of criteria based on the sample of alternatives.
4. Calculation of the Gaussian factor for each criterion.
5. Weighting of the decision matrix.
6. Normalisation of results.
7. Obtaining the new ranking.

For items 2, 3, and 4, in the listed steps, the following formulas are considered (i), (ii) and (iii):

(i) Arithmetic mean: $\overline{x} = \frac{\sum_{i=1}^{n} x_i}{n}$

(ii) Standard deviation: $s = \sqrt{\frac{(x_i - \overline{x})^2}{n-1}}$

(iii) Gaussian factor: $\frac{s}{\overline{x}}$.

It is important to highlight that one of the main advantages of this method lies in its ability to dispense with the use of pairwise comparisons or Saaty's scale to establish the criterion weights in a multicriteria context. These weights are calculated based on quantitative information regarding the analysis criteria, "eliminating the need for subjectivity" on the decision maker's part and reducing biases associated with weight generation. It is important to note that the method is applicable only to problems that have cardinal inputs for the criteria; and the decision-maker doesn't feel comfortable assigning weights to the criteria.

Definition of Criteria and Alternatives

The first step in applying the algorithms involved identifying qualitative criteria with the assistance of a tourism expert. These criteria can be observed in Table 1.

The next step involved defining alternatives for the problem, which was also done with the help of a specialist in the field. A survey conducted by the Data.rio portal of the Pereira Passos Institute (PPI) shows that in the last quarter of 2022, the main tourist attraction in Rio de Janeiro is its coastal region, both for domestic and international demand, as shown in Figs. 2 and 3, respectively.

Thus, opening the new branch in a neighbourhood that includes a beach can be considered a strategic choice since the predominant motivation for tourists in Rio de Janeiro is to enjoy the sun and the beach. By capitalizing on this preference for the natural attractions of the coastline, the company could meet tourists' needs and expectations, increasing its competitiveness and chances of success. Therefore, only

Table 1 Variables and criteria

Variable	Criteria
Price m^2	C_1
Per Capita income	C_2
Urban mobility (bus and subway lines)	C_3
Hotel occupancy rate	C_4
Robberies	C_5
Thefts	C_6
Number of occurrence records	C_7
Beach Water Quality Index	C_8

Source Authors (2023)

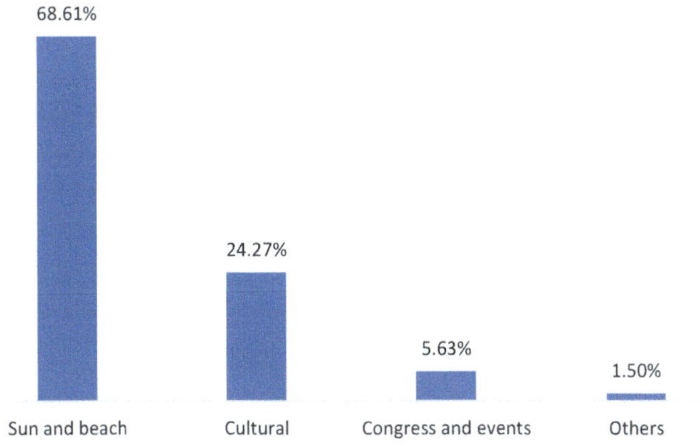

Fig. 2 National travel motivation—4th quarter of 2022. *Source* Authors (2023)

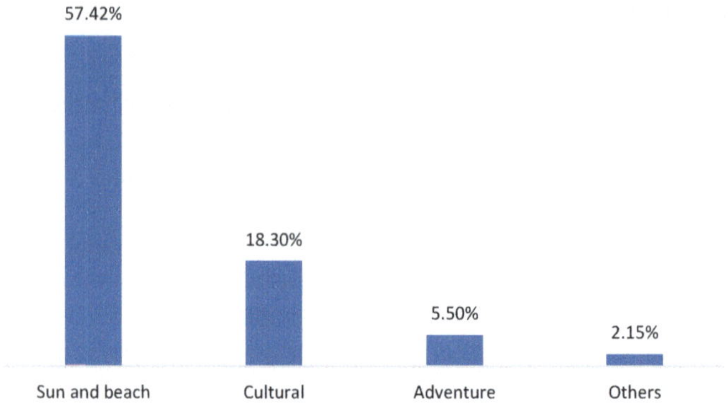

Fig. 3 International travel motivation—4th quarter of 2022. *Source* Authors (2023)

Table 2 Matrix of criteria importance

Criteria	C_1	C_2	C_3	C_4	C_5	C_6	C_7	C_8
C_1		1/3	5	3	1/3	1/3	1/3	1/5
C_2	3		5	5	1	1	1	1/3
C_3	1/5	1/5		1	1/5	1/5	1/5	1/9
C_4	1/3	1/5	1		1/3	1/3	1/3	1/7
C_5	3	1	5	3		3	3	1/3
C_6	3	1	5	3	1/3		1	1/3
C_7	3	1	5	3	1/3	1		1/3
C_8	5	3	9	7	3	3	3	

Source Authors (2023)

neighbourhoods with access to a beach were considered as alternatives: Barra da Tijuca, Copacabana, Ipanema, Leblon, Leme, and São Conrado.

Another important step is defining the weights for each criterion. However, in this step, one of the main differences between the AHP and AHP-Gaussian methods becomes apparent. In the latter, the weights will be calculated according to the score of each alternative on each of the criteria, eliminating the need for this step. Thus, a matrix was constructed, as shown in Table 2.

Classification of Alternatives in Each Criterion

During the stage of classifying alternatives from the perspective of each criterion, the data collection process is carried out. At this phase, all necessary information for the analysis is gathered, considering the previously established criteria. Each alternative is carefully assessed concerning each criterion, and the corresponding data is recorded. This can be done through surveys, questionnaires, interviews, observations, or reviewing existing data. Ensuring the accuracy and reliability of the collected data is essential for proper analysis and informed decision-making. In the case under study, data were collected from various sources, as observed in the Table 3.

From the mentioned sources, it was possible to gather the necessary information to fill in the classification of each alternative in each criterion, as shown in Table 4.

Table 3 Data source

Variable	Criteria	Source
Price m^2	C_1	Loft Brasil Tecnologia
Per Capita income	C_2	SIDRA (IBGE Census 2010)
Urban mobility (bus and subway lines)	C_3	Mobility App Moovit
Hotel occupancy rate	C_4	Data.rio (Pereira Passos Institute)
Robberies	C_5	State Public Security Institute
Thefts	C_6	State Public Security Institute
Number of occurrence records	C_7	State Public Security Institute
Beach Water Quality Index	C_8	INEA

Source authors (2023)

Table 4 Classification of each alternative in each criterion

		Criteria		
Alternative	*Price* m^2 *(reais R\$)*	*Per Capita income (reais R\$)*	*Urban mobility (bus and subway lines)*	*Hotel occupancy rate (%)*
Leme	11408	4392	3,0	64
Copacabana	10955	4173	9,0	64
Ipanema	15328	6550	8,0	62
Leblon	16928	6861	8,0	62
São Conrado	10402	6925	5,0	41
Barra da Tijuca	11038	6855	8,0	41
Alternative	*Robberies*	*Thefts*	*Number of occurrence records*	*Beach Water Quality Index (%)*
Leme	73,0	365	937	98
Copacabana	92,0	507	1358	99
Ipanema	40,0	371	824	83
Leblon	40,0	371	824	49
São Conrado	25,0	161	502	74
Barra da Tijuca	94,0	483	1594	74

Source Authors (2023)

Application of Algorithms

With criteria, alternatives, and rankings defined, the next step was to apply the algorithms of the two multicriteria analysis methods, AHP and AHP-Gaussian. In this phase, calculations and quantitative analyses are performed to establish the priorities and relative weights of criteria and alternatives. The algorithms of the methods utilize the information collected previously, such as the comparison matrices between criteria and alternatives, to carry out the necessary calculations.

Both AHP and AHP-G methods follow a mathematical approach to establish the relative weights of criteria and the priorities of alternatives. This approach involves decomposing the comparison matrix into eigenvalues and eigenvectors, which are utilized to calculate the relative weights of criteria. Then, the required multiplications and summations are performed to obtain the priorities of alternatives.

In this study the MS Excel tool was used to apply the method (Baldini et al., 2021). The results obtained for AHP and AHP-G, represented as AHP-G, are presented in Table 5.

The methods enabled a detailed quantitative analysis based on concrete data, culminating in the choice of the most suitable alternative.

The AHP and AHP-G algorithms recommend São Conrado as the best neighbourhood for the firm, with Ipanema coming in second. However, the AHP-Gaussian algorithm placed Leblon in third place, whereas the AHP ranks Leme as the third-best choice.

As a result, it was clear that no substantial differences existed between the approaches used. However, significant modifications in each model's scores were detected. This is critical since just determining the order of priority may not capture major nuances and inconsistencies between the alternatives under consideration. This provides a more thorough and detailed perspective of the choice landscape, as the method quantifies the relative differences between alternatives, delivering significant insights into priorities and relative relevance of alternatives based on the defined criteria.

Table 5 Algorithm results

Alternative	AHP score	Rank AHP	AHP-G score	AHP-G rank
Barra da Tijuca	0.137	6	0.130	6
Copacabana	0.149	5	0.139	5
Ipanema	0.175	2	0.176	2
Leblon	0.152	4	0.170	3
Leme	0.159	3	0.141	4
São Conrado	0.227	1	0.245	1

Source Authors (2023)

Conclusions

The primary goal of this study was to provide knowledge to aid the strategic decision-making process for a tourism-related corporation looking for a location for a new branch. The use of multicriteria analysis methodologies in the context of tourism in Rio de Janeiro enabled the discovery of a few ongoing difficulties that the city faces, affecting its development and growth. Security was one of the most critical challenges, influencing the decision to make a new investment and presenting high values in the collected data. Another significant difficulty is the management of natural resource preservation, particularly beaches, whose water quality index did not obtain maximum ratings in any of the trend scenarios.

It is crucial to acknowledge that this study has several limitations. The findings are based on assumptions and facts available at the time of the study, which may be vulnerable to changes in the economic, social, and political context. The effectiveness of the procedures is dependent on the quality of the information used, as well as the accurate formulation of the criteria and weights. Given these constraints, subsequent research could be guided by examining the situation through the lens of various multicriteria methods that incorporate quantitative and qualitative values, as well as broadening the scope of the study to include alternatives in other regions of the municipality and state. Furthermore, periodic updates and reviews of such prospective studies are encouraged, considering changes in the context and the availability of fresh data.

In conclusion, this study underscored the significance of employing multicriteria analysis methods for strategic decision-making in the tourism sector. The combination of techniques provides a more comprehensive and well-founded analysis, considering uncertainties and opportunities present in the environment. The findings from this study are anticipated to enhance sustainable tourism development in the municipality of Rio de Janeiro and serve as a foundation for future research and practical applications in this field.

Acknowledgements This work was supported by national funds through FCT/MCTES (PIDDAC): UNIAG, UIDB/04752/2020 (DOI https://doi.org/10.54499/UIDB/04752/2020) and UIDP/04752/2020 (DOI https://doi.org/10.54499/UIDP/04752/2020).

References

Baldini, F., Santos, M., Coelho, L. S., Mariani, V. C. (2021). AHP-GAUSSIANO em VBA, v. 1. Retrieved March 18, 2023, from https://fabiobaldini.com.br/ahp-gaussiano-no-excel/. (Online)

Bhushan, N., & Rai, K. (2004). *Strategic decision making*. Springer.

Božić, S., Vujicic, M. D., Kennell, J., Besermenji, S., & Solarević, M. (2018). Sun, sea and shrines: Application of analytic hierarchy process (AHP) to assess the attractiveness of six cultural heritage sites in Phuket: Thailand. *Geographica Pannonica, 22*(2), 121–138. https://doi.org/10.5937/22-16983.

Costa, H. G. (2016). Modelo para webibliomining: Proposta e caso de aplicação. *Revista da FAE, 13*(1), 115–126. Retrieved from https://revistafae.fae.edu/revistafae/article/view/226.

Crouch, G. I., & Ritchie, J. B. (2005). Application of the analytic hierarchy process to tourism choice and decision making: A review and illustration applied to destination competitiveness. *Tourism Analysis, 10*(1), 17–25.

Da Silva, G. B., Costa, H. G., & de Barros, M. D. (2015). Entrepreneurship in engineering education: A literature review. *International Journal of Engineering Education, 31*(6), 1701–1710.

Da Silva, R. C., Fonseca, M. A. P., & Borges, A. L. M. (2021). Tourism and management of public investments: an analysis of the tourist regions of Rio Grande do Norte–Brazil 2003/2016. *Turismo-Visão e Ação, 23*(1), 26–47. https://doi.org/10.14210/rtva.

Dodevska, Z., Radovanović, S., Petrović, A., & Delibašić, B. (2023). When fairness meets consistency in AHP pairwise comparisons. *Mathematics, 11*(3), 604. https://doi.org/10.3390/math11030604.

Farida, I., & Setiawan, D. (2022). Business strategies and competitive advantage: The role of performance and innovation. *Journal of Open Innovation: Technology, Market, and Complexity, 8*(3), 163. https://doi.org/10.3390/joitmc8030163.

Fratucci, A. C. (2014). Turismo e território: Relações e complexidades. *Caderno Virtual De Turismo, 14*(1), 87–96.

Goeldner, C. A., & Brent Ritchie, J. R. (2016). *Tourism: Principles, practices, philosophies* (10th ed.). Wiley and Sons.

Kumari, S., Behera, M. D., & Tewari, H. R. (2010). Identification of potential ecotourism sites in West District, Sikkim using geospatial tools. *Tropical Ecology, 51*(1), 75–85.

Oliveira, J. L. S., Santos, L. O. da S., & Netto, A. P. (2022). *Teoria do turismo: interfaces, educação e práticas*. Edições EACH: São Paulo.

Ozkaya, G., & Demirhan, A. (2022). Multi-criteria analysis of sustainable travel and tourism competitiveness in Europe and Eurasia. *Sustainability, 14*, 15396. https://doi.org/10.3390/su142215396.

Raha S., Mondal M., & Gayen S. K. (2021). Ecotourism potential zone mapping by using analytic hierarchy process (AHP) and weighted linear algorithm: A study on West Bengal, India. *Journal of Geographical Studies, 5*, 44–64. https://doi.org/10.21523/gcj5.21050201.

Saaty, R. W. (1987). The analytic hierarchy process: What it is and how it is used. *Mathematical Modelling, 9*(3–5), 161–176.

Santos, M. dos, Costa, I. P. de A., Gomes, C. F. S. (2021). Multicriteria decision-making in the selection of warships: A new approach to the AHP method. *International Journal of the Analytic Hierarchy Process, 13*(1). https://doi.org/10.13033/ijahp.v13i1.833.

Spadaro, I., Pirlone, F., Bruno, F., Saba, G., Poggio, B., & Bruzzone, S. (2023). Stakeholder participation in planning of a sustainable and competitive tourism destination: The Genoa integrated action plan. *Sustainability, 15*(6), 5005. https://doi.org/10.3390/su15065005.

Walton, S., O'Kane, P., & Ruwhiu, D. (2019). Developing a theory of plausibility in scenario building: Designing plausible scenarios. *Futures, 111*, 42–56.

Exploring the Impact of Classical Music Festivals on National Tourism—A Case Study of the "George Enescu" Festival

Oana Bălan-Budoiu

Abstract The impact of classical music festivals on international tourism can be significant, and the "George Enescu" Festival in Bucharest, Romania, is a relevant case study in this regard. This Music Festival is one of the most important cultural events in Romania and Europe, attracting a significant number of international tourists with each edition. It provides a platform for world-renowned artists and outstanding performances of classical music, contributing to the promotion and appreciation of the work of the Romanian composers. The "George Enescu" Festival captivates visitors coming from diverse corners of the globe, all eager to immerse themselves in the enchanting concerts and exclusive events meticulously orchestrated during the festival. The number of tourists notably bolsters the influx of visitors in Romania, throughout the festival's duration. These types of events serve as a pivotal opportunity to showcase the cities as a vibrant hub of cultural significance within Europe, thus amplifying their presence on the global cultural map. In economic terms, the festivals wield considerable influence, fostering a palpable economic uplift within the cities and their surrounding regions. Tourists who converge to the festival inject substantial capital into the local economy by financially supporting hotels, restaurants, shops, and various other establishments, thereby fostering robust economic growth and development within the community. The festivals also encourage exploration of the country's diverse cultural and historical treasures, beyond the realms of musical performance, serving as a dynamic platform for cultural exchange, fostering dialogue and interaction between artists, the public, and both the local and international communities. This vibrant exchange of cultural perspectives not only enriches the tourist experience but also serves as a catalyst for promoting cultural diversity and understanding among different communities. This study will pivot towards analyzing statistical data concerning the festival's audience dynamics over the past few years. Through empirical evidence and insightful analysis, we will shed light on the festival's role in attracting visitors, boosting economic activity, and promoting cultural exchange.

O. Bălan-Budoiu (✉)
Academia Națională de Muzică „Gh. Dima", Cluj-Napoca, Romania
e-mail: oana.balan@amgd.ro

A. L. Negrușa et al. (eds.), *Sustainable and Resilient Businesses in the Global Economy*, Springer Proceedings in Business and Economics,
https://doi.org/10.1007/978-3-031-75883-6_3

Keywords Cultural visibility · Tourism development · Community engagement · Economic growth

Introduction

In many countries, music is presented as a vital element of the connection to the community, while it contributes to regional development and to people's active involvement in the cultural phenomenon (Gibson & Connell, 2012). Besides these important social benefits, large-scale events, such as international classical music festivals, contribute to national economic developments (Andersson & Andersson, 2006), while providing an important opportunity for various regions to present their cultural uniqueness and diversity.

The "George Enescu" Festival in Romania has become an important landmark on Europe's cultural map, attracting an impressive number of international tourists with each edition. Due to the exceptional quality of the musical renditions and to the platform it offers to world-famous artists, this festival brings a significant contribution to the promotion and appreciation of the music written by Romanian composers, having an essential role in amplifying Romania's cultural presence on the world stage. The festival not only gathers music lovers from all over the world, but it also serves as an important catalyst for a vibrant cultural exchange, consolidated by the interaction among the artists, the large audience, and the local and international communities. This exchange enriches the touristic experience and cultivates an enhanced understanding and appreciation of cultural diversity (Long & Robinson, 2004). Moreover, the substantial international presence in the festival stimulates local economy, as visitors contribute significantly to the revenues of hotels, restaurants, stores, and other local businesses.

The role of classical music festivals, such as the "George Enescu" Festival, transcends the traditional framework of entertainment, bringing valuable contributions to the development of tourism and economy. Within this study, we shall explore the dynamics of the festival's audience during the past few years, analyzing the data provided by ARTEXIM (a public institution subordinated to the Romanian Ministry of Culture) in order to illustrate its direct and indirect impact on tourism. It is essential to understand how these major cultural events not only attract visitors, but also influence the international perception about Romania as a top travel destination.

Besides bringing direct economic benefits, the classical music festivals, including the "George Enescu" Festival, play a pivotal role in revitalizing and promoting the host regions. They transform cities into centres of cultural and tourist attraction, offering unique experience opportunities for visitors and amplifying the cultural visibility of the destinations. In this context, the festival not only enriches Bucharest's cultural offer, but also contributes to a feeling of pride and cultural belonging for the local community.

This study will approach the long-term impact of the "George Enescu" International Festival on tourism, emphasizing the way in which large-scale cultural events can serve as engines of sustainable growth in the tourism sector. We shall investigate the way the festival influences not only the local economy, but also the way in which Romania is perceived on an international level, arguing in favour of the fact that investments in high-quality culture are essential for the development and sustainability of national and international tourism. By means of this analysis, we hope to offer a comprehensive perspective on the strategic value of integrating performing arts in tourism development strategies.

The structure of the paper follows key sections mean to highlight the role of classical music festivals, using both empirical data and comprehensive analysis. It focuses on the impact of these festivals in promoting cultural exchange, boosting tourism, and fostering economic growth. After discussing the existing research and theoretical background, we will deeply analyse a case study of the „George Enescu" International Festival. This includes data collections from the period 2017–2021, representing the most recent iterations of the festival, including post-pandemic insights. The objective of the paper is to present findings over different years and to discuss the effectiveness of communication strategies in attracting tourists and stimulating local economies. Additionally, we will evaluate how festivals stimulate economic growth and advocate for the integration of high-qualitaty cultural events in tourism development.

Literature Review

Special attention has been paid over the past few years to the development of tourism with the help of festivals. Classical music festivals play a significant role in tourism dynamics, influencing both the appeal of the travel destinations, and the local economy. These events transform host cities in centres of cultural and touristic attraction, offering unique experience opportunities for visitors.

There are several key themes in literature that widely recognizes the significant economic benefits of festivals. The studies emphasize that classical music festivals enhance the attractiveness of their host cities, transforming the locations into tourist destinations.

The study of festivals, such as the festival in Lucerne, shoes how they can revitalize entire regions, transforming them into points of attraction for cultural tourism. For instance, the Lucerne festival not only attracts classical music lovers from all over the world, but also stimulates investments in infrastructure and local services, highlighting the symbiotic relationship between culture and development (Smith, 2016).

The literature of the recent years has analyzed the impact that festivals can have on regions. The works authored by Raj (2004), Nurse (2001), Egresi and Kara (2014) offer detailed perspectives and comprehensive analyses in touristic and economic terms. Raj (2004), known for his studies in the field of event management, including music festivals, analyses in his paper the economic impact of festivals, highlighting the way in which they stimulate local economies by attracting tourists who spend money in various sectors, from accommodation to entertainment and gastronomy.

Raj also highlights how festivals can improve the image of a destination and can contribute to cultural branding.

Nurse (2001) focuses on the intersection of tourism, economy, and culture. He explores the concept of festivals as part of an "experimental" economy, where visitors are attracted not merely by the musical performance, but also by the opportunity to experience local culture in an authentic way. Nurse discusses how music festivals can serve as a catalyst for cultural development and how they facilitate cultural exchanges.

Egresi and Kara (2014) analyse the ecologic impact of music festivals, an increasingly important aspect in the context of the growing global awareness with respect to sustainability. Jones offers analyses on how festival organizers can implement sustainable practices, such as waste management, using renewable energy and reducing carbon footprint. His work proposes solutions to turn festivals into green events that contribute positively to the environment as well as to the local communities.

Lee (2017) explores the cultural capital dynamics in a study that discusses the perception of the visitors who attend a festival, namely the level of satisfaction and the extension of their knowledge from a tourist's perspective, during and after the attendance at the musical events. This research stresses the fact that the level of satisfaction that combines the cultural and the touristic experiences has a significant influence on the quality of an individual's life, emphasizing that an experience of this type can also subjectively shape how tourist services from a certain region are perceived.

Picard and Robinson (2006) emphasize the cultural transformations that festivals inevitably exert on communities since the very first editions. The fact that they bring together aspects of global and local cultures, while concomitantly highlighting the identity of the host people, produces interactions which remodel a travel destination. In this respect, festivals act as catalysts of social evolution by promoting and encouraging society's diversity and welfare as a whole. Like in most of the studies that have approached this topic, Picard and Robinson (2006) analyse the contribution brought by this combination in the economic area, as festivals have a capacity to generate income in the sector of hospitality, trade, and tourism, stressing the complexity of these relationships.

In the study "Cultural Tourism: The Partnership Between Tourism and Cultural Heritage Management", McKercher (2002) shows how the touristic experience can be improved through interaction with elements of the cultural heritage, attracting tourists and satisfying their curiosity through multiple information. Case studies are conducted about information with touristic potential such as archaeological sites, museums, events presenting local traditions etc., a collaborative approach that integrates all the aspects that define certain regions and can reinforce the image of the location in question.

Kruger and Trandafoiu (2014) closely discusses the influence of globalization processes, namely the manner in which migration, supported by large-scale events (such as festivals) facilitates the interaction of various local communities. By the nature of these art events, the phenomenon of cultural mobility, assisted by artists, leads to an exchange of cultural identity and influence in intercultural relationships. Music becomes an instrument of tourist attraction and a method of promoting cultural destinations, an ample discussion on the complexity of the touristic phenomenon due to this cultural and social richness that can be provided by such events as the international festivals.

The recent literature (Myunghee, 2020) has focused on the development of national industries as an impact of the large-scale cultural events, showing how the long-term effects on the local economy can be traced in the revenue reported by local businesses. Therefore, these actions are not mere elements of entertainment, but notably important engines for the development of tourism due to the stimulation of the regional economies and of potential tourists' interest, having a clear positive impact on the consolidation of local communities, the enrichment and highlighting of the local heritage.

From a social viewpoint, it has been acknowledged that festivals contribute to the wellbeing of various communities due to the intensity of the activities and the new ambiance they bring to regions, namely a feeling of cohesion that intensifies among community members. Also, an important component that develops in relationship to these actions is the administrative one, as many of the locations where actions take place are hyper-activated during the festival days. This is how, for the life of the local community, the benefits are apparent in many spheres, making fewer known locations profitable, developing new destinations for tourists and transforming host cities in vibrant points of attraction.

Festivals not only revitalize regions by stimulating investments in infrastructure and services, but also by promoting an improved image of the respective destinations. Economic aspects are highlighted through the generation of diverse types of income, from accommodation to food, and through the provision of jobs, thus contributing to local development in the long run. Beside the economic impact, festivals also facilitate cultural exchanges, enriching community life and offering a platform for the expression of cultural diversity. Thus, classical music festivals prove to be more than mere entertaining events, acting as powerful catalysts of touristic and cultural development, with positive effects on the local communities and the international perception.

As we see in the literature, the classical music festivals play a multifaced role in economic and cultural development, serving as vital platforms for social interactions. It is clearly demonstrate that they enhance the cultural visibility of their host cities, contributing to their reputation and recognizing the need for sustainable practices to maximize their positive impact.

The Global Perspectives of Classical Music Festivals: Examining the „George Enescu" Festival's Role in Promoting Cultural Tourism

The multifunctional role and the profound impact of classical music festivals, which is not limited to aspects of entertainment, as they also significantly extend their contributions to the development of community, culture, and economy, proves to be very important for connecting communities through music, facilitating active involvement and cultural education, which not only enrich individual experiences, but also consolidate a sense of belongingness among the various audience categories, from amateurs to connoisseurs. Besides the social benefits, the economic contributions are noteworthy, as they generate substantial revenues for tourism and the local economy, creating temporary employment and increasing demand for local products and services. This economic dynamics is complemented by the attraction of an international audience, due to the presence of renowned artists and the diversity of the offered repertoire, which transforms these festivals into genuine travel destinations. The classical music festivals harmoniously mingle art with tangible benefits for the community and the economy, thus consolidating their status as cultural phenomena of global importance.

The research of the past few years shows that music festivals can contribute significantly to generating revenues for the national tourism and economy, by selling tickets, creating temporary employment opportunities, increasing demand for local products and services, boosting hotels' and restaurants' profits, improving transport infrastructure, etc. Nevertheless, the experience offered by a classical music festival can be different from one consumer to the next, being influenced by many factors, among which we mention only: the interest in a particular music genre, the artistic quality (the musicians' reputation), the educational and explorative component which can increase the consumer's satisfaction, the social factor of bonding to the "elite" community, the individual experience by which each consumer can experience the artistic act per se.

McGuire (2017) wrote in a study on the dynamics of British classical music festivals that they tend to define audiences, namely the way in which the audience appreciate and invest in a cultural event by means of their level of education and the quality of the education, splitting consumers into: "amateur audience" and "connoisseur audience"; in other words, the uninformed audience, also known as the "public at large" and the informed audience (who have either frequently attended classical music concerts, or have a music education), while each of the two groups bonds to their own consumption dynamics. Because these categories relate to the artistic act differently, other types of relationships implicitly arise with regard to financial investments (buying tickets, longer periods of time spent at the concert venues, consumption of local services, etc.).

In the same area of touristic influence grow the roots of tradition, heritage, history, which enhance the attractiveness of a music festival.

Other arguments on the capacity of music festivals to be an important tourist attraction are high-quality cultural experiences, which can be appreciated by:

- The presence of internationally acclaimed performers who, due to their name and reputation, can attract tourists from all over the world;
- The rich variety of repertoires that can turn the event into a touristic destination for music lovers with diverse tastes and preferences;
- The context and location of the festival which can become a destination for a cultural trip in its own right;
- The diversity of accommodation and transportation options which can encourage an increasing dynamics of the cultural tourism.

The UNESCO study "Festival Statistics: Key Concepts and Current Practices" (2009), highlights the importance of festivals and the necessity of better instruments for the measurement of their diversified impact. As in many other publications, it emphasizes the fact that classical music festivals have lately become global phenomena that attract millions of tourists, evolving from modest gatherings to high-impact events. Beyond their entertainment facet, they represent a vibrant manifestation of society, and have played a major role in attracting tourists in recent years.

For the past 65 years Romania has been the host of one of the grand festivals of the world of classical music, the "George Enescu" Festival, which reached its edition number XXVI in 2023. Internationally acclaimed soloists, world-famous conductors, exceptional premieres—all these have combined throughout the impressive history of this festival. Being designed for a wide range of consumers, with cycles of concerts, recitals, competitions, high-level scientific sessions, the "George Enescu" Festival brings together great artists, with impressive techniques and repertoires, who attract a large number of foreign tourists.

Being one of the most important festivals in the world, situated in the first ranks of the cultural life, alongside other grand events of the same type, such as the ones in London, Edinburgh, and Lucerne, the event that honours the memory and personality of the great Romanian musician George Enescu gathers hundreds of musicians who come to Romania together with the best orchestras of the world for the concerts given in 6 cities of our country in autumn.

The event makes our country a first-class cultural destination, attracting people from all over the world, enhancing the interest of tourists in Romania, who thus implicitly get into contact with the Romanian cultural identity. Being a highly prestigious event, the "George Enescu" Festival improves and consolidates the international image, showing that our country has a cultural richness and diversity that we can be proud of.

From the perspective of the touristic impact, the festival promotes the preservation of Romanian classical music, which it promotes on a high level of the international network consisting of highly prestigious artists who are invited to each edition to Romania.

The "George Enescu" Festival significantly influences the cultural tourism in Romania, particularly in Bucharest, by improving the city's branding and its visibility in the mass-media. This major cultural event attracts thousands of international and local tourists and is considered the most important cultural event in Romania, having a positive impact on the local infrastructure and branding (Alexe & Tapardel, 2013). Also, the festival serves as a key example of cultural tourism, contributing economically and culturally to the region by attracting a specific audience interested in classical music and cultural experiences (Bunghez & Ştefănescu, 2017).

Research Methodology

In order to evaluate the impact of the "George Enescu" Festival on the national tourism of Romania, we propose a methodological approach in several stages. Based on the reports of the 2017, 2019, and 2021 editions, provided directly by the Festival organizers, we shall use a multifaceted methodology, focusing on measuring the strategic impact of communication campaigns on tourism and cultural visibility. The study of the strategic impact of the communication campaigns on the tourism and the cultural visibility of a classical music festival can be considered a valid research method, because it offers valuable data and perspectives on the way in which communication influences the visitors' perceptions and behaviours. Before evaluating the impact, it is crucial to understand the specific goals of the festival's communication campaigns. Also, determining the target groups of the campaign, which can include international tourists, local visitors, classical music lovers, and cultural critics, provides important data as well.

For the scope of this study, we shall analyze:

1. The Advertising value equivalency (AVE) that will measure the impact of the communication campaign by equalizing it with the cost that would have been incurred by obtaining the same visibility employing paid advertising methods (Barra & Zotti, 2016). It is a common method to evaluate the efficiency and the profit of investing in public relations. Its relationship to tourism is assessed by its high advertising value, which can indicate a high visibility and implicitly attract more visitors, improving the reputation of an event or a location, stimulating tourism indirectly.
2. The total number of readers and potential contacts, measurements that quantify the direct and indirect audience who had access to promotional materials, offering a perspective on the penetration degree of the campaign into the target market. High numbers seen here suggest an extensive coverage and increase the likelihood of attracting visitors from various demographic areas, which is essential for events which target a large audience.

3. The media and the impact of the advertising methods, namely the analysis of the various media channels (TV, written media, social media) meant to determine which channel had the biggest impact with regard to AVE and the audience's engagement. Understanding the most efficient channel can help organizers optimize marketing strategies so as to attract tourists, concentrating resources in places where the impact is maximal. Comparing the efficiency of posts on various social media platforms in order to determine which generates the most impressions and interactions, as well as optimizing communication channels can improve the visibility of the events and can attract more segments of the public, including international tourists.

By using these methodologies, the "George Enescu" International Festival has managed to maximize its impact on tourism and to stand out as an important cultural landmark, thus contributing to the promotion of culture and tourism in the region. These methodological strategies offer a solid basis for the continuous evaluation and optimization of communication efforts in the context of large-scale cultural events.

Results and Discussions

The "George Enescu" International Festival, am emblematic event of the Romanian cultural scene, has been, throughout its numerous editions, a barometer of the dynamics and evolution of the area of cultural tourism and communication management. In the context of globalization and accelerated digitalization, the festival has constantly adapted, showing an impressive capacity to use communication means strategically in order to maximize its visibility and economic and cultural impact. The statistic reports of the 2017, 2019, and 2021 editions provide data that reflect not only the success of the festival in reaching a large and diversified audience, but also the efficiency of its communication campaigns in promoting a musical heritage of universal value. These reports highlight a notable growth of the generated materials and of the interaction in the social media, indicating a constant evolution of the communication strategies adopted by the organizers. The comparative analysis of these editions offer a valuable perspective of the way in which the "George Enescu" Festival has succeeded in establishing itself as a nucleus of touristic attraction and a catalyst of cultural and economic development in Romania.

Based on the report of the 2017 edition, the statistical analysis highlights success in the strategic use of communication for:

- The Advertising value equivalency (AVE) for the campaign, which was 2,342,021 Euro, reflecting the extended impact of the communication actions on various media channels;
- The total number of readers of the materials generated by the PR and advertising campaign was 8,668,096;
- The number of potential contacts (people who saw the campaign's messages several times) was 92,392,063;

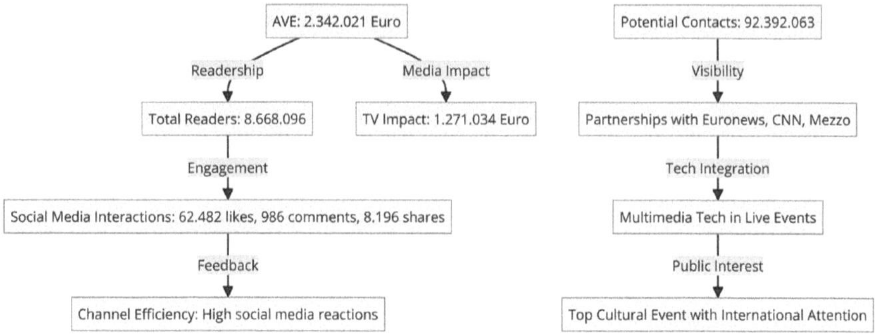

Fig. 1 The impact analysis of communication strategies for "George Enescu" Festival 2017

- Media and Channel Impact—television: had the highest AVE value of 1,271,034 Euro;
- Social media activity: 62,482 likes, 986 comments and 8.196 shares on Facebook, indicating a high level of interaction with the festival's content;
- The efficiency of the communication channels which show that the social-media posts were efficient, having the most numerous reactions compared to other social platforms;
- The partnerships with Euronews, CNN, and Mezzo considerably extended the festival's international visibility;
- The integration of multimedia technologies and projections at live events added value to the participants' experience and attracted the attention of the media.

The festival continued to consolidate its position as a first rank cultural event, attracting international attention and increasing the public's interest (Fig. 1).

The 2019 edition offered a fertile ground for the analysis of the impact of press releases and the active social media presence, highlighting the evolution and the continuous adaptation to the new media tendencies and the public's preferences, a remarkable evolution in terms of media impact and social media presence compared to the 2017 edition. The statistic analysis of the press releases and of the social network activity highlights a range of aspects worth analyzing. The number of materials generated by the press has increased significantly: Whereas in 2017 it released 616 materials, in 2019 their number reached 810. This indicator shows an impressive growth of the impact of the press releases issued by the organizers. With regard to the efficiency of press releases, the data show a clear improvement of the performance. Whereas in 2017, each press release generated an average of 56 materials, in 2019 the average number was 81 materials per release, thus highlighting a more efficient use of this communication resource.

The distribution of materials according to media types was a subject of interest as well. Comparing the way in which the materials were distributed on channels such as Facebook, Twitter, and YouTube in 2017 as compared to 2019, we can notice changes in the communication strategy, adapted to the tendencies of the digital media consumption. The presence of the festival on the social media also evinced a

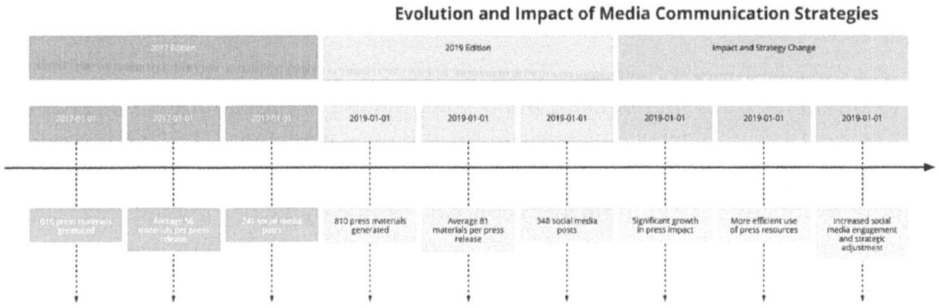

Fig. 2 Evolution and impact of media communication strategies between the 2017 and 2019 editions of "George Enescu" Festival

significant growth. Whereas 241 posts were recorded in 2017, their number grew to 348 in 2019, reflecting a more aggressive strategy with a better engagement in the online dialogue with the public (Fig. 2).

The 2021 edition of the "George Enescu" International Festival managed to impress due to the number of attendees, to the extended media coverage as well as to the remarkable advertising impact. The event attracted the attention of a large audience, both in the concert halls, and online, demonstrating its adaptability in a changing global context.

The festival occupied in 2021 an impressive number of 54,000 seats in the concert halls, and the online presence was also noteworthy, with 20,358 views on the official website, a digital attendance that highlighted the growing interest of the audience, as well as the efficiency of the online advertising strategies. The advertising of the festival in 2021 was vast, with 8,749 media appearances during the campaign, reaching an estimated audience of 214 million contacts. This level of visibility reflects not only the popularity of the festival, but also the efficiency of its public relations tactics. The value of the advertising impact was estimated at 4.54 million Euros, indicating a significant added value and a strong Return on Investment (ROI) for the festival's sponsors and partners. In the context of the "George Enescu" International Festival, mentioning a strong ROI suggests that the investments made in organizing and promoting the festival generated substantial benefits, in terms of finances, visibility, and media impact. In 2021, the festival's website counted 135,486 unique users and was accessed 2,099,331 times, which proves a deep engagement and a constant interest in its contents. On the social networks, the interaction was vibrant too, with approximately 1.4 million people involved and 387,400 interactions during the course of the festival. These numbers highlight an efficient digital communication strategy and an active online community. Analyzing the data in the previous reports, we notice a continuous growth in terms of visibility and impact.

Compared to the previous editions, the festival demonstrated an impressive expansion in all of its dimensions, from attendance at concert halls to media impact and digital presence, which proves its cultural value and its growing relevance in the world's artistic landscape.

Conclusions

The results of the study highlight the importance of integrating cultural events, such as the "George Enescu" Festival, in touristic development strategies. The classical music festivals can be used as strategic instruments to attract tourists and improve travel offers, consolidating the cultural and economic profile of the destination. Therefore, the local and national authorities should support and invest in these festivals as an integral part of tourism policy.

Based on the performed analysis, it is recommendable that tourism authorities collaborate closely to the organizers of classical music festivals in order to maximize their impact on tourism. It is essential to develop a touristic infrastructure that can support the visitors' influx, to improve local services and to promote cultural events in a sustainable manner. Also, marketing strategies should include efficient communication campaigns that can highlight the uniqueness and the added value of such cultural events as the "George Enescu" Festival.

The relationship between classical music festivals and national tourism is undisputable and offers valuable opportunities for Romania's touristic and cultural development. The "George Enescu" Festival represents an eloquent example in this respect, demonstrating the capacity of cultural events to transform the touristic landscape and to contribute to a nation's economic and cultural prosperity.

Beside the direct benefits brought by the increase in the visitors' numbers, the "George Enescu" Festival contributes significantly to the local development. The investments in infrastructure necessary to support the event, such as improving public transportation and modernizing accommodation facilities have long-term effects on the community. These improvements not only serve the immediate needs of the festival, but also remain as a permanent benefit for residents and future visitors, thus stimulating tourism in the long term.

From the perspective of increasing international visibility, the "George Enescu" Festival plays a crucial role in promoting Romania on the world's cultural stage. By attracting prestigious names from the world of classical music and an international audience, the festival strengthens Romania's international profile as a first-class travel destination. The international media often covers the event, offering a valuable visibility that transcends the traditional boundaries of touristic advertising. Thus, the touristic marketing efforts capitalize on this coverage in order to attract new categories of tourists.

With a view to creating a touristic strategy, the collaborations between the cultural and the commercial sectors, such as the partnerships with local hotels, restaurants, and other businesses can amplify the economic impact of the festival. The travel packages that include tickets to the festival, accommodation and thematic gastronomic experiences can offer visitors an enhanced experience and an increased source of income for the local businesses. This synergic approach of culture and business can lead to innovative developments, attracting a larger, more diversified audience.

Therefore the "George Enescu" Festival is not just a prestigious cultural event, but also a platform that promotes Romania on the international stage, attracting visitors from all over the world. With the help of this event, the national tourism is stimulated, as visitors are attracted not only by the quality of the musical performances, but also by Romania's cultural and historic heritage.

From the economic point of view, the "George Enescu" Festival generates multiple benefits for the local financial dynamics, by increasing the number of tourists who spend money in the hotel sector, restaurants, transportation, and other services. The intensification of this economic flow is crucial for the sustainable development of the region, creating jobs and stimulating local businesses.

The festival also promotes cultural diversity and strengthens bonds between various regions with the help of international artists and of the diverse audience who attend the event. This cultural exchange enriches the touristic experience and enhances intercultural understanding and appreciation.

The results indicate that the „George Enescu" Festival scientifically boosts the local economy by attracting tourists who spend on hotels, restaurants and other services. This alignment suggests that the event shares common economic benefits with other festivals, as indicated by Raj (2004) and Andersson and Andersson (2006), who discuss the role of cultural festival in the development of economic benefits and cultural visibility. The findings also indicate that „George Enescu" Festival enhance Bucharest's international image, attracting diverse audiences enhancing tourist experiences. The study also shows how interaction among international artists and the local community foster a sense of belonging. We can see how communication campaigns adds depth to the general tourism development, providing a nuanced presentation of how the promoting channels may contribute to a festival's success.

Acknowledgements We thank the Artexim institute and its director Mihai Constantinescu for the generous support they offered by providing us with the essential data which facilitated the analyses presented in this research. The information we received was essential in understanding the impact of the "George Enescu" Festival on the cultural tourism in Romania. We are also grateful to everyone involved in issuing the reports of the "George Enescu" Festival, for the dedication and the professionalism with which they completed these materials.

References

Alexe, F. A., & Tapardel, A. C. (2013). Case study: The impact of organizing the George Enescu international festival on the branding and promotion of Bucharest City. *International Journal of Economics, Finance and Management Sciences, 1*(5), 241. https://doi.org/10.11648/j.ijefm.20130105.15.

Andersson, A., & Andresson E. (2006). *The economics of experiences, the arts of entertainement* (pp. 59–116). Eduard Elgar Publishing House.

Barra, C., & Zotti, R. (2016). Measuring efficiency in higher education: an empirical study using a bootstrapped data development analysus. *International Advance in Economic Research*, 11–33.

Bunghez, C. L., & Ştefănescu, V. (2017). Cultural Tourism. *Revista Economică, 69*(1), 29.

Egresi, I., & Kara, F. (2014). *Economic and tourism impact of small events: The case of small-scale festivals in Istanbul.* Studia Universitatis Babes-Bolyai, Geographia.

Gibson, C., & Connell, J. (2016). *Music festivals and regional development in Austria.* Routledge.

Kruger, S., & Trandafoiu, R. (2014). *The globalization of musics in transit: Music migration and tourism*. Routledge.
Lee, K. -J. (2017). *Cultural capital of visitors, classical music festival satisfaction and quality of life*. Global Business and Finance Review.
Long, P., & Robinson, M. (2004). *Festival and tourism: Marketing, management and evaluation*. Business Publishers Ltd.
McGuire, C. (2017). Of programs and prima donnas: Investigating British music with the musical festivals database. *Notes, 73*(3), 432–472.
McKercher, B. (2002). *Cultural tourism: The partnership between tourism and cultural heritage management*. Routledge.
Myunghee, J. (2020). *Impacts of festivals and events*. Salem State University.
Nurse, K. (2001). Festival tourism in the Caribbean: An economic impact assessment. *Journal of Management*, 223–230. (Cognizant Communication Corporation)
Picard, D., & Robinson, M. (2006). *Festivals*. Remaking Worlds. Channel View Publications.
Raj, R. (2004). *The impact of cultural festivals on tourism*. Tourism Today.
Smith, W. (2016). *Making money in the music business: A comprehensive guide to being an entrepreneur and making a living in the music industry*, CreateSpace Independent Publishing Platform.
UNESCO. (2009) Festival statistics: Key concepts and current practices. UNESCO Institute for Statistics, https://uis.unesco.org/sites/default/files/documents/festival-statistics-key-concepts-and-current-practices-handbook-3-2015-en.pdf.

Bicycle Travel as a Sustainable Strategy for Improving Green Tourism Development: Theories and Practices

Le Thanh Tung and Radoslaw Dolecki

Abstract Green tourism has been highlighted as an important trend for developing tourism industry in the future. One of the potential ways of improving the green tourism industry is bicycle travel, which can be considered as a sustainable strategy for this industry. However, the concept of bicycle travel and the contribution of it for supporting green tourism industry has not received enough attention from academics. This paper aims to conduct a summary of theories related to the bicycle travel concept and suggest a theoretical framework which presents possible effects of this phenomenon on the green tourism industry. This study has used multiple methods for modelling the bicycle travel concept including the systematic review method and the graphing technique. This study brings its contributions through several points. First, we present the theories of green tourism development. Second, the study, provides a literature review which focuses only on bicycle travel and highlights the role of bicycle travel on green tourism industry. Third, the study introduces a theoretical framework for conducting the bicycle travel as a variable of green tourism industry. Fourth, we deliver some theoretical and practical implications for enhancing bicycle travel in countries. These results will promote green tourism development in the future. In addition, this study will provide some practical implications for applying this concept to sustainable development in the tourism industry.

Keywords Bicycle travel · Green tourism · Sustainable development · Systematic review · Tourism development

L. T. Tung
Policy and Applied Economics Research Group, Faculty of Economics and Public Management, Ho Chi Minh City Open University, Ho Chi Minh City, Vietnam
e-mail: tung.lt@ou.edu.vn

R. Dolecki (✉)
Faculty of Social Science, Vincent Pol University, Lublin, Poland
e-mail: rdolecki@wssp.edu.pl

A. L. Negrușa et al. (eds.), *Sustainable and Resilient Businesses in the Global Economy*, Springer Proceedings in Business and Economics,
https://doi.org/10.1007/978-3-031-75883-6_4

Introduction

The post pandemic state of economies has placed tremendous stress on the tourism industry. Growing costs of natural resources as well as energy prices made it difficult for many travellers to continue the tourism habits they had been used to. The experience of lockdowns as well as administrative difficulties and limitations on foreign travel also contributed to changing behaviours of travellers. This led to growing popularity of such phenomena as staycation as well as domestic tourism and a rediscovery of local attractions. In the post-pandemic COVID-19 years, some tourism development strategies have mentioned and discussed in academic field (Lee & Iwasa, 2020; Sharpley, 2022; Han et al., 2023). The resignation and shift away from package travel allows tourists to modify the choices and options available to them also in terms of mode of transportation, which may induce some more ecologically conscious and oriented tourists incline towards eco-friendly solutions.

Green tourism which is understood as a more ecologically responsible choice of recreation activities is the industry's response to this growing demand for sustainable offers. At the same time, green tourism is also a trend that is being consciously shaped and promoted by central authorities, local administration and stakeholders alike across popular tourist destinations of the globe. The rise in green tourism is accompanied by the growing popularity of bicycle travel as the two trends complement one another when bicycles are the preferred mode of transportation both in terms of urban as well as rural tourism (Lamont, 2014). Bicycles allow for slow paced forms of tourism, they are also low cost and environmentally friendly (Han et al., 2020). Hence, they are easily integrated in tourism development strategies both as an independent means of transport and as an element in multimodal concepts of transportation in urban or rural settings. The promotion of cycling is viewed as a way of counterbalancing the problem of exceeded tourist capacity of popular destinations as cyclists eagerly engage in activities where the road and the time spent on travelling is an attraction in itself and the goal is not limited only to reaching the final destination and maximizing the time spent there. However, the tourists' behaviour needs to be defined because the diversity in study results is related to this kind of tourism transportation (Han et al., 2016).

This study brings its contributions through several points. First, we present the theories of green tourism development. Second, the study provides a literature review which focuses only on bicycle travel and highlights the role of bicycle travel in green tourism industry. Third, the study introduces a theoretical framework for conducting the bicycle travel as a variable of green tourism industry. We explore how bicycle tourism can be effectively implemented as a sustainable strategy for the development of green tourism in the future. Fourth, we deliver some theoretical and practical implications for enhancing bicycle travel in countries. These activities will promote green tourism development in the future.

Literature Review

Theories of Green Tourism Development

Green tourism, also referred to and often used interchangeably with ecotourism and sustainable tourism, involves the type of environment-responsible travel. It can be associated with travel to destinations that opt for a balanced approach to tourism development as well as use of resources, and minimize the environmental and cultural impact of the tourism movement (Furqan et al., 2010). Fennell (2008) defines ecotourism as a low-impact, non-consumptive, locally focused, ethically managed type of nature-based tourism that is sustainable, non-invasive, and largely focused on learning about nature first hand and is meant to aid in the preservation of natural places. Green tourism combines it with equal attention being paid to the conservation of cultural diversity and urban tourism assets (Furqan et al., 2010).

Green tourism is as much an element in tourism development strategies adopted at local or central government levels as it is a matter of personal choices of persons engaging in travel activities, such as going to lengths so as to minimise or offset the carbon trace of the travel experience or efforts at limiting the environmental impact of the transportation part alone. Support to investments which help foster greener and low-carbon economy, mobility and urban development is at the heart of and constitutes one of the funding priorities under the 2021–2027 European Regional Development Fund. The shared nature of responsibility for the management of EU funding distributed under this framework and exercised by the European Commission and authorities of member states means that the priorities directly translate to compatible local policies and development strategies. For this reason, EU-co-funded programmes and projects include among their specific goals efforts at supporting the development of a green, low-carbon economy relying on clean energy, green investments, and sustainable, multimodal urban mobility, which will eventually lead to further construction of cycling infrastructure, bicycle tourism trails as well as necessary accompanying facilities. Sustainable transport solutions integrate on an increasingly wider scale cycle networks in cities as well as the country while special multi-user tourism trails connect entire regions rich in tourism assets, a path that was once pioneered by such countries as Denmark, the Netherlands or Switzerland (Lumsdon, 2000). Both in terms of research, conceptual studies as well as actual strategic and implementation work, one can observe a bias and increased activity in Western countries as well as Australia in favour of Asia and other popular tourism destinations (Sabri et al., 2019).

Through conscious planning decisions on the part of stakeholders and the destination policymakers, the negligible demand generated by most enthusiastic amateurs of cycling holidays has gradually started to evolve into a substantial trend in modern tourism industry. In the tourism industry, a shift in demand is most likely to follow the prerequired supply of necessary facilities and services. Investment, often government-backed, needs to precede promotional campaigns and policies aimed at generating the interest in green travel and cycling. This course of action is achieved more easily in highly-urbanised areas, where bicycle-oriented solutions can be integrated with the existent motor and pedestrian traffic infrastructure in a relatively cost-efficient manner and, more-importantly, they will be shared and intensely used by bicycle tourists and the large local population alike. For those reasons, in some destinations urban tourism has the upper hand compared to rural tourism (for example, Nilsson, 2019).

Bicycle Travel as a Sustainable Strategy for Improving the Green Tourism Development

Sabri et al. (2019) have executed a systematic review on bicycle tourism and concluded that there is a general consensus in the literature regarding the positive economic, social and environmental impact of bicycle tourism on destinations, however, this form of niche tourism needs to be fostered with effective marketing campaigns as well as proper governmental policies and good management of infrastructure. Sabri et al. conclude that safety of tourists, quality and directness of cycling routes, inclusion of necessary facilities and parking spaces, correct signage, as well as the integration of the network in a cohesive way while avoiding unnecessary contact with vehicular traffic are among key factors in developing bicycle tourism.

Bicycle travel and tourism are viewed as an indispensable element in the development strategies focused on green and smart growth as well as sustainable mobility, as evidenced by examples of forerunners and leaders in this trend, such as the city of Amsterdam or the Greater Copenhagen Region (Nilsson, 2019). The development of urban tourism in these types of destinations is harmonious and synergic with the necessary construction of bike lanes, passes and similar infrastructure primarily intended for local users. At the same time, bicycle as a mode of transport for tourists considerably reduces the burden on municipal road infrastructure and parking space during the high seasons. However, as observed by many researchers (see Sabri et al., 2019), this harmonious growth ceases to exist at city boundaries with local rural authorities lagging behind in terms of cycling infrastructural projects. This is understandable as they involve larger distances entailing higher costs as well as significantly lower user rates due to small rural populations. Also in Poland, the regions with the highest levels of urbanisation and metropolitan areas characterised by large populations and high standards of economic development can boast the most dense and long networks of bicycle lanes (Roman & Roman, 2014), while frequent and unavoidable repair works to intensely used city roads allow for their swifter modifications and adaptations to growing bicycle traffic.

In order to overcome the aforementioned negative tendencies and the lack of balance in the availability of bicycle lanes, local municipal authorities should better coordinate their efforts with surrounding rural communes under metropolitan area cooperation frameworks, especially in destinations with observable suburbanisation tendencies further boosted by the technological revolution in multiple industries (e.g., distributed teams, home office options). Extended cycling networks would allow commuters from the growing suburban populations to escape inevitable car-dependence, especially in the light of increasingly widespread use of long-distance electric bicycles. The bicycle lane grid would of course also serve tourists and would contribute to the growth of green tourism in the area. den Hoed (2019) advocates the development of tourism closer-to-home as well as its efficient integration with active mobility patterns to benefit from the multi-modal nature of everyday travel of local populations. Bicycle tourism experience in such cases often becomes one in the numerous purposes of the trip, it does not have to be the sole goal but rather a factor among other social causes and lifestyle patterns. It can mean a moment of recreation to a local commuter or an important reason for the trip and a source of motivation to a tourist. Importantly, the extended network of infrastructure will allow to draw tourists away from the most popular attractions in areas at the brink of their tourist carrying capacity and to distribute them more evenly between other nearby destinations enjoying less public attention.

In more remote rural areas, the development of the said bicycle tourism infrastructure requires sustained support and coordination from authorities at various levels of administration. An example of such initiatives is the Eastern Poland Cycling Trail Green Velo, which spans across five different Polish regions and has a total length of 2078 kms with 228 newly-constructed stations along its length. The entire project, which also included adoption, reconstruction and development of new bicycle roads, required close cooperation of authorities from all concerned destinations for the purpose of coordinating the endeavour, which, however, would not have been possible without the support from state budget and especially co-funding from the European Regional Development Fund exceeding EUR 60 million. However, as of now the Green Velo trail connects destinations and sites of significant natural, cultural and historic value and it is a destination in itself, according to a study from 2018—to a population exceeding 65,000 tourists per month (415 483 persons, May to September that year, see: Dziedzic & Trojanowski, 2021). Interestingly, the study proved 99.5% of that group to be domestic tourists, despite promotional campaigns also targeting international audiences. This fact would support the hypothesis that bicycle tourism is favoured particularly in alternative and domestic forms of tourism, such as slow tourism, staycation and green tourism, and is in line with studies which demonstrate the decisive preponderance of leisure day trippers and domestic holidaymakers over foreign cycling holidaymakers (Ritchie, 1998).

Empirical Studies Regarding Bicycle Travel and Green Tourism Development

Green development of tourism industry has been highlighted as a trend for conducting strategy development plans in many countries. This strategy creates a tourism sector in a specific location but helps increase the industry in as sustainable a manner as possible. The green tourism development can improve the industry in two dimensions, first one is the efficiencies in attaining the business targets, and the second one is the preservation of the environment, natural wildlife, local culture, and spillover social economic benefits for local people. However, green tourism development is not easy because of the limitations of resources in destinations.

We have checked the current literature of bicycle tourism and green tourism industry and saw that the number of empirical studies have been significantly increased in many countries. It is the evidence for the important role of this kind of transportation in modern tourism. Bicycle travel contributes to the sustainable tourism development, therefore, it enhances green tourism industry (Lumsdon, 2000). Bicycle travel contributes to the sustainable tourism development. Therefore, this kind of travel has received some empirical studies in recent years. Meng and Han (2016) demonstrated that decision-making processes of bike-travelling should be investigated by further empirical studies. Obviously, bicycle tourism is a good choice as a leisure-time in China. The study shows that two incorporated constructs were the robustest influencing the bicyclers' decision process. Moreover, other variables include attitude, subjective norm, perceived behavioural control, positive anticipated emotion affecting tourists' desire, therefore, they influence the behavioural intentions. Asan (2021) discussed that bicycle travellers have increased as a convenience way for Turkey's tourists and this kind of transportation would enhance the sustainable development of tourism industry. In the context of consideration for selection the transportation, the bicycle travellers were affected by the comments of travel influencers. There are four categories which found by the study including informative effects, motivating effects, effects as a role model, and communal effects. Buning and Lamont (2021) use a systematic review and highlight that bicycle travel is a trend in tourism in Australia. However, there is a gap regarding the methodologies for investigating the bicycle travel in the current literature.

A renaissance of bicycling in transportation has operated in the urban areas in western Europe (Nilsson, 2019). Bicycle tourism has improved as sustainable pathways and helped to develop the tourism sector by advantaged infrastructure and modern facilities in cities. Bicycle tourism has been considered as an innovative trend in mobility systems between destinations. However, the supporting activities are required for enhancing this kind of travel. Lamont (2009) confirmed that bicycle tourism has been located a recreational, leisure and sporting activity in developed countries such as European region. The study developed an academic definition of bicycle tourism which helped exact measurements this definition in the practical environment. Sann et al. (2024) applied the theory of planned behaviour (TPB) for cycling tourism in Thailand. By analysing the sample including 395 bicycle tourists, the study confirmed that attitude towards behaviour, subjective norm, perceived behavioural control, behavioural intention, actual behaviour had effects on behavioural intention on the actual behaviour of cycling travellers. The finding provided the knowledge for understanding the intention formation of Thai cyclists in tourism. Xu et al. (2019) described the gradual rise in bicycle tourism and recreation in Hainan, China. The study investigated the relationship between cycling tourism and well-being of cycling people. The study suggested that the human–bicycle relationship contributed to enhancing the well-being of cycling tourists.

Khajehshahkoohi et al. (2022) shown that bicycle tourism was an emerging and growing type of tourism in Gorgan, Iran. By a survey having 846 tourists, the results implied that the perceived behavioural control and attitude had significant direct effect on behavioural intention of cycling in tourism. However, the infrastructure and facilities have an important role for developing bicycle travel in this country. Lee and Huang (2014) conducted a study for identifying bicycle tourism as a variable of attractiveness of Taiwan's destination. The study confirmed that comfortable climate, suitable infrastructure and facilities were important determinants of attractiveness for tourists as bicycle tourism destination. The finding suggested that bicycle tourism in Taiwan got a promising view in the future. Daniele et al. (2023) identified the effect of public transportation on supporting city bike tourism in Italy. The study concentrated on bike tourists and their decision making process based on the convenience of using public transportation to connect within-destination places during holidays. Therefore, the connection of public transportation with cycling system could reduce the cost of travelling. However, bicycle travel is good pathway for enhancing the green tourism development in destinations.

Research Methodology

This study has used multiple methods for modelling the bicycle travel concept. Firstly, a short literature review of theories is applied in order to conduct the concept. We employ the systematic review method for all previous theories and empirical studies regarding bicycle travel for summarising the current literature of this concept (Harris et al., 2014). Secondly, a graphing technique (Mann, 2006) will be employed for suggesting the effect of bicycle travel on green tourism development.

Results and Discussions

Based on the green tourism model (Furqan et al., 2010), we put bicycle tourism as an element for supporting the green tourism development. In detail, first, bicycle tourism helps to ensuring economic vitality and sustainability (in local economic vitality aspect), second, protecting, conserving, and enhancing nature and the physical (in environmental responsibility aspect), third, participation in, involvement with, nature, people, places and cultures (in experiential richness aspect), fourth, ensure the continued well-being of local or host cultures (in cultural diversity aspect). Therefore, bicycle travel is a sustainable strategy for improving the green tourism development (Fig. 1).

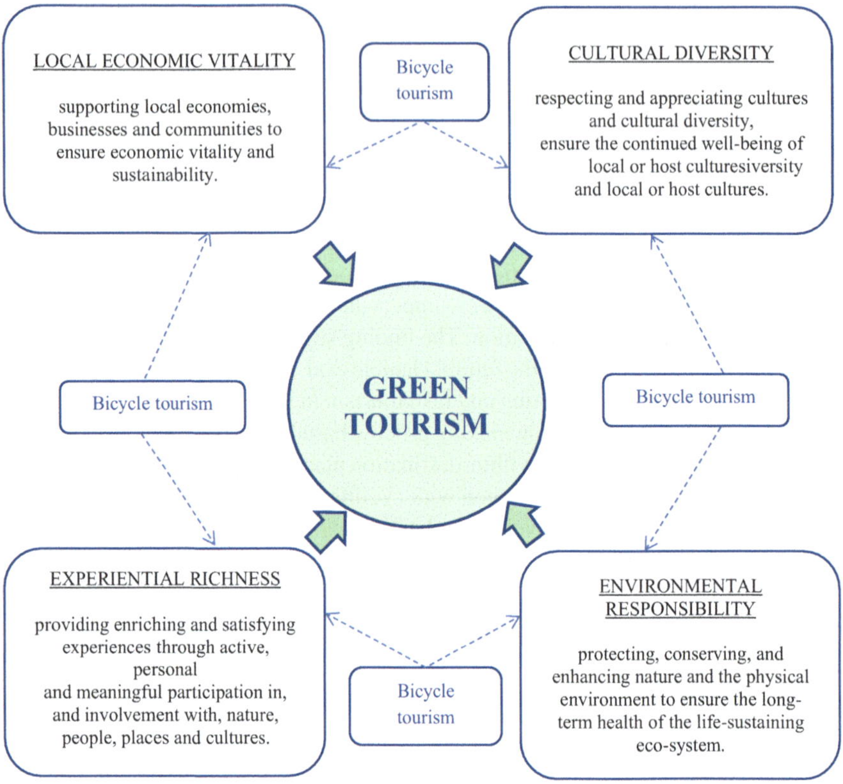

Fig. 1 Bicycle tourism in Green tourism model, authors developed from Furqan et al. (2010)

Bicycle Tourism and Environmental Responsibility

Because it does not use any natural energy, bicycle tourism is popularly recognized as a factor which contributes to green tourism development in urban, as well as rural areas and nature conservation (Hannam et al., 2021). When tourists arrive at destinations by bicycles, it means they bring net-zero emission of transportation. Therefore, air pollution is not a problem in destinations.

In many well known international destinations, air pollution is one of the greatest environmental risks to local people's health. Outdoor air pollution is a huge environmental health problem affecting people everywhere, especially in areas having a lot of people such as tourism destinations. The more cars and buses bring tourists to destinations, the more vehicle emissions put out there. The bicycle tourism helps reducing air pollution levels, then local people can reduce health problem, for example, the burden of lung cancer or chronic and acute respiratory diseases.

Besides, private cars and buses require park areas larger than bicycles. In the case of cars and buses, bigger land areas will be lost for construction of parking areas. And traffic jams can increase if all tourists come by private cars and buses. Obviously, bicycle tourism supports green tourism by protecting, conserving, and enhancing nature and the physical environment to ensure the long-term health of the life-sustaining eco-system in destinations.

Bicycle Tourism and Local Economic Vitality

Tourism is a "no-smoke" industry, it provides many benefits for local societies not only environment but also in economic aspect, supporting local economies, businesses and communities to ensure economic vitality and sustainability. Cycling tourists engage in many activities supporting the local economic development in destinations. Cycling allows freedom, flexible, and sightseeing potential in many styles of destinations that is not made possible by other transportations. There are some helpful ways that are through their demands and their consumptions of locally-based private services, for example, local foods (street foods), local accommodations (the home stay service) or local transportation (the private taxi service). On the other hand, bicycle travellers stay longer in regions and spend more than the average typical tourists. Because bicycle tourists spend on a variety of goods and services, they can help to reduce poverty and income inequality in destinations. The benefits of bicycle tourism on local economic vitality are clear and they are evidences for the connecting between bicycle tourism and green tourism development in destinations.

Bicycle Tourism and Cultural Diversity

Bicycle tourism is complementary with the slow tourism trend. The tourists using this kind of transportation always have shorter distance trips between destinations (Wang, 2023). Therefore, they make less environmental effect, spent more time in sightseeing cultural attractions. After the trips, they can share the experiences and knowledge for other people and expand the cultural diversity. Bicycle tourism supports cultural diversity in destinations and it helps to develop green tourism in these areas.

Bicycle Tourism and Experiential Richness

Bicycle tourism evolves the incorporation of cultural and culinary experiences. The travellers will achieve a higher level of experiential richness compared to the cases when they use other transportation. Tourists come and stay longer in destinations and their trips are maybe the same with adventures seamlessly blending with huge opportunities to explore local culture, enjoy heritage sites and regional cuisines. The trips of bicycle tourists have a holistic and enriching travel experience. Then the relationship between bicycle tourism and experiential richness helps to improve green tourism development in destinations.

Implications for Destinations in Enhancing Bicycle Travel and Green Tourism Development

In the light of the above, it is clear that destinations should include and give great importance to bicycle tourism as one of the efficient paths leading to development of green, sustainable tourism movement in a given area. In order to contribute to the growth of the said mode of travel and foster an increase in bicycle tourism, destinations need to take into account the above-discussed environmental, economic, cultural and experiential factors. In order to enhancing the bicycle travel and green tourism development in destinations, we provides some practical implications for policy-makers as well as business people in the tourism sector.

Despite the potential environmental impact entailed by the necessity to construct cycling infrastructure, destinations can greatly improve the sustainability of tourism by promoting bicycle tourism, and the negative impact of construction works can be decreased by integrating the cycling infrastructure with existing road and road-servicing network or pedestrian routes. Eventually, the investment will reduce the strain on both natural environment as well as the local community. It will also benefit the destination both by promoting sustainable tourism behaviour as well as promoting environmentally-responsible behaviour among the local population, especially in areas of high density and large population. As Veerappampalayam Easwaramoorthy et al. (2020) state, with urban mobility accounting for approximately 64% of the mileage travelled globally, some destinations opt for righteous scenarios, facilitating mobility at a fair cost and low carbon discharge; in Seoul, for example, bike rental system aims to alleviate traffic congestion, oil prices, and air pollution in the city while creating a healthier trend for the society and improving their quality of life. Similarly, in destinations which manifest particular environmental awareness, such as Scandinavia or Finland, the municipal networks smoothly blend with bicycle routes reaching out to suburban and even remote rural destinations, thus creating in a community a cycling culture. In Finnish Oulu, which is located just 100 km south of the Arctic Circle but still self-defines as the unofficial capital of year-round cycling, maintaining the cycling routes clear of snow or ice throughout the entire winter is of utmost priority. The routes are also well-lit and in many locations light signage is additionally projected onto their snowy surfaces. The fact that most routes are segregated and offer efficient shortcuts circumventing traffic and obstacles such as crossroads or traffic lights additionally helps to promote them as a viable green alternative to vehicular transport both among local residents as well as tourists.

The positive economic impact of bicycle tourism consists both in the fact that tourists using this mode of transport are able to reach and frequent locations unavailable easily to motorists as well as their tendency to consume attractions at a leisurely pace while spending more time on a series of short stop-overs. Since the ride is at least as important as being at a given destination, destinations can greatly profit, also economically, by making more attractions available and reachable by safe routes. This will help improve the tourist carrying capacity of the destination but most importantly distribute tourist income across communities fostering a more sustainable and balanced growth of the area. Sabri et al. (2019) conclude that the key recommendation lies in providing information and appropriate promotion of alternative routes, also prior to travel, however matched by correct preparation of safe, quality infrastructure for riding and parking, which proves essential to tourist perceptions of the trip. One needs to agree with this conclusion as it not only allows the destination to avoid post-purchase dissonance effect but also to induce the cycling tourists to visit more locations scattered across the area.

In order for destinations to harness bicycle tourism to support their cultural diversity, it is necessary to ensure the tourists are well-aware of the attractions and confidently navigate their ways along the network laid out for them. Sabri et al. (2019) advocate the dissemination of cycle maps for on and off-road infrastructure in order to increase the willingness of tourists to visit more locations. The maps certainly prove useful in terms of promoting specifically designed trails linking particular types of attractions curated according to a theme or criterion. However, similarly to bike computers or smartphone bike navigation applications, they are all not necessarily convenient in some weather conditions, require good lighting or occasional charging and overall distract the tourist from keeping eyes on the route or enjoying the surroundings. That is why proper signage and marking of trails is still essential in raising the awareness of cyclists as to the available options and possible directions. The aforementioned city of Oulu uses an extensive system of boards with distances, maps and numbered and coloured cycle routes both to help cyclists navigate but also to motivate users to explore and arouse their curiosity regarding other locations across the city and beyond its limits.

High experiential richness facilitated by the bicycle tourism context positively contributes to green tourism development. Destinations can improve this experience by facilitating the tourists' immersion in local culture as well as nature in a variety of environmentally-responsible ways. In the words of Ashraf et al. (2020), "one possible solution is to design memorable experiences for tourists that comply with their values", which means designing experiences in an engaging manner and with a broad range of on-site sensations and activities contributing to the conservation of local environment and community. A possible obstacle lies in the necessity to reconcile the disparate interests of cycling tourists and everyday commuters co-sharing a large portion of the cycling infrastructure. Their needs differ as commuters tend to favour speed and directness over experiencing sensations and gathering memories. However, these needs and wants are not mutually-exclusive and can be accommodated through careful planning, the more so since local commuters can often become tourists in their own home destinations, exploring attractions and enjoying the assets of the area in their free time. By promoting the growth of healthy lifestyle in a community as well as the use of alternative modes of transport, authorities and stakeholders alike can foster the creation of a cycling culture which will be welcoming towards bicycle tourists and will allow their better integration and involvement with the local nature and society.

Conclusions

Bicycle tourism has received much interest from people worldwide. There are more and more tourists engaging in this mode of travel in their trips. This paper aims to present a literature review of bicycle travel as a strategy for supporting green tourism development. Based on the theories and previous studies, we develop a research map which merges bicycle travel and green tourism development. The relationship between bicycle tourism has been discussed in terms of some aspects such as bicycle tourism and environmental responsibility, bicycle tourism and local economic vitality, bicycle tourism and cultural diversity, bicycle tourism and experiential richness. The connection between bicycle tourism and green tourism development is clear, with implications for enhancing the bicycle travel as important activities toward green tourism development. Implications for destinations for enhancing bicycle travel is provided. Suitable policies will promote the green tourism development in future.

Acknowledgements There is no conflict of interest between authors and other parties.

References

Asan, K. (2021). Measuring the impacts of travel influencers on bicycle travellers. *Current Issues in Tourism, 25*(6), 978–994.

Ashraf, M. S., Hou, F., Kim, W. G., Ahmad, W., & Ashraf, R. U. (2020). Modeling tourists' visiting intentions toward ecofriendly destinations: Implications for sustainable tourism operators. *Business Strategy and the Environment, 29*(1), 54–71.

Buning, R. J., & Lamont, M. (2021). Mountain bike tourism economic impacts: A critical analysis of academic and practitioner studies. *Tourism Economics, 27*(3), 500–509.

Daniele, C., Elena, M., & Evangelia, P. (2023). Assessing the role of public transportation to foster city bike tourism. The case of Italy. *Case Studies on Transport Policy, 12*, 101015.

den Hoed, W. (2019). Where everyday mobility meets tourism: An age-friendly perspective on cycling in the Netherlands and the UK. *Journal of Sustainable Tourism, 28*(2), 185–203.

Dziedzic, U., & Trojanowski, A. (2021). Utrzymanie i Wykorzystanie Wschodniego Szlaku Rowerowego Green Velo (Maintenance and Use of the Eastern Poland Cycling Trail Green Velo. In Polish). *Kontrola Państwowa, 3*, 141–156.

Fennell, D. (2008). *Ecotourism* (3rd ed.). Routledge.

Furqan, A., Mat Som, A. P., & Hussin, R. (2010). Promoting green tourism for future sustainability. *Theoretical and Empirical Researches in Urban Management, 8*(17), 64–74.

Han, H., Meng, B., & Kim, W. (2016). Emerging bicycle tourism and the theory of planned behavior. *Journal of Sustainable Tourism, 25*(2), 292–309.

Han, H., Lho, L. H., Al-Ansi, A., & Yu, J. (2020). Cycling tourism: A perspective article. *Tourism Review, 75*(1), 162–164.

Han, S., Ramkissoon, H., You, E., & Kim, M. J. (2023). Support of residents for sustainable tourism development in nature-based destinations: Applying theories of social exchange and bottom-up spillover. *Journal of Outdoor Recreation and Tourism, 43*, 100643.

Hannam, K., Butler, G., Witte, A., & Zuev, D. (2021). Tourist's mobilities: Walking, cycling, driving and waiting. *Tourist Studies, 21*(1), 57–69.

Harris, J. D., Quatman, C. E., Manring, M. M., Siston, R. A., & Flanigan, D. C. (2014). How to write a systematic review. *The American Journal of Sports Medicine, 42*(11), 2761–2768.

Khajehshahkoohi, M., Davoodi, S. R., & Shaaban, K. (2022). Factors affecting the behavioral intention of tourists on the use of bike sharing in tourism areas. *Research in Transportation Business & Management, 43*, 100742.

Lamont, M. (2009). Reinventing the wheel: A definitional discussion of bicycle tourism. *Journal of Sport & Tourism, 14*(1), 5–23.

Lamont, M. (2014). Introduction: Cycling and tourism. *Tourism Review International, 18*(1), 1–7.

Lee, C.-L., & Huang, H.-I. (2014). The attractiveness of Taiwan as a bicycle tourism destination: A supply-side approach. *Asia Pacific Journal of Tourism Research, 19*(3), 273–299.

Lee, J. H., & Iwasa, Y. (2020). Ecotourism development and the heterogeneity of tourists. *Theoretical Ecology, 13*, 371–383.

Lumsdon, L. (2000). Transport and tourism: Cycle tourism–a model for sustainable development? *Journal of Sustainable Tourism, 8*(5), 361–377.

Mann, P. S. (2006). *Introductory statistics* (6th ed.). Wiley.

Meng, B., & Han, H. (2016). Effect of environmental perceptions on bicycle travelers' decision-making process: Developing an extended model of goal-directed behavior. *Asia Pacific Journal of Tourism Research, 21*(11), 1184–1197.

Nilsson, J. H. (2019). Urban bicycle tourism: Path dependencies and innovation in Greater Copenhagen. *Journal of Sustainable Tourism, 27*(11), 1648–1662.

Ritchie, B. W. (1998). Bicycle tourism in the South Island of New Zealand: Planning and management issues. *Tourism Management, 19*(6), 567–582.

Roman, M., & Roman, M. (2014). Bicycle transport as an opportunity to develop urban tourism-Warsaw example. *Procedia-Social and Behavioral Sciences, 151*, 295–301.

Rus, K. A., Dezsi, Ș., & Ciascai, O. R. (2023). Transformative experiences in cycling tourism: A conceptual framework. *Sustainability, 15*(20), 15124.

Sabri, N. A. M., Anuar, F. I., Adib, M. A., & Azahar, N. (2019). A systematic review on bicycle tourism: Concept, issues, and future directions. *TEAM Journal of Hospitality and Tourism, 16*(1), 49–66.

Sann, R., Jansom, S., & Muennaburan, T. (2024). An extension of the theory of planned behaviour in Thailand cycling tourism: The mediating role of attractiveness of sustainable alternatives. *Leisure Studies, 43*(1), 47–61.

Sharpley, R. (2022). Tourism and development theory: Which way now? *Tourism Planning & Development, 19*(1), 1–12.

Veerappampalayam Easwaramoorthy, S., Park, J., & Cho, Y. (2020). Seoul bike trip duration prediction using data mining techniques. *IET Intelligent Transport Systems., 14*(11), 1351–1506.

Wang, J. (2023). The role of attributes of cycling tourism and enduring involvement in developing tourists' place attachment. *International Journal of Tourism Cities, 9*(4), 1063–1078.

Xu, H., Yuan, M., & Li., J. (2019). Well-being of cycling tourists in Hainan, China. *Sport in Society, 22*(8), 1449–1465.

Engagement Dynamics of Individuals with Intellectual Disabilities in Touristic Activities: A Case Study of Best Practices

Sandra Rebelo and Mafalda Patuleia

Abstract This article investigates the involvement of individuals with intellectual disabilities in tourism activities, focusing on inclusive dynamics to ensure effective participation. Despite the extensive literature on accessible tourism for physical disabilities, there is a notable gap regarding intellectual disabilities. This study aims to address this gap by analyzing barriers and facilitators of participation for individuals with intellectual disabilities, aligning with human rights principles to promote accessible tourism for all. The research utilizes a qualitative method, examining six national and six European best practices (from Spain, France, and Italy) in tourism activities. The findings highlight the predominant use of Augmentative and Alternative Communication (AAC) techniques, specifically pictograms, alongside sensory and therapeutic experiences. Notably, museums exhibit a higher adoption of inclusive strategies. The study underscores the importance of Universal Design and the need for awareness, training, and education for tourism professionals to enhance communication and attitudes toward individuals with intellectual disabilities. The analysis concludes that inclusive practices, although varied, demonstrate the potential for improved and satisfactory tourism experiences for individuals with intellectual disabilities, emphasizing a multifaceted approach to effective inclusion.

Keywords Intellectual disability · Accessible and inclusive tourism · Tourism activities

---

S. Rebelo
Institute of Geography and Spatial Planning of the University of Lisbon (IGOT-ULisboa), in collaboration with the School of Hospitality and Tourism of Estoril (ESHTE), Lisboa, Portugal

M. Patuleia (✉)
CETRAD| Intrepid Lab, Lusofona University, Lisboa, Portugal
e-mail: p4998@ulusofona.pt

A. L. Negrușa et al. (eds.), *Sustainable and Resilient Businesses in the Global Economy*, Springer Proceedings in Business and Economics,
https://doi.org/10.1007/978-3-031-75883-6_5

Introduction

This article explores the nature of the involvement of individuals with intellectual disabilities in tourism activities, considering the inclusive dynamics of these activities, aiming to understand and analyze practices adopted to ensure the effective participation of individuals with intellectual disabilities in different tourism contexts. The existing literature review indicates a lack of studies addressing inclusive tourism focusing on intellectual disabilities and other non-motor limitations, as research on this topic generally focuses on physical disabilities and mobility. This suggests the need to promote more accessible tourism activities for everyone, not excluding the needs of individuals with intellectual disabilities, as a priority parameter to be developed in line with the basic principles of human rights. Reflecting on tourism and human rights involves discussing its basic status as a human practice and its relationship with meeting human needs, including leisure, mobility, cultural appreciation, emotional growth, and identity expression (Ferraz & Serpa, 2021). Tourism activities should prioritize inclusive dynamics, providing effective participation for all.

The main objective of this article is to investigate and understand the dynamics of the involvement of individuals with intellectual disabilities in tourism activities, highlighting best practices that promote the effective inclusion of individuals with intellectual disabilities. This objective is subdivided into two secondary objectives: to analyze the barriers faced by individuals with intellectual disabilities when participating in tourism activities, identifying the main challenges and obstacles that negatively impact their participation, and to investigate the facilitating factors of participation and inclusion of individuals with intellectual disabilities in tourism activities, seeking to understand the practices that promote more effective and positive participation.

Thus, the organization of this text begins with the definition of disability and specifically intellectual disability. Next, an approach to inclusive and accessible tourism is presented, discussing indicators of inclusion of individuals with intellectual disabilities at the European level. The next step is to problematize the nature of the relationship between individuals with intellectual disabilities and tourism activities; this is a fundamental issue for discussing the relationship between tourism activities and individuals with intellectual disabilities. Moving on to the analysis method used a qualitative method, where through a case study, six good practices at the national level and six good practices in the following European Union countries, Spain, France, and Italy, were analyzed. Finally, a discussion of the results obtained will be conducted, as well as conclusions, aiming to debate and answer the research question of this study—To what extent can the dynamics of involvement of individuals with intellectual disabilities in tourism activities be understood and improved to promote a more accessible and satisfying tourism experience?

Literature Review

Definition of Disability and Intellectual Disability

According to Crow (2010), there is ambiguity in defining ***disability***, which encompasses three distinct perspectives: medical, functional, and social. From a social perspective, individuals are considered to live with a unique lifestyle, where disability is viewed as being created by environmental and attitudinal barriers. From a medical perspective, the World Health Organization (WHO) and the Directorate-General of Health (2004) define disability as a problem in bodily functions, including physiological functions of organic systems (including psychological functions) or body structure. The Assembly of the Portuguese Republic corroborates and adds in Article 2 of Law No. 38/2004, of August 18, that a person with a disability "is one who, due to loss or anomaly, congenital or acquired, of body functions or structures, including psychological functions, presents specific difficulties that, in conjunction with environmental factors, may limit or hinder activity and participation under conditions of equality with other persons" (Assembly of the Portuguese Republic, 2004; 5232). Thus, considering the different definitions presented above, the WHO (2021) adds that disability is categorized into four types: intellectual, physical/motor, hearing, and visual, as stated by the United Nations Convention, which states that persons with disabilities are those who have long-term physical, mental, intellectual, or sensory impairments that may hinder their full interaction and participation in society (Leonardi et al., 2006).

There is considerable confusion in defining ***intellectual disability***, often related to the variety of terms used to describe it. However, it has been characterized by visible limitations in intellectual functioning and behavior, affecting conceptual, social, and practical adaptive capabilities (Patterson & Pegg, 2009). According to the WHO (1992), intellectual disability refers to a condition of arrested or incomplete development of the mind, characterized especially by diminished cognitive, linguistic, motor, and social abilities, manifested primarily during an individual's developmental period. The American Association on Mental Retardation (AAMR) (1992) further adds that intellectual disability manifests before the age of eighteen and is characterized by significantly below-average intellectual functioning, limiting two or more of the following areas: communication, self-care, home life, social skills, community interaction, health, leisure, and work. The AAMR (2002) added in a more recent study that intellectual disability is lifelong and characterized by significant limitations both in intellect and in conceptual, social, and practical adaptive skills.

Inclusive and Accessible Tourism

The term ***inclusive tourism*** is often referred to as accessible tourism or even tourism for people with disabilities (Münch & Ulrich, 2011). However, according to Scheyvens and Riddulph (2017), accessible and inclusive tourism, despite their symbiosis, have slightly different definitions, considering that inclusive tourism, compared to accessible tourism, is more concerned with access to tourism consumption and production for all marginalized groups. Duarte and Umbelino (2016) further elaborate on this distinction, stating that accessible tourism focuses on access to infrastructure, while inclusive tourism focuses on specific service delivery needs. However, both definitions are interconnected and aim to make tourism available to everyone, regardless of individuals' characteristics, ensuring full access to all tourism services (Darcy & Dickson, 2009; Turismo de Portugal, 2014).

Indicators of Inclusion of People with Intellectual Disability at the European Level

In a study by Inclusion Europe (2023), an assessment of inclusion was conducted across 29 European countries for people with intellectual disabilities. As represented in Fig. 1, the level of inclusion practiced was divided into seven categories, where categories 1–2 represent the least inclusive countries and categories 9–10 are the most inclusive. It is observed that no country falls into the last category; however, Portugal, France, Belgium, the Netherlands, the Czech Republic, Lithuania, Ukraine, Romania, Bulgaria, Greece, Albania, Croatia, and Slovenia are in the category immediately above 3–4, representing a low level of inclusion of people with intellectual disabilities. On the other hand, Norway, Sweden, and Finland stand out for their high level of inclusion.

The Relationship Between Individuals with Intellectual Disabilities and Tourism Activities

The Participation of Individuals with Intellectual Disabilities in Tourism Activities

The participation of individuals with intellectual disabilities in tourism activities is influenced and limited by three factors: financial, emotional, and social means (Hunter-Jones, 2010; McKercher & Darcy, 2018). Recognizing these factors, it is essential to develop inclusive tourism products, services, and experiences through accessible information, the use of simple language, and the incorporation of augmentative and alternative communication (AAC) techniques. Thus, it is crucial to undertake awareness-raising actions, training, and education to reduce stigma against individuals with intellectual disabilities, cultivate empathy, and enable quality participation in tourism activities (McIntosh, 2020). The participation of individuals with intellectual disabilities in tourism activities can yield various positive outcomes at both intrapersonal and interpersonal levels, such as psychological and physical health benefits, personal development, and social inclusion (Innes et al., 2016), thereby

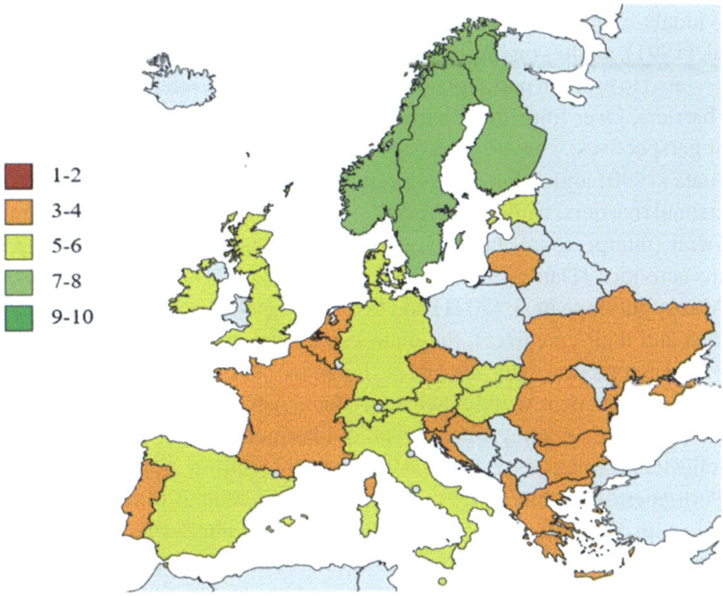

Fig. 1 Analysis of the level of inclusion in 29 European countries. *Source* Inclusion Europe (2023)

enhancing the quality of life and satisfaction of individuals with this type of disability (McCabe & Johnson, 2013). According to Gillovic and McIntosh (2020), there is a significant lack of understanding regarding the participation of individuals with disabilities in tourism, particularly, as highlighted in a study conducted in 2021, individuals with intellectual disabilities (Gillovic et al., 2021). In the European Union, in recent years, the market of tourists with intellectual disabilities has been growing. However, tourism companies and professional training agencies lack training on the needs of visitors with intellectual disabilities, often leading tourism professionals to adopt attitudes of overprotection or impatience, resulting in poor quality tourism services (Kronenberg & Szczecinska, 2013). Therefore, greater attention to social inclusion concerning the participation of individuals with intellectual disabilities in tourism activities becomes indispensable, as they have diverse consumption patterns and constitute a significant segment for the tourism market (Cohen et al., 2014).

Barriers of Tourism Activities for Individuals with Intellectual Disabilities

Social policies and service improvements to facilitate access and inclusion of people with disabilities in the community have made progress (Milner & Mirfin-Veitch, 2013); however, according to Backman (2010), individuals with intellectual disabilities continue to face barriers and challenges, which, as stated by Popay et al. (2006), increase social isolation, poor health, and lower life satisfaction. It is a fact that barriers to access and participation in tourism experiences and activities may be felt by many tourists without any limitations; however, these tend to be amplified for individuals with disabilities (McKercher & Darcy, 2018). According to Crawford et al. (1991), studies on tourism for people with disabilities define three types of barriers related to tourism participation, namely: structural, interpersonal, and intrapersonal barriers. Over time, these barriers for people with disabilities have taken on different perspectives. That is, older studies like those of Smith (1987) and Murray and Sproats (1990) argue that the most significant obstacles are related to intrinsic (intrapersonal) barriers, while more recent studies argue that structural barriers, especially social (interpersonal) barriers, are the ones that pose the greatest hindrance to leisure activities (Daniels et al., 2005; Daruwalla & Darcy, 2005; Innes et al., 2016). Mastrogiuseppe et al. (2021) add that specifically for people with intellectual disabilities and their caregivers, the main barriers encountered include: perception and physical interaction with resources; language and symbols; understanding of content; and engagement with knowledge. Mactavish et al. (2007) further adds that main barriers include reading signs and other interpretative information. Individuals with intellectual limitations require easily accessible information to strengthen their skills (Portuguese Republic, 2019). Therefore, it is essential to reduce the barriers that people with disabilities, especially individuals with intellectual disabilities, face, preventing them from participating in an equitable community (National Advisory Committee on Health & Disability, 2003). To reduce the identified barriers above, it is essential that service provision and tourism activities are conceived and planned based on the principles of Universal Design (Darcy et al., 2010).

Research Methodology

Research Problem

According to the UN (2021), worldwide, at least one billion people have some form of physical or intellectual disability, with studies such as Krahn et al. (2006) and Lennox (2007) specifically adding that individuals with intellectual disabilities represent about 2% of the population. In Portugal, according to Pordata (2001), there are approximately seventy thousand nine hundred ninety-four people with this type of disability. Many people with disabilities enjoy an active life filled with travel (Yau et al., 2004); however, tourism is developed for people without disabilities (Figueiredo et al., 2012). The lack of understanding within the tourism sector regarding the participation of people with disabilities, especially intellectual disabilities, and other non-motor disabilities (Gillovic et al., 2021), leads to a widespread misconception that people with these disabilities cannot or do not want to travel (Porto et al., 2019). Therefore, it is necessary to promote more accessible tourism activities for all, avoiding the marginalization of people with intellectual disabilities and their caregivers during their tourist trips (O'Grady et al., 2004), thus giving "voice" to these individuals (Correia et al., 2017), who continue to receive little attention in the field of tourism (Boxall et al., 2018), as well as in research, since studies on inclusive tourism generally focus on physical disabilities and mobility (Gillovic & McIntosh, (2020), with research on tourists with intellectual disabilities still in its infancy (Devile & Kastenholz, 2018) and even, as stated by Mactavish et al. (2007), neglected. Thus, this study aims, on the one hand, to conduct research based on equity and inclusion, promoting inclusive tourism strategies for individuals with intellectual disabilities, based on creating a dignified, fair, and inclusive industry (Richards et al., 2010), and on the other hand, to meet the research needs mentioned by Devile (2014), who recommends the research line of his doctoral dissertation—Dynamics of involvement of people with disabilities in tourism activities—adapted or replicated to other dimensions of disability, namely intellectual disability.

Study Objectives and Research Question

This article aims is to investigate and understand the dynamics of involvement of individuals with intellectual disabilities in tourism activities, highlighting best practices that promote effective inclusion of individuals with intellectual disabilities. This objective is subdivided into two secondary objectives: to analyze the barriers faced by individuals with intellectual disabilities when participating in tourism activities, identifying the main challenges and obstacles that negatively impact their participation, and to investigate the facilitating factors of participation and inclusion of individuals with intellectual disabilities in tourism activities, seeking to understand the practices that promote more effective and positive participation.

Therefore, the starting question was defined as follows: to what extent can the dynamics of involvement of individuals with intellectual disabilities in tourism activities be understood and improved to promote a more accessible and satisfying tourism experience?—which, considering the research conducted and the ideologies advocated in the theoretical framework of reference, aims to be answered through an analysis model based on the theoretical idea of inclusive tourism and people with intellectual disabilities, considering dimensions that encompass inclusion techniques for this group of people.

Research Method

The primary data of the study included a qualitative research method, where a case study was used to analyze and describe six best practices in tourism activities for people with intellectual disabilities, implemented in Portugal, and two best practices carried out in Spain, France, and Italy, respectively. Thus, emphasis was placed on national-level practices; however, to conduct a comparative study, tourist companies from European Union countries geographically closer to Portugal were analyzed. Case studies regarding inclusive tourism, are scarce, contributing to the lack of support in planning and inclusive practices by tourist companies (Darcy et al.,

2010). Therefore, this methodological procedure is intended to increase and improve research on inclusive tourism specifically for people with intellectual disabilities, analyzing some successful practices to meet the study by Corcoran et al. (2004), who argue that case studies can be used as a mechanism to transform and improve the practice under study. The collection of information for the development of this case study was based on the investigation of tourist projects that were based on Easy-to-read techniques; augmentative and alternative communication (AAC), such as easy-to-understand symbology—pictograms, and therapeutic sessions.

Easy-To-Read

Easy-to-read is a project of Inclusion Europe, which is part of the European Union, promoting a set of rules to facilitate and understand written information, ensuring the participation of people with intellectual disabilities. This method can be used by everyone and should have communication that must be read and understood by the elderly, people with poor mastery of the native language, and people with intellectual disabilities. When a text complies with the guidelines of easy-to-read, as mentioned in Table 1, its symbol is represented (Inclusion Europe, 2020, 2021).

Table 1 Guidelines a text must meet to be considered easy-to-read

Numbers	Numbered pages
	All pages numbered except the cover
	Numbering in the bottom right corner of each page
	Numbers used as digits, not words
Writing	Arial font, size 14 or larger
	Text aligned to the left
Photos	Images placed alongside the text for easy interpretation
	Images easy to understand
Titles	Clear and easily understandable titles
	Titles in bold
Sentences and words	Each sentence starts on a new line
	Sentences are no longer than one or two lines
	Words are simple
	Most sentences are positive
	Difficult words are always explained
	Avoid splitting a word into two lines
Information	Information easy to find
	Examples to explain information

Source Inclusion Europe (2020, 2021)

Augmentative and Alternative Communication (AAC)

To understand the concept of AAC, it is important to distinguish the terms speech, communication, and language, where communication is defined as intentional or unintentional, involving conventional or non-conventional signals, which may occur through speech or other means. Communication can be used in various ways, and its effectiveness and efficiency vary due to numerous individual and environmental factors (American Speech-Language-Hearing Association—ASHA, 2004). Von Tetzchener and Martinsen (2000) define alternative communication as communication other than speech that is used in a face-to-face context, while augmentative communication is a type of complementary communication that aims to promote and support speech and ensure there is a means of communication. According to Ferreira et al. (2020), AAC techniques are associated with Augmentative and Alternative Communication Systems (AAC), which are an "integrated set of techniques, supports, strategies, and skills that a person with communication difficulties uses to communicate" (p. 21), and Assistive Technologies (AT).

Pictograms

Pictograms are characterized using easy-to-understand symbology (Turismo de Portugal, 2014), which helps people with intellectual disabilities understand content. A pictogram is a symbol that, through illustration, represents a concept or object (The International Federation of Library Associations & Institutions, 2012).

Results and Discussions

The analysis of the results of the present study considered the identification of best practices for people with intellectual disabilities, through the implementation of easy-to-read methods, AAC, namely pictograms, and sensory and therapeutic sessions. During the present investigation, it was noted that most websites for different tourist activities focus their inclusive offerings on people with motor limitations. In Table 2, some of the best practices for people with intellectual disabilities implemented especially at the national level and generally in Spain, France, and Italy are observed.

The study is centered on Portugal, and analyzing more Portuguese cities allows for a deeper understanding of the national context. This is crucial for identifying best practices, specific challenges, and particular needs of individuals with intellectual disabilities within the country. Furthermore, inclusion policies and support structures vary significantly between countries. By focusing more on Portugal, the study can offer recommendations and solutions that are more specific and applicable to the Portuguese context, considering the legislative, cultural, and social particularities of

Table 2 Successful practices, considering easy-to-read techniques; CAA; sensory and therapeutic experiences

Portugal		
Country/city	Local	Practices
Batalha	Museum of the Batalha Municipality Community	Utilizes AAC strategies. Texts in easy-to-read format, games and exploration activities, texts with pictorial support, tactile experiences—replicas, models, objects to touch, guided tours (MCCB, 2020)
Lisbon	MAAT	Open to All Program. Integrated program of creative actions aimed at the population with disabilities and/or mental illness, through the creation of spaces for dialogue, experimentation, and sharing, namely therapeutic sessions (MAAT, 2023)
Lisbon	Calouste Gulbenkian Museum	It is part of the I Am—Inclusive and Accessible Museums project, where specifically for intellectual disabilities, it created tools for sensory exploration in exhibition rooms in an inclusive and accessible manner. Additionally, it developed an activity called "curator for a day" through a virtual classroom (Calouste Gulbenkian Museum, 2023)
Mafra	Quinta dos Machados (Accommodation)	Sensory Path composed of five stations inspired by the five senses distributed throughout the route (Quinta dos Machados, 2023). This accommodation is featured on the website of Portuguese Tourism, All for All Program, as one of the good practices of inclusive tourism in Portugal
Porto	Caves Calém	– Informative and directional signage adapted for people with special needs, such as the existence of informative pictogram pieces (Turismo de Portugal, 2022)
Mira	Tourist Information Office and Ethnographic Museum of Mira	Communicational Accessibility: Use of simple language and writing; System of pictographic symbols; Professional training in specialized and inclusive customer service (Turismo de Portugal, 2022); This building is featured on the website of Portuguese Tourism, All for All Program, as one of the good practices of inclusive tourism in Portugal

(continued)

Table 2 (continued)

Portugal		
Country/city	Local	Practices
Other countries		
Spain Madrid	Ico Museum	Easy-to-read method, where the informative brochure is placed at the museum reception (available on the museum's website). During the exhibition, the easy-to-read logo is represented in various locations (Fundación ICO, 2013)
Spain Sevilha	Virtual itineraries in the city and territory	Creating a 360º video allows for a heritage route, through a system of easy-to-read and pictograms (University of Seville, 2019)
France—La Croix en Touraine	Association Bibliochouette. La Croix en Touraine	The Ministry of Culture, the Association of Librarians of France (ABF), the National Agency for Combating Illiteracy (ANLCI), and the Interregional Federation of Books and Speaking implement the easy-to-read reading offerings in French territory (Ministère de la Culture, 2023)
France Paris	Paris Zoological Garden	Quality of facilities for visitors with psychiatric and learning disabilities. Guided tours and various illustrations for people with intellectual disabilities, providing a special way to learn about animals and vegetation. Pictograms widely used to facilitate the orientation of people with intellectual disabilities (Parc Zoologique Paris, n.d.)
Italy Rome	National Roman Museum	It is part of the Museu per tutti project, where the participation of people with intellectual disabilities is facilitated through sharing experiences with other participants with the same disability. Independent or small group guided tours (Museu per tutti, n.d.)
Italy—Milan	Museum BeGo	Adaptation of rooms and respective exhibitions accessible to people with intellectual disabilities, acquiring the "Museum for All" project developed by Associazione l'Abilità Onlus in cooperation with the De Agostini Foundation. It has a trained team to welcome people with intellectual disabilities. Ensures correct relational and communication methods through provided materials (easy-to-read sheets, social stories, and AAC cards). This museum complies with the easy-read method (Museo for all, 2018)

the country. Therefore, the emphasis on more Portuguese cities compared to other cities in the European Union ensures that the study provides valuable and practical insights for promoting truly inclusive tourism in Portugal, with a direct and immediate impact on the national reality.

Considering the information obtained in Table 2, it was possible to observe that, among all tourist activities, museums have shown a greater concern for people with intellectual disabilities, adopting specific strategies, although one accommodation, one tourist office, one virtual tourist itinerary, and one library were identified.

Analysis of Practices Accessible to People with Intellectual Disabilities Analyzing national practices, six tourist activities were observed, where it was found that the most commonly used methods are sensory and therapeutic experiences and the AAC method. On the other hand, it was found that none of the tourist activities used the easy-to-read symbol on their official websites, although the museum in Batalha and the Tourist Office and Ethnographic Museum of Mira mentioned easy-to-interpret writing and consequently easy reading.

As observed in Table 2, of the four countries mentioned, the easy-to-read technique is used in Spain, France, and Italy (one company from each), not being represented in the Portuguese context, which focuses on AAC techniques and sensory and therapeutic sessions. The AAC technique was used by companies from all four observed countries, showing a greater emphasis (six companies), represented specifically through pictograms. Sensory and therapeutic sessions were observed in companies from only two countries, Portugal, and Italy, thus demonstrating a greater representativeness in the Portuguese context.

Overall, considering the twelve companies observed, AAC and specifically pictograms are the most used practices, followed by sensory and therapeutic sessions, and finally the Easy-to-read technique (Table 3).

Table 3 Techniques used per tourist activity

Inclusive techniques	Tourism activities
Easy-to-read	SP—Ico Museum FR—Association Bibliochouette. La Croix en Touraine IT—MuBeGo Museum
AAC	PT—Batalha Municipal Community Museum PT—Calém Caves PT—Mira Tourist Office and Ethnographic Museum SP—Virtual itineraries in the city (Seville) and in the territory FR—Paris Zoological Garden IT—BeGo Museum
Sensory and therapeutic sessions	PT—MAAT; PT—Calouste Gulbenkian Museum PT—Quinta dos Machados (Accommodation) IT—National Roman Museum

Final Reflection and Conclusions

This study highlights the importance of inclusive tourism with an emphasis on intellectual disabilities, aiming to investigate and understand the dynamics of involvement of people with intellectual disabilities in tourist activities, highlighting best practices that promote effective inclusion of people with intellectual disabilities. Specifically, it aims to analyze the barriers faced by people with intellectual disabilities when participating in tourist activities, as well as investigate the facilitating factors of participation and inclusion. The results obtained demonstrate coherence between the conceptions of the theoretical framework and the practices implemented in the analyzed tourist activities.

Thus, to meet the mentioned objectives, through the theoretical framework, it was possible to verify that currently the major barriers faced by people with intellectual disabilities in tourist activities are related to structural and interpersonal barriers, making the implementation of Universal Design extremely important. Furthermore, the importance of training, awareness, and education actions for tourism professionals is emphasized to develop communication and attitudes towards people with intellectual disabilities.

Regarding the developed case study, through the analysis of successful best practices in Portugal and in the European Union countries, Spain, France, and Italy, various approaches were highlighted to make tourism more accessible to people with intellectual disabilities, such as easy-to-read techniques, AAC, specifically pictograms, and sensory and therapeutic experiences.

When compared, these countries reveal differences regarding the adopted approaches; however, overall, the AAC technique is highlighted as the most used technique, followed by sensory and therapeutic experiences, and finally the easy-to-read technique, which is not represented in Portugal. However, it is extremely important to mention that the sample of analyzed practices is not fully representative and may vary in different regions and companies in these countries.

The identification of these best practices, based on inclusive methods for people with intellectual disabilities, highlights the need for multifaceted approaches to promote effective inclusion. Furthermore, museums emerge as spaces that frequently adopt inclusive strategies.

Thus, through the theoretical framework and the identification of the above-mentioned best practices, it was possible to answer the defined starting question—How can the dynamics of involvement of people with intellectual disabilities in tourist activities be understood and improved to promote a more accessible and satisfactory tourist experience?

It was concluded that these dynamics of involvement can be understood and improved with the adoption of inclusive techniques, such as easy-to-read methods, AAC techniques, such as pictograms, and sensory and therapeutic experiences.

In summary, it is intended that this analysis of best practices provides valuable insights for promoting effective inclusion and contributes to the development of strategies to make tourism an accessible and satisfactory experience for people with intellectual disabilities.

References

AAMR. (1992). *Mental retardation: Definition, classification, and systems of supports* (9th ed.). American Association on Mental Retardation.

AMMR. (2002). *Definition, classification, and systems of support* (10th ed.). American Association on Mental Retardation.

ASLHA. (2004). Roles and responsibilities of speech-language pathologists with respect to augmentative and alternative communication: technical report (Technical Report). Retrieved December 13, 2023 from www.asha.org/policy.

Assembly of Portuguese Republic. (2004). Law No 38/2004-The general principles of the legal framework for the prevention, habilitation, rehabilitation, and participation of persons with disabilities are defined.

Backman, C. L. (2010). Occupational balance and well-being. *Introduction to occupation: The art and science of living*, 231–249.

Boxall, K., Nyanjom, J., & Slaven, J. (2018). Disability, hospitality, and the new sharing economy. *International Journal of Contemporary Hospitality Management, 30*(1), 539–556.

Cohen, S. A., Prayag, G., & Moital, M. (2014). Consumer behaviour in tourism: Concepts, influences, and opportunities. *Current Issues in Tourism, 17*(10), 872–909.

Corcoran, P. B., Walker, K., E., & Wals, A. E. (2004). Case studies, make-your-case studies, and case stories: A critique of case-study methodology in sustainability in higher education. *Environmental Education Research, 10*(1), 7–21.

Correia, R. A., Seabra-Santos, M. J., Campos Pinto, P., & Brown, I. (2017). Giving voice to persons with intellectual disabilities about family quality of life. *Journal of Policy and Practice in Intellectual Disability, 14*(1), 59–67.

Crawford, D. W., Jackson, E. L., & Godbey, G. (1991). A hierarchical model of leisure constraints. *Leisure Sciences, 13*, 309–320.

Crow, L. (2010). Including all of our lives: Renewing the social model of disability. In *Equality, participation and inclusion* (Vol. 1, pp. 136–152). Routledge.

Daniels, M., Rodgers, E., & Wiggins, B. (2005). Travel tales: An interpretive analysis of constraints and negotiations to pleasure travel as experienced by persons with physical disabilities. *Tourism Management, 26*(6), 919–930.

Darcy, S., & Dickson, T. (2009). A whole-of-life approach to tourism: The case for accessible tourism experiences. *Journal of Hospitality and Tourism Management, 16*(1), 32–44.

Darcy, S., Cameron, B., & Pegg, S. (2010). Accessible tourism and sustainability: A discussion and case study. *Journal of Sustainable Tourism, 18*(4), 515–537.

Daruwalla, P., & Darcy, S. (2005). Personal and societal attitudes to disability. *Annals of Tourism Research, 32*(3), 549–570.

Devile, E., & Kastenholz, E. (2018). Accessible tourism experiences: The voice of people with visual disabilities. *Journal of Policy Research in Tourism, Leisure and Events, 10*(3), 1–21.

Devile, E. (2014). *Dynamics of involvement of people with disabilities in tourism activities*. Doctoral thesis, University of Aveiro.

Duarte, A., & Umbelino, J. (2016). *Accessible and inclusive tourism "Conceptual and methodological framework". Bredait Project*. Building a Regional Network for the Development of Accessible and Inclusive Tourism.

Ferraz, J., & Serpa, S. (2021). Tourism and human rights: A complex relationship. *Academic Journal of Interdisciplinary Studies, 10*(2), 75.

Ferreira, M., Ponte, M., & Azevedo, L. (2020). *Curricular innovation in the implementation of alternative means of communication in children with severe neuromotor disabilities*. SNRIPD.

Figueiredo, E., Eusébio, C., & Kastenholz, E. (2012). How diverse are tourists with disabilities? A pilot study on accessible leisure tourism experiences in Portugal. *International Journal of Tourism Research, 14*, 531–550.

Fundación ICO. (2013). Your visit to the ICO museum. Retrieved December 18, 2023 from https://www.fundacionico.es/web/guest/arte/inclusion-y-accesibilidad.

Gillovic, B., & McIntosh, A. (2020). Accessibility and inclusive tourism development: Current state and future agenda. *Sustainability, 12*(22), 9722.

Gillovic, B., McIntosh, A., Cockburn-Wootten, C., & Darcy, S. (2021). Experiences of tourists with intellectual disabilities: A phenomenological approach. *Journal of Hospitality and Tourism Management, 48*, 155–162.

Hunter-Jones, P. (2010). Consumer vulnerability and exclusion: A study of carers in the tourism marketplace. *Journal of Marketing Management, 27*(1–2), 165–180.

Inclusion Europe. (2020). Easy-to-read. Easy-to-read information is easier to understand. Retrieved December 18, 2023 from https://www.inclusion-europe.eu/easy-to-read/.

Inlcusion Europe. (2021). Annual report 2021. Retrieved December 18, 2023 from https://www.inclusion-europe.eu/wp-content/uploads/2022/06/IE_annual_report_2021.pdf.

Inclusion Europe. (2023). Inclusion indicators 2023. In *Rights and inclusion of people with intellectual disabilities in 29 European vountries.*

Innes, A., Page, S. J., & Cutler, C. (2016). Barriers to leisure participation for people with dementia and their carers: An exploratory analysis of carer and people with dementia's experiences. *Dementia, 15*(6), 1643–1665.

Krahn, G. L., Hammond, L., & Turner, A. (2006). *A Cascade of Disparities: Health and Health Care Access for People with Intellectual Disabilities, 12*(1), 70–82. https://doi.org/10.1002/mrdd.20098.

Kronenberg, M., & Szczecinska, J. (2013). *Overall desk research on situation of tourist guides in context of working with people with intellectual disabilities and/or learning difficulties*. T-GUIDE Tourist Guides for People with Intellectual Disability and/or Learning Difficulties in Europe.

Lennox, N., Bain, C., Rey-Conde, T., Purdie, D., Bush, R., & Pandeya, N. (2007). Effects of a comprehensive health assessment programme for Australian adults with intellectual disability: A cluster randomized trial. *International Journal of Epidemiology, 36*(1), 139–146.

Leonardi, M., Bickenbach, J., Ustun, T. B., Kostanjsek, N., & Chatterji, S. (2006). The definition of disability: What is in a name? *The Lancet, 368*(9543), 1219–1221.

MAAT. (2023). MAAT-museum open to all: Specific educational needs program. Retrieved December 18, 2023 from https://maat.pt/pt/event/maat-museu-aberto-todos-programa-necessidades-educativas-especificas.

Mactavish, J., McKay, K. J., Iwasaki, Y., & Betteridge, D. (2007). Family caregivers of individuals with intellectual disability: Perspectives on life quality and the role of vacations. *Journal of Leisure Research, 39*(1), 127–155.

Mastrogiuseppe, M., Span, S., & Bortolotti, E. (2021). Improving accessibility to cultural heritage for people with intellectual disabilities: A tool for observing the obstacles and facilitators for the access to knowledge. *Alter, 15*(2), 113–123.

McCabe, S., & Johnson, S. (2013). The happiness factor in tourism: Subjective wellbeing and social tourism. *Annals of Tourism Research, 41*, 42–65.

MCCB. (2020). Inclusive solutions. Retrieved December 17, 2023 from https://museubatalha.com/pt/recursos-de-acessibilidade-solucoes-inclusivas.

McIntosh, A. (2020). The hidden side of travel: Epilepsy and tourism. *Annals of Tourism Research, 81*, 1–10.

McKercher, B., & Darcy, S. (2018). Re-conceptualizing barriers to travel by people with disabilities. *Tourism Management Perspectives, 26*, 59–66.

Milner, P., & Mirfin-Veitch, B. (2013). I am here: The Article 19 project plain English version. Dunedin: Donald Beasley Institute.

Ministère de la Culture. (2023). Easy-to-read reading in libraries and mediation. Retrieved December 18, 2023 from https://www.culture.gouv.fr/en/Thematic/Book-and-Reading/Public-libraries/Easy-to-read.

Münch, H., & Ulrich, R. (2011). Inclusive tourism: Accessible tourism. *The long tail of tourism: Holiday niches and their impact on mainstream tourism* (pp. 159–169). Gabler.

Murray, M., & Sproats, J. (1990). The disabled traveler: Tourism and disability in Australia. *Journal of Tourism Studies, 1*(1), 9–14.

Museo for all. (2018). Museo for everyone. Retrieved December 20, 2023 from http://www.museobenozzogozzoli.it/en_GB/museo-per-tutti.html.

Museo per Tutti. (n.d.). What is Museo per Tutti. Retrieved December 19, 2023 from https://www.museopertutti.org/en/che-cose-museo-per-tutti-eng/.

Museu Calouste Gulebenkian. (2023). Cultural participation of young people with specific educational needs. The role of the museum. Retrieved December 18, 2023 from https://gulbenkian.pt/museu/read-watch-listen/a-participacao-cultural-dos-jovens-com-necessidades-educativas-especificas-o-papel-do-museu/.

National Advisory Committee on Health & Disability. (2003). To have an 'Ordinary' life: Community membership for adults with an intellectual disability. A report to the Minister of Health and the Minister for Disability Issues.

O'Grady, A., Pleasence, P., Balmer., N. J., Buck, A., & Genn, H. (2004). Disability social exclusion and the consequential experience of justiciable problems. *Disability & Society, 19*, 259–271.

Parc Zoologique Paris (n.d). Mentally impaired visitors. Retrieved December 19, 2023 from https://www.parczoologiquedeparis.fr/en/mentally-impaired-visitors-2051.

Patterson, I., & Pegg, S. (2009). Serious leisure and people with intellectual disabilities: Benefits and opportunities. *Leisure Studies, 28*(4), 387–402.

Popay, J., Enoch, E., Johnston, H., & Rispel, L. (2006). *Social exclusion knowledge network (SEKN): Scoping of SEKN and proposed approach.*

PORDATA. (2001). Resident population with disabilities according to the Census: total and by type of disability (2001). Retrieved November 19, 2023 from https://www.pordata.pt/portugal/populacao+residente+com+deficiencia+segundo+os+censos+total+e+por+tipo+de+deficiencia+(2001)-1239.

Porto, N., Rucci, A. C., Darcy, S., Garbero, N., & Almond, B. (2019). Critical elements in accessible tourism for destination competitiveness and comparison: Principal component analysis from Oceania and South America. *Tourism Management, 75.*

Portuguese Republic. (2019). *Practical guide to the rights of people with disabilities in Portugal.* Simplex.

Quinta dos Machados. (2023). Experiences. Here you feel nature. Retrieved December 18, 2023 from https://www.quintamachados.com/experiencias.html.

Richards, V., Pritchard, A., & Morgan, N. J. (2010). (Re)envisioning tourism and visual impairment. *Annals of Tourism Research, 37*(4), 1097–1116.

Scheyvens, R., & Biddulph, R. (2017). Inclusive tourism development. *Geography of Tourism,* 1–21. (Unapei. (2019). Falc quality reference. The reference of easy to read and understand services. Editing-Transcription. France.)

Smith, R. W. (1987). Leisure of disabled tourists: Barriers to participation. *Annals of Tourism Research, 14*(3), 376–389.

Von Tetzchner, S., Martinsen, H. (2000). Introduction to augmentative and alternative communication. *Special Education Collection, Porto, Porto Editora.*

The International Federation of Library Associations and institutions. (2012). *Guidelines for easy-to-read materials.*

Netherlands Turismo de Portugal (2014). *Good practices guides for accessibility-active tourism.*

Turismo de Portugal. (2022). Caves Calém. Retrieved December 18, 2023 from https://business.turismodeportugal.pt/pt/crescer/Historias_Sucesso/Paginas/caves-calem.aspx.

Universidade de Sevilha. (2019). Virtual itineraries in the city (Seville) and in the territory. Retrieved December 18, 2023 from https://ra.sav.us.es/index.php/realidad-virtual/2-itinerario-por-sevilla.

WHO & Direção Geral da Saúde. (2004). Classificação internacional de funcionalidade, incapacidade e saúde: CIF. Lisboa. OMS (2004). CIF–Classificação Internacional de Funcionalidade, Incapacidade e Saúde (International Classification of Functionality, Incapacity and Health). Lisbon: Direcção Geral da Saúde.

WHO. (1992). International classification of diseases (ICD-10) Geneva, from which the WHO's international classification of impairments. Disabilities and Handicaps: A Manual of Classification relating to the Consequences of Disease (1980) was developed.

WHO. (2021). 1 Billion people with disabilities among those most impacted by the pandemic. Retrieved December 11, 2023 from https://news.un.org/pt/story/2021/12/1772482.

Yau, M., McKercher, B., & Packer, T. (2004). Traveling with a disability: More than an access issue. *Annals of Tourism Research, 31*(4), 946–960.

Improving Tourist Interpretation in the Castelo Neighborhood in Lamego: Evaluation and Proposals for the Enhancement of Cultural and Historical Heritage

Isabel Vieira, Didiana Fernandes, Liliana Índio, and Ana Rodrigues

Abstract This study aims to understand visitors' perspectives on the importance of interpretation in enhancing heritage offerings and the tourist experience. We chose the Castelo Neighbourhood in Lamego, Portugal, as our case study. Methodologically, a quantitative approach was employed, using a questionnaire administered to 106 visitors, with descriptive analysis performed using the SPSS software. The results provided a deeper understanding of visitors in terms of their sociodemographic profile and shed light on which interpretation methods they value most. Based on the collected data, we offered suggestions for improvements to enhance the facilities in the Castelo Neighborhood. This knowledge is essential for aligning offerings with demand in a tourism development context that aims to be sustainable.

Keywords Interpretation · Visitors · Sustainable tourism · Heritage · Historic quarters

I. Vieira (✉) · D. Fernandes
CIDEI-Centre for Studies in Education and Innovation, Polytechnic Institute of Viseu, Lamego School of Technology and Management, Lamego, Portugal
e-mail: ivieira@estgl.ipv.pt

D. Fernandes
e-mail: dfernandes@estgl.ipv.pt

L. Índio
Polytechnic Institute of Viseu, Lamego School of Technology and Management, Lamego, Portugal

A. Rodrigues
Centre for Transdisciplinary Studies for Development–CETRAD, University of Trás-Os-Montes and Alto Douro (UTAD), Vila Real, Portugal
e-mail: anarodri@utad.pt

A. L. Negrușa et al. (eds.), *Sustainable and Resilient Businesses in the Global Economy*, Springer Proceedings in Business and Economics,
https://doi.org/10.1007/978-3-031-75883-6_6

Introduction

Over the past 30 years, heritage tourism has experienced significant development, with growing demands from tourists for destinations that offer excellent quality attracting more visitors (García-Delgado et al., 2020). It is within this context that interpretation applied to heritage resources emerges, aiding in the formation of tourist products by addressing various issues in heritage management and tourism (Hart-Robertson, 2015). Interpretation is increasingly recognized as an essential tool for understanding, promoting, and preserving the history of each location, enhancing visitor interaction and appreciation (Hams, 2016; Miranda, 2001).

Since empirical studies on this subject are scarce, the present study seeks to bridge that gap and enhance the existing knowledge. Thus, for the purpose of studying how interpretation helps in reviving historical neighbourhoods, we selected Lamego's Castelo Neighbourhood as a case in point. Oftentimes, our definition of a theme depends upon where we value it such as Lamego city. The city has a rich cultural background and historical heritage, with the Castelo Neighbourhood being at its heart. This research seeks to determine how tourist interpretation can help improve amenities in this area so that it becomes more meaningful to people living here. In this regard, the following key questions were posed:

1. How can tourist interpretation be utilized as a fundamental tool for the presentation and enhancement of the heritage in the Castle Neighborhood?
2. What interpretative instruments and means can be developed to improve the tourist experience and the interaction between visitors and the monuments in the Castle Neighborhood?
3. In what way can tourist interpretation contribute to the improvement of facilities and services in the Castle Neighborhood, promoting a self-sustaining and attractive environment for visitors?

To better understand the perceptions and needs of the visitors to the Castle Neighborhood in Lamego, several hypotheses were formulated to guide the investigation. Firstly, it was assumed that on present days there are no sufficient interpretative and directional signs installed in the territory to fulfill the audiences' expectations. Secondly, it was presupposed that there is an insufficiency of the detailed informational-orientation materials in different languages, and it forms an unsatisfactory tourist experience. Furthermore, it was noted that such aspects as the lack of specialized guides and interactive games decrease the chances for visitors to engage with the heritage. To achieve this, we highlight the following:

a. Characterize the socio-demographic profile of visitors.
b. Investigate the sources of information used.
c. Evaluate the visitor experience during their visit.
d. Understand how visitors assess heritage, management, and the tourist experience in the Castelo Neighbourhood.
e. Identify the most valued forms of interpretation.
f. Develop a draft interpretation plan for the facilities in the Castelo Neighbourhood.

Therefore, this article is structured into five sections, in addition to this introduction. The first section presents the theoretical framework, elucidating concepts related to tourism, tourists, historic cities, historic neighbourhoods, cultural heritage interpretation, and new technologies applied to interpretation. The second section provides a brief characterization of Lamego, its Castelo Neighbourhood, existing facilities, and available interpretation methods. The third section outlines the methodology used for data collection and analysis in this study, along with the predefined objectives. The fourth section presents and analyses the study's case data, proposing improvements aimed at enhancing the Castelo Neighbourhood's facilities through the application of new interpretation methods. Finally, the conclusions, limitations, and suggestions for future research will be discussed.

Tourism, Historic Cities, and Heritage Interpretation

Tourism and Historic Cities

Cities emerge as a culturally and physically constructed product of their societies, dating back to ancient times. In antiquity, cities evolved from small settlements and, over time, culminated in some cases in large metropolises (Salgueiro, 2006).

From a general perspective, we understand that cities convey messages through their dynamics. Some cities have an urban culture that supersedes their historical identity. A city is an urbanised area that differentiates itself from towns and other urban spaces through various criteria: population, infrastructure, organisations, and services, among others. A city is characterised by the particular lifestyle of its inhabitants, urbanisation, and the concentration of economic activities (Salgueiro, 2006). On the other hand, historic cities are linked to the past of a people or culture. They are among the main tourist attractions today, encompassing valuable cultural heritage and serving as an important tourist resource. Cultural heritage (both material and intangible) forms the essence of the cultural identity of historic cities and similarly underpins their primary appeal to tourists (Richards, 2016).

In Europe, tourist demand and visits to historic cities represent one of the most attractive options for travel in the contemporary world (Calle-Vaquero 2006).

The main allure of historic cities is not the individual elements considered in isolation (monumental heritage, urban fabric, museums, religious and cultural events, lifestyles, etc.) but rather the unique character of each historic city, materialised in a singular and unrepeatable sense of place that emerges not from a mere collection of cultural elements for tourist consumption. The development of tourism has been essential in adapting various historic cities for tourism purposes. Most historic cities are increasingly leveraging their historical heritage to create tourist products. This heritage is a vital resource for tourism development. On the one hand, it fosters the creation and development of cultural tourism itineraries, while on the other, it enhances and revitalises pre-existing offerings. Indeed, tourism is a fundamental instrument for rehabilitating heritage in historic cities (Almeida et al., 2017).

The growing interest in cultural heritage, be it material or intangible, and its protection and preservation stems from an increasingly informed and demanding society (Benhamou, 2013). According to this premise, a whole process emerges that leads to various communication and management tools that contribute to meeting these citizens' demands. It is in this context that Interpretation techniques emerge, facilitating a symbiosis between heritage and its enjoyment (Ham, 2016).

Interpretation of Cultural Heritage

Tilden (2009) describes interpretation as an educational activity aimed at establishing relationships and revealing meanings through the use of original objects, through direct contact with the resource through illustrative means, not confined merely to providing information.

Interpretation is an act of communication. One could say that interpreting is the art of communicating messages and emotions from a text, a musical score, a work of art, an environment, or a cultural expression (Murta & Albano, 2005). Aldridge, cited by Miranda, defined "interpretation as the art of explaining the place of Man in his environment, to increase the visitor's awareness of the importance of this interaction, arousing in him the desire to contribute to the conservation of the environment" (Albrige, 1981, p. 33, cited by Miranda, 2001).

Although these authors have defined the concept of interpretation in different ways, it has existed since the first travellers recorded their impressions in diaries and the first commemorative monuments were erected in public spaces (Ham, 2016). Indeed, the personal desire to speak about one's place, historical past, recent events, or even current dilemmas, as well as the collection of personal evidence of history, are fundamental elements for the interpretation and valorisation process. As we know, in all cultures, personal memories and past experiences, faded photographs, and records of family events are milestones of individual lives and are of great value to the interpretation process. Interpreting is more than just informing; it is revealing meanings, evoking emotions, stimulating curiosity, entertaining, and inspiring new attitudes, and providing a memorable and high-quality experience for visitors.

However, the rush to promote cities and regions as tourist destinations, investing in the restoration, interpretation, and presentation of their heritage, as well as diversifying their attractions, led the Northern Hemisphere to make mistakes, which were felt only in the 1990s. Overall, the main mistake was excluding the local population from the planning process without effective creative collaboration between preservation and tourism bodies, resulting in the creation of fossilised historical attractions for quick tourist consumption, lacking the vitality of local community cultural practices (Murta & Goodey, 2005). Therefore, at the ECO92 conference in Rio de Janeiro, the issue of sustainability gained conceptual and concrete forms, leading to new conceptions of how to develop tourism and its related activities sustainably.

Due to the rapid growth of cultural and nature tourism in various forms, interpretation, as a communication tool, has become even more crucial for instilling preservation attitudes in host communities, visitors, and tourism entrepreneurs. The problems and perspectives of interpretation practice have been debated by several authors and recorded in a book edited by Uzell and Ballantyne, published in 1998.

In June 1999, the International Heritage Interpretation Conference in Europe was held at Bournemouth University, England, aimed at evaluating current interpretation practice, as well as trends and prospects for the future. However, there were some complications at the end of the millennium; humanity was not winning the fight for preservation, no matter how great the effort was to classify historic ensembles and restore them or to delimit natural conservation and protection areas, development and behavioural standards have been more destructive than a preservationist. On the other hand, the only way to reverse this situation is to bring preservation to the hearts and minds of people, dealing with their emotions to develop in them an affection for the environment that surrounds them, for public space, nature, and historic sites (Murta & Goodey, 2005, pp. 16–17).

The interpretation medium requires extensive knowledge of the interpretation process, but must also consider theoretical and philosophical issues about communication and basic tools to convey a message. In the late 1960s, most knowledge about interpretation practice was gained more through experience than through research. Today, interpretation means are applied in the development of communication services for spaces with heritage value (Ham, 2009).

All interpretive services must be equipped according to certain conditions, both for visitors and for the resource. Regarding visitor variables, we can mention the time available, their attractive interests, and the money they intend to spend. As for the resource, we can underline access to its features and its degree of difficulty, available energy sources, and the site's carrying capacity.

New technologies emerged in the Information Revolution, and gradually developed since the second half of the 1970s and especially in the 1990s. Most of these new technologies are characterised by agility, facilitating communication through digitisation and network transmission.

The use of various technologies in presentation and interpretation has been occurring for decades. The first use of voice-guided navigation technology in cultural spaces was in 1952 by the Stedelijk Museum **in Amsterdam, when it provided historical context and exhibition context to visitors through portable guides (Loic, 2008). In this way, several documents for the protection of international and national cultural heritage highlight the importance of technological integration in heritage interpretation at different levels. Since the 1990s, technologies have been widely used in the context of cultural heritage, which has become a support for interpretation and heritage interpretation (Rahaman & Kiang, 2017). In particular, the integration of reality creation technologies, such as virtual reality (VR), augmented reality (AR), and mixed reality (MR), which provide various content formats (video, 3D animation, multimedia, etc.) and abundant interaction methods (voice interaction, touch-screen interaction, etc.), has changed the way heritage history is displayed and explained from expressed images, textual descriptions, or other limited aesthetic settings. These reality-creation technologies have increased the authenticity, experience, and interactivity of presentations (Griffin et al., 2017; Guerra et al., 2015) and enable new interactions and connections between heritage and the public (Vieira et al., 2022). It is important to note that new technologies in interpretation should not be considered merely information distribution platforms or an alternative to traditional means of presenting information about the surrounding area (museums, monuments, etc.), but should be considered a way to connect and engage visitors with objects, collections, and exhibitions (Liu, 2020).

Characterisation of the study area

Lamego is an ancient city. During the reign of the Visigothic King Sisebuto from 612 to 621, a coin was minted, that attests to the commercial, cultural, and historical importance of Lamego. This was followed by alternating periods of peace and war. Sometimes Christian, sometimes Arab, it changed hands several times until Fernando Magno de Leão definitively conquered the city in 1057 (Pinto, 2022). In 1835, the districts were instituted by a reform of Mouzinho da Silveira; at this time, Lamego was initially planned as the district capital, but in the same year, the district headquarters were moved to the city of Viseu due to its more central position (Monterey, n.d., pp. 63–88).

Lamego is the seat of the diocese, and it is the only Portuguese diocese that does not correspond to a district capital; therefore, religious monuments abound, among which we highlight the Cathedral See, the Church of São Pedro de Balsemão, and the Sanctuary of Nossa Senhora dos Remédios, which also gives its name to the annual pilgrimage, with the main day being September 8, which is a municipal holiday (Monterey, n.d., pp. 63–88).

However, two important events entirely changed the economic and social life of this region: the conquest of Granada, which expelled the last Moors from the Peninsula, taking with them commercial vigour, and the discovery of the maritime route to India, weakening all trade in Lamego.

The economy of this city is essentially filled by the following activities: services, commerce, and agriculture, with the latter representing a source of wealth, mainly originating from the wine sector, as the municipality is integrated into the Douro Demarcated Region. In addition to the production of the famous port wine, there is also a clear focus on the Controlled Designation of Origin (DOC) and the production of sparkling wines, which are known as an important promotional showcase nationally and internationally (CML, 2019).

The municipality of Lamego is located in the district of Viseu, covering an area of 165.42 km^2. The city is the second-largest city in the district, with 24,348 inhabitants (Censos, 2021).

The Castle Neighbourhood

The General Directorate of Buildings and National Monuments classified Lamego Castle and the Cistern (in 1910) along with its walls (in 1951), and established the perimeter of its protected zone (see Fig. 1). In 1964 and 1960, efforts were made to restore the designated properties within that area. This included the restoration of the Castle tower and walls.

Nowadays, the Castle Neighbourhood constitutes an urban ensemble delimited by the ancient city walls (medieval core), serving as its principal historical nucleus. Given its unique geographical location, forming a festoon line, the walled nucleus of the castle acts as the constructed boundary of the upper zone of the city to the

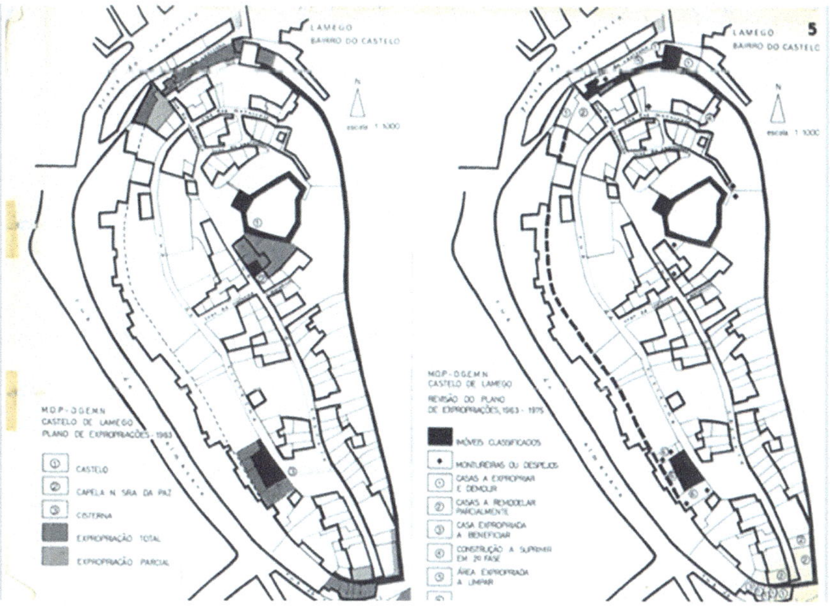

Fig. 1 Protection zone. *Source* DGEMN (1978)

East. For this reason, and due to some buildings in ruins or an advanced state of degradation, efforts began in 2007 to recover and requalify the Castle Neighbourhood through the programme—"Valorisation and Urban Integration of the Historic Centre of Lamego", which was a submission to the City Policy Programme/Partnerships for Urban Regeneration, in which the Castle Neighbourhood was included.

Methodology

This article was developed based on research employing a quantitative approach, specifically a descriptive and cross-sectional study. This is a commonly used research method in studies of this nature, aiming to obtain a profile of visitors and their opinions on the interpretation of the site (Lynch et al., 2011). To achieve the objectives of this study and test the formulated hypotheses, an approach was chosen that allows for an objective and systematic analysis of the collected data. This method was selected for several reasons:

- Descriptive and Cross-Sectional: Descriptive and cross-sectional research is suitable for obtaining a detailed profile of visitors to the Castelo Neighbourhood in Lamego, capturing their perceptions and experiences at a specific point. This is crucial for identifying areas for improvement in heritage interpretation.

- Structured Questionnaires: The use of structured questionnaires allowed for the collection of data from a broad and diverse sample of visitors. This method is effective for obtaining standardized responses and facilitating statistical analysis and comparison of results. The questionnaires included closed-ended questions and Likert scales to assess visitor satisfaction and identify critical areas.
- Random Sampling: Random sampling was employed to ensure that the sample was representative of the visitor population, including different ages, nationalities, and levels of tourist experience. This increases the external validity of the results, allowing for more accurate generalizations.
- Statistical Analysis: The collected data were analyzed using statistical techniques such as frequency analysis, means, and standard deviation to identify significant trends and correlations. This quantitative approach provides a solid foundation for the proposed recommendations, ensuring they are based on empirical evidence.
- Data Triangulation: In addition to the questionnaires, secondary sources such as tourism reports and previous studies were analyzed to complement and validate the primary data. Data triangulation increases the robustness of the results and provides a more comprehensive view of the issue.

The choice of these methods was based on the need to obtain a detailed and accurate understanding of visitors' perceptions, allowing for the identification of specific areas that require improvement. The quantitative approach, combined with rigorous statistical analysis, ensures that the conclusions and recommendations are based on concrete and reliable data.

The questionnaire design was preceded by consulting other questionnaires conducted in the same context (Cesar, 2016; Sol et al., 2017), with the main source being the study conducted by Duarte (2016). Thus, this study served as the basis for formulating some questions, always seeking to adapt them to the specific case under analysis.

Regarding the method of application of this research, we opted to conduct personal interviews supported by a structured questionnaire directed at visitors to the historic city of Lamego. The administration of the questionnaire involved the participation of four members of the tourism office and the Castelo de Lamego team. The questionnaire structure is divided into groups of questions:

Group A: Data on visitor profile in terms of consumed product: The first part of the questionnaire consists of a set of questions aimed at understanding the visitor profile regarding the consumed product, covering dimensions such as familiarity with the destination; information sources used to plan the visit; reasons for visiting the region and country; the importance of factors affecting the choice of the city.

Group B: Data on the evaluation of Castle Neighbourhood facilities and monument quality, conditions of various sites for receiving visitors, and aspects that should be improved (signage, brochures, maps, guides), management, and tourist experience.

Group C: General data about the respondent: In the final section, we aimed to obtain personal information regarding the socio-economic characterization of the respondents (gender, age, nationality, country of residence, education level, occupation, and monthly income).

The surveys were conducted in person, with an average duration of 15 min per applied questionnaire, and were available in four languages: Portuguese, English, French, and Spanish. From the obtained responses, some issues were identified. Accordingly, we made corrections, mainly regarding the format and sequence of questions, and based on the criticisms and suggestions received, some questions were reformulated, some were eliminated, and others were added. Changes were also made to the answer options for multiple-choice questions and the scales used.

Primary data collection was conducted in September, October, November, and December 2022, with 120 questionnaires applied to visitors based on a non-probabilistic convenience sample, of which 106 were validated. The validated sample size and the selected sampling method were due to budget limitations for this study. To validate the sample size, we considered the two criteria outlined by Hill and Hill (2012): the "minimum effort criterion", whereby the sample is sized to be as large as possible within the limits of available resources, and the "Trumb rule criterion", which stipulates that the minimum sample size should allow for adequate statistical analysis according to the objectives.

Regarding the analysis of the surveys, conducted in January 2023, the Statistical Package for the Social Sciences (SPSS V26) and Microsoft Excel 2016 statistical program were used for data elaboration and/or presentation of results.

Finally, the collected data were analyzed quantitatively to identify patterns and trends in the visitors' responses. The analysis focused on evaluating visitor satisfaction with the tourist interpretation and identifying specific areas that require improvement.

Analysis and Discussion of Results

Considering the methodological approach developed in the scope of the empirical study, we aim, at this point, to analyse the responses to understand the profile of visitors to Lamego Castle Neighbourhood in terms of the consumed product and their evaluation of heritage, the assessment of the Castle Neighbourhood facilities and monument quality, the conditions of various locations to accommodate visitors, and aspects that should be improved, as well as the socio-demographic characterization and associated travel type.

Socio-demographic characterization of surveyed visitors

Table 1 presents the main information used to define the socio-demographic profile of the visitor.

Some interdependencies can be drawn from the data in Table 1 regarding the socio-demographic profile of visitors to the Castelo Neighborhood:

Table 1 Visitor profile

Characterization of the respondent	
Gender	Men (57.0%); Women (41.5%)
Age brackets	18–25 years old (15.1%); 26–35 years old (21.7%); 36–45 years old (28.3%); 56–65 years old (5.7%); >65 years old (2.8%)
Nationality	Portugal (28.3%): Lisbon and Porto represent (43.5%) of domestic demand Foreign (28.9%): Spain (21.7%), USA and France (14.2% each), Brazil (10.4%), United Kingdom (3.8%), Italy (2.8%)
Qualifications	Basic education (3.7%); Secondary education (28.3%); Higher education (50.9%)
Employment status and occupation	Unemployed (9.4%); Student (14.2%); Retired (1.9%); Homemaker (1.9%); Self-employed (15.1%); Employed by others (40.6%)

Source Own elaboration

- The majority of visitors are in the age groups of 26–35 years (28.3%) and 36–45 years (24.5%), which are also the ages when people are generally most active in the labor market. This is corroborated by the fact that 40.6% of visitors are employed by others and 15.1% are self-employed.
- The high percentage of visitors with higher education (50.9%) may be related to the diversity of nationalities, especially from countries such as the USA, France, and the United Kingdom, where higher education is quite common. This suggests that the Castelo Neighborhood attracts an educated and international audience.
- Foreign visitors (28.9%) from countries such as Spain, the USA, France, and Brazil may be more concentrated in the age groups of 26–35 and 36–45 years, which are typical ages for international travel, whether for leisure or work.
- The majority of visitors employed by others (40.6%) and self-employed (15.1%) may include both national and international visitors, suggesting that the Castelo Neighborhood is an attractive destination for professionals who may be visiting for work or leisure.
- The age group of 26–35 years, which is the most represented, is also the age when many people complete higher education and begin their careers. This may explain the high percentage of visitors with higher education (50.9%).

These correlations help to better understand the profile of visitors and can guide marketing strategies and improvements in the facilities and services of the Castelo Neighborhood to better meet the needs and expectations of this diverse audience.

Visitor Profile: Motivation and Information Sources

One of the specific objectives defined was to understand the visitor profile concerning the consumed product. In the research instrument, we formulated a series of questions based on a literature review, which seemed relevant for understanding this topic (see Table 2).

The analysis of the data reveals several key insights into the visitor profile, sources of information, motivations, and perceptions regarding the Castelo Neighbourhood in Lamego.

- Visitor Travel Frequency: The majority of visitors travel 2–3 times per year, followed by those who travel more than 3 times per year. This suggests that the Castelo Neighbourhood attracts both frequent travellers and those who travel occasionally.
- Sources of Information: Visitors use various sources to plan their trips, with travel agencies, the internet, and personal recommendations being the most prominent. Specifically, 25.5% of visitors received information through travel agencies or tour operators, which contradicts the trend identified by the Institute for Tourism Planning and Development (IPDT, 2022) states that 90% of tourists seek information on websites. This discrepancy may reflect the specificities of tourism demand in Lamego, greatly influenced by river tourism in the Douro region. The internet accounts for 17% of information sources, gaining increasing importance, while direct communication through "word of mouth" from acquaintances and family ranks third at 16.0%. Social media platforms such as Facebook, Instagram, and Twitter account for 10.4%, being one of the most influential sources for tourists today. Repeat visitors also play a significant role at 10.4%. Lastly, TripAdvisor awards account for 5.7%, and posters, flyers, and billboards account for 2.8%.
- Motivations for Visiting: Religious tourism and cultural heritage are the main motivations for visiting, followed by the exploration of new places and local traditions. The majority of respondents (43.4%) associate Lamego with being a "city of churches," followed by 28.3% who mentioned gastronomic quality, 11.3% mentioned Castelo Neighbourhood, and 16% mentioned other attributes such as

Table 2 Main motivations and sources of information

Frequency of travel	Less than twice a year (17%) 2 to 3 times a year (45.3%) More than 3 times a year (35.8%)
Sources of information used in trip planning	Indication by Travel Agencies and Tour Operators (25.5%); Internet (17%); Recommendation from friends and family (16%); Previous visit (10.4%); Social Media (10.4%); Awards—TripAdvisor Certificates of Excellence (5.7%); Posters, Flyers, and Outdoor advertisements (2.8%)
Reason for the visit	Religious Tourism (15.5%); Cultural Heritage (14.9%); Exploring a new place (14.9%); Local traditions (12.2%); Culture (12.8%); Escaping routine (11.5%); Gastronomy (10.8%); Visiting family/friends (7.4%)
Attributes associated with the city of Lamego	Gastronomic Quality (28.3%); City of Churches (43.4%); Castle Neighbourhood (11.3%); Others (16%)

Source Own elaboration

Historic City, Romaria dos Remédios, and Culture. Lamego's image is strongly linked to its ecclesiastical heritage and its renowned gastronomy, mainly related to smoked meats.

Understanding the intricate correlations derived from visitor data is crucial for enhancing the overall tourist experience in the Castelo Neighbourhood. By analysing patterns and trends, we can identify key factors that influence visitor satisfaction and preferences. For instance, the high percentage of visitors in the 26–35 age group, many of whom have completed higher education and are beginning their careers, correlates with the significant number of visitors with higher education (50.9%). This suggests that local cuisine is a strong attraction for visitors, as evidenced by the perception of the city as a destination of high gastronomic quality (28.3%). The motivation to explore new places (14.9%) is often associated with cultural interest (12.8%). Visitors seeking to escape routine (11.5%) may also be looking for enriching cultural experiences. Although TripAdvisor awards (5.7%) and external advertising (2.8%) have a smaller percentage, they still play a role in attracting visitors interested in quality and recognition, which may be related to the pursuit of high-quality gastronomic and cultural experiences. Regarding the evaluation of BC facilities and equipment conservation, and the conditions of various places to welcome visitors.

Heritage keeps the memory of other peoples or civilizations alive. Castelo Neighbourhood is one such example. Preserving its heritage has allowed its legacy and history to reach our people. Individuals were asked about what they consider to be the advantages of preserving cultural/archaeological heritage. Upon analysing Table 3, it was found that 25.5% of visitors believe that preserving cultural/archaeological heritage supports the tourism development of a region, 20.8% believe that it preserves local identity and tradition, 19.8% responded that it is important for understanding local history, 17.9% consider it important for tourist attraction, and 16% did not respond to this question.

Regarding the most visited monuments, it is noted that the Castle of Lamego and the Cistern are the most visited monuments by 29.1% of respondents. The Archaeological Site of Porta dos Figos (27.8%) is the second most visited location, which is a very positive result considering this facility opened less than 5 years ago. Lastly, 13.9% visited three facilities within the Castelo Neighbourhood. The valorization and preservation of heritage are essential for the development of tourism activity. Only by preserving heritage can we preserve memories of the past and its values. In this regard, from the total number of individuals, 45.3% responded that the preservation and valorization work carried out in the Castelo Neighbourhood facilities is good; 30.2% consider this work satisfactory; 2.8% consider it very good, and 8.5% consider it unsatisfactory.

Respondents were asked if they thought the Castelo Neighbourhood facilities are adapted to receive various types of visitors, considering different age groups—children, youth, adults, and seniors. Some visitors responded affirmatively (45.9%), but 50.9% did not share the same opinion, while 5% did not respond to the question. Regarding the question: "Is the information provided easily understandable by diverse audiences?" 61.3% of respondents believe that the information available

Table 3 Evaluation of the facilities in Castelo Neighbourhood (BC)

Advantages of preserving cultural/archaeological heritage	Conservation facilitates the tourism development of a region (25.5%), preserves local identity (20.8%), important for understanding the local past (19.8%), important for increasing tourist attraction (17.9%), no response (16%)
Most visited monuments in BC	Castelo de Lamego (29.1%), Cisterna (29.1%), Núcleo Arqueológico da Porta dos Figos (27.8%), All (13.9%)
Preservation and valorization of heritage	Unsatisfactory (8.5%), Satisfactory (30.2%), Good (45.3%), and Very good (2.8%)
Cultural facilities in BC adapted to different age groups (children, youth, adults, seniors)	Yes (45.9%), No (50.9%), No response (5%)
Information provided in BC is adapted to different audiences	Yes (61.3%), No (29.2%), Don't know, no response (9.4%)
Staff at BC facilities are well-trained	Yes (70.8%); No (13.2%)
The information available at BC facilities	Yes (12.3%); No (64.2%)
Conditions that allow interaction between visitors and the monument	Yes (26.4%); No (55.7%); No response (17.9%)

Source Own elaboration

at the Castelo Neighbourhood facilities is easily understood by the diverse public, 29.2% responded negatively, and 9.4% did not respond to this question. Concerning the question: "Do you think the staff at the places you visited are properly trained to provide assistance/information?" 70.8% of respondents said the staff is adequately trained, while 13.2% of respondents believe otherwise, and 16% did not respond to this question.

Analysing the question "Do you consider the information at the Castelo Neighbourhood facilities sufficient?" it is observed that 64.2% of respondents believe that the information about the facilities is not sufficient, 23.6% consider it sufficient, and 12.3% did not respond. We can conclude that the information provided at the Castelo Neighbourhood facilities is not sufficient. Regarding the question "Do you think sufficient conditions have been created to establish interaction between the visitor and the monument?" 26.4% of respondents answered affirmatively, while 55.7% answered negatively due to the reasons mentioned above, and 17.9% did not respond to this question.

There are significant areas that need improvement. Investing in signage, information in multiple languages, staff training, and interactive activities can substantially enhance the visitor experience and better meet their needs and expectations. Tourist interpretation helps visitors develop a deep awareness and understanding of the place they are visiting and also provides appropriate information services for various types of tourists/visitors. However, this is not happening at the Castelo Neighbourhood facilities. By analysing Table 4, we see that 32.1% of respondents rated the tourist interpretation at the Castelo Neighbourhood facilities as unsatisfactory, and 19.8% rated it as poor, meaning that more than half of the visitors (51.9%) are not satisfied with the tourist interpretation available at the Castelo Neighbourhood facilities, which causes visitors to lose interest in these monuments. On the other hand, some visitors responded that the tourist interpretation at these facilities is satisfactory (26.4%) and good (4.7%).

Regarding aspects related to interpretation that should be improved in Castelo Neighbourhood, 18.7% of respondents consider interpretive signs and posters to be the most important aspect to be improved, followed by directional signage (signs and markers) at 17.9%. This is followed by brochures available in more languages (14.3%); brochures with the history of Castelo Neighbourhood facilities (13.3%), and publications (13%). Additionally, visitors believe that automatic audiovisuals (11%) should be improved; interactive games (7.2%), and specialized guides (4.6%). These data indicate that the overall dissatisfaction with interpretive means is strongly linked to the need to improve signage and the availability of information in multiple languages.

The analysis of Table 5, which cross-references the questions "Which aspects of the Castelo Neighborhood facilities should be improved" and "Do you consider the information available in the Castelo Neighborhood to be sufficient," reveals important correlations.

The majority of visitors who consider the information insufficient also highlight the need for improvements in various areas. For example, 50 visitors who find the information insufficient point out the need to improve interpretive signs and directional signage. Additionally, 35 visitors who consider the information insufficient mention the need to improve publications, such as brochures, guides, and maps, while 42 visitors suggest the inclusion of brochures with the history of the facilities.

Other aspects that need improvement include interactive games (21 visitors), specialized guides (16 visitors), automatic audiovisuals (32 visitors), and the availability of brochures in multiple languages (43 visitors). In contrast, among those who consider the information sufficient, the suggested improvements are less frequent,

Table 4 Aspects that should be improved in the interpretation of heritage in the Castelo Neighbourhood

Evaluation of personal and non-personal interpretative means Areas for improvement in BC interpretation	Poor (19.8%); Unsatisfactory (32.1%); Satisfactory (26.4%); Good (4.7%)
Evaluation of personal and non-personal interpretative means	Interpretative signs and posters (18.7%); directional signage (signs and marks) (17.9%); Brochures available in more languages (14.3%); brochures with the history of Castelo Neighbourhood facilities (13.3%); Publications (13%); Automatic audiovisuals (11%); interactive games (7.2%); Specialized guides (4.6%)

Source Own elaboration

Table 5 Cross-tabulation "Which aspects of the Castelo Neighbourhood facilities should be improved"/"Do you consider the information available in Castelo Neighbourhood to be sufficient"

		Interpretative signs and posters	Signage (Signs and Marks)	Publications (brochures, guides, and maps)	Brochures with equipment history	Interactive games	Specialized guides	Automatic audiovisuals	Brochures available in multiple languages	Total
Do you consider the information available in the Castle Neighborhood to be sufficient?	Yes	19	17	14	9	5	1	10	11	24
	No	50	50	35	42	21	16	32	43	65
Total		69	67	49	51	26	17	42	54	89

Source Own elaboration

with 19 visitors mentioning interpretive signs, 17 directional signages, 14 publications, 9 brochures with the history of the facilities, 5 interactive games, 1 specialized guide, 10 automatic audiovisuals, and 11 brochures in multiple languages.

This analysis indicates that the perception of insufficient information is strongly associated with the need for improvements in various interpretive and signage areas.

Proposals

After collecting data and evaluating interpretive methods, we understand that regarding tourist interpretation at Castelo Neighbourhood facilities, there is a long way to go. It is crucial to develop an interpretation plan to contribute to a better qualification of tourist interpretation.

An interpretation plan for enhancing a village, city, or region is important for a tourist destination as it provides a strategy for action for municipalities, the community, and entrepreneurs, among other stakeholders (Murta & Albano, 2005).

An interpretive plan aims to establish a network of discoveries and enjoyment for residents, visitors, and tourists, enhancing the potential for tourism and cultural projects. It should also include alternatives for facility management and promotion, as well as a strategy for monitoring, evaluation, and updating, to sustain quality standards in preserving and promoting self-sustainable environments and attractions.

Moreover, the plan should consider the entire socio-environmental context, where customer service aspects such as parking, security, signage, restaurants, and restrooms cannot be neglected, as they can leave visitors with a negative impression of the place, even if the interpretation itself is excellent. Additionally, it guides the cleaning and discovery of original facades; the harmony of signage and design of signs and plaques compatible with the "spirit" of the place; and urban trails that "decode" the city and informative signs that enhance visitors' environmental perception (Biesek, 2004).

Given the study conducted through the analysis of primary and secondary sources, we suggest the following proposals aimed at improving the interpretation of Castelo Neighbourhood:

Improvements to signposting and interpretive information
External and internal signalling:
Improve external signage: Implement external signage on the streets and in the respective facilities of the Castelo neighborhood, according to established standards
Improve internal signage: Implement descriptions in each facility in several languages (French, Spanish, and English) and develop a project so that this signage complies with current standards and is applied effectively and efficiently
Publications and leaflets:
Develop brochures: Present the history of the neighborhood and its monuments
Making publications available in multiple languages: This will not only fulfil a need expressed by visitors but will also enrich the tourist experience
Interactive technologies and resources
Automatic Audiovisuals and Interactive Games:

(continued)

(continued)

Improve automatic audiovisual resources: Introducing interactive games can make the visit more engaging, especially for younger audiences. These resources can provide information in a dynamic and accessible way
Specialized Guides:
The presence of specialised guides: This can significantly improve the visitor experience by offering detailed information and answering questions in real-time. Investing in ongoing training for these guides is crucial
Accessibility and inclusion
Accessibility for People with Reduced Mobility:
Install removable ramps: Other facilities for people with reduced mobility can make the Castelo neighbourhood more accessible to all visitors, regardless of their physical abilities
Information in multiple languages:
Making information available in multiple languages: This includes not only leaflets and publications but also signage and digital resources
Infrastructure and services
Hygiene and sanitation:
Improving hygiene and sanitation conditions: Including the installation of public toilets can significantly improve the visitor experience. Keeping the streets and facilities clean is also essential
Interaction between visitors and monuments:
Create conditions for greater interaction: Practical activities and workshops can increase visitor interest and satisfaction
Marketing and promotion strategies
Marketing campaigns:
Develop marketing campaigns: Highlight improvements and new attractions to attract more visitors. This includes promoting cultural and religious events, as well as local gastronomy
Continuous Monitoring and Evaluation:
Implement a continuous monitoring and evaluation strategy: Ensure that improvements meet visitors' needs and identify new areas for improvement

Source Own elaboration

Upon analysing the data, it is clear that several areas require attention to improve the visitor experience in the Castelo Neighborhood. Investing in signage, interpretive resources, accessibility, infrastructure, and marketing can increase visitor satisfaction and promote the sustainable development of tourism in the region.

To expand on the management implications of the proposals presented in the study, it is essential to adopt a broader, comparative approach. A comparative analysis between Lamego Castle and other historic cities of similar size and characteristics can reveal commonalities and differences that influence the implementation of the proposals. Case studies of other cities that have implemented similar improvements to their heritage interpretation practices can provide practical examples and lessons learned (Benhamou, 2013; Calle Vaquero, 2006).

Flexibility and adaptation of proposals are crucial for different cultural, geographical, and socio-economic contexts. For instance, the need for multilingual signage can vary depending on the linguistic diversity of visitors in different locations (Ham, 2016). Additionally, integrating new technologies such as augmented and virtual reality can significantly enhance the visitor experience (Mantzou et al., 2023).

The benefits and impacts of improvements in heritage interpretation should also be analyzed. Enhanced interpretation can positively impact the local economy by attracting more tourists and extending their stays (Hart-Robertson, 2015). Furthermore, effective interpretation practices can contribute to the sustainability of tourism by helping to preserve cultural and natural heritage (Murta & Albano, 2005).

For the implementation and management of the proposals, a strategic planning model that other cities can follow is recommended. Capacity building and continuous training of staff involved in heritage management and visitor care are fundamental to the success of the initiatives (Hill & Hill, 2012). Community involvement is another vital aspect. Engaging the local community in improving heritage interpretation can increase local support and ensure that initiatives are culturally sensitive and sustainable (Lynch et al., 2011). Methods for collecting and analyzing visitor feedback on an ongoing basis are also essential to allow for constant adjustments and improvements (Miranda, 2001). Additionally, public policies that encourage the improvement of heritage interpretation in historic cities should be recommended. Partnerships between local governments, tourism organizations, academic institutions, and the private sector can facilitate the implementation of proposals.

Finally, the monitoring and evaluation of initiatives are crucial. Defining clear indicators to evaluate the success of heritage interpretation initiatives is essential. Creating periodic reports to share the results and best practices with other cities and stakeholders can promote transparency and the dissemination of knowledge (Hill & Hill, 2012). By recognizing these limitations and suggesting directions for future research, we hope to contribute to a more robust and applicable understanding of heritage interpretation in diverse tourism contexts.

Conclusions

Interpretation plays a crucial role in communicating ideas, serving as a tool to reveal underlying meanings of a location, often not immediately perceptible. In addition to uncovering these unknown aspects, it has the power to promote environmental knowledge and safeguard heritage, fostering attitudes of respect and protection.

Therefore, interpretation emerges as an essential resource for understanding the historical relevance of a location and the reasons that give it value. Additionally, it plays a crucial role in the tourism context, serving as a vital component for valorization, preservation, and understanding of heritage. It is important to highlight that interpretation not only allows visitors to know and appreciate places better but can also positively influence their experience, extending their stay and even encouraging future visits.

The case study selected for this research was Castelo Neighbourhood, located in Lamego, Portugal. This choice was motivated firstly by the growing interest in the tourism industry in the city, and then by the perception that there is significant potential to be explored in terms of existing tourist resources.

As support, we conducted a survey among visitors to Castelo Neighbourhood in Lamego. The survey aimed to understand visitors' profiles as well as their opinions on available facilities, local heritage, the city of Lamego, and other relevant aspects. The results revealed that the main reasons for visiting are related to religious tourism and cultural and archaeological heritage. Approximately a quarter of respondents were Portuguese nationals, while the remaining nationalities were distributed among Spaniards, French, Americans, Brazilians, English, and Italians, reflecting a diversification of nationalities in line with new territorial dynamics.

As evidenced in this research, the city of Lamego is primarily noted for its distinct religious heritage, often referred to as the "City of Churches." In the context of Castelo Neighbourhood facilities, participants identified Lamego Castle and the Cistern as the most frequented spaces. Regarding aspects that could be improved, respondents emphasized the need for enhancements in signs and interpretive posters, the implementation of more comprehensive signage, and intensified efforts in producing publications such as brochures detailing the history of the facilities, as well as making them available in multiple languages.

Regarding future perspectives, we propose the development of an interpretation plan to improve certain aspects such as missing signage (many visitors have to ask locals or individuals familiar with the city for information); improve facilities for people with reduced mobility, such as installing removable ramps at monuments; enhance existing interpretive means at the facilities to enhance tourist attraction at Castelo Neighbourhood and provide an unforgettable tourist experience to visitors.

Regarding the contributions of this study to heritage tourism management, the conceptual work addressing crucial issues in heritage interpretation stands out. Understanding the profile of visitors and their perspective on interpretive techniques is essential for aligning supply with demand and promoting a vision of sustainable development. In this context, the presented results can be valuable for all involved in the (sustainable) tourism development of this locality, particularly for local policymakers, as the obtained results can help conceptualize a heritage interpretation plan more effectively. It can also be useful for heritage managers, as this study can assist them in planning and outlining strategies, especially regarding interpretation techniques and dissemination and promotion methods.

While this research offers valuable insights into heritage interpretation and visitor experiences at Lamego Castle, it's crucial to acknowledge certain constraints. The study used a limited, non-probabilistic sample of 106 visitors, which may not accurately represent the broader tourist population. In more populous areas, a probabilistic sampling approach might be necessary to ensure data representativeness. Moreover, as the study was conducted in a small town, its findings may not apply to larger cities or places with different features. Bigger urban centers, with their diverse visitor base and complex attractions, might require more comprehensive research methods and a larger pool of respondents.

Another consideration is the variability in visitor preferences. These can differ significantly across locations and cultural contexts. Thus, this study's findings might not directly apply to other historic cities with different visitor profiles and tourism infrastructure. Furthermore, larger cities often have more resources and advanced facilities to implement heritage interpretation technologies. This could impact the effectiveness of the suggestions made in this study, which were based on the specific conditions of a castle in a small town.

The data collection methods also have limitations. Questionnaires were used, which may not capture all the subtleties of visitors' experiences. In larger venues, additional techniques like in-depth interviews and focus groups might be necessary to gain a more comprehensive understanding of visitors' needs and expectations.

To address these limitations, future studies should consider using larger, more representative samples, especially in bigger cities, to ensure the findings can be generalized. It's also recommended to use a mix of quantitative and qualitative methods to capture a wider range of visitor perceptions and experiences. Conducting comparative studies across different cities and historic sites could help identify variations in visitor preferences and heritage interpretation practices. Longitudinal studies are also important to assess the impact of improvements in heritage interpretation over time. By recognizing these limitations and suggesting future research directions, we aim to contribute to a more robust and applicable understanding of heritage interpretation in diverse tourism contexts.

In conclusion, the continuous improvement of interpretive methods is essential for enhancing the visitor experience and promoting the sustainable development of heritage tourism in the Castelo Neighbourhood. By addressing the identified areas for improvement and implementing the recommended strategies, the region can better preserve its rich heritage while providing an unforgettable experience for its visitors.

Acknowledgements This work is funded by National Funds through FCT—Foundation for Science and Technology, IP, under the project reference UIDB/05507/2020 and DOI identifier https://doi.org/https://doi.org/10.54499/UIDB/05507/2020. Additionally, we would like to thank the Centre for Studies in Education and Innovation (Ci&DEI) and the Polytechnic Institute of Viseu for their support.

References

Almeida, J., Mello, M., & Costa, A. R. (2017). Avaliação do impacto do turismo na reabilitação e uso do edificado dos centros históricos: O caso da Baixa Pombalina. *Journal of Tourism & Development (27/28)*, 93–507. https://doi.org/10.34624/rtd.v1i27/28.8549.

Benhamou, F. (2013). *Public intervention for cultural heritage: Normative issues and tools*. Edward Elgar Publishing.

Calle Vaquero, M. (2006). *La cuidad histórica como destino turistico*. Ariel.

Censos. (2021). Obtido de Plataforma de divulgação dos Censos 2021. https://censos.ine.pt/scripts/db_censos_2021.html.

CML. (2019). *Cidade com História. Obtido de cidade com um património e cultura com séculos de história*. https://www.cm-lamego.pt/municipio/historia.

Da Silva, N., & De Francisco, A. (2008). *Segmentação de mercado: turismo religioso como elemento motivador do ser humano.* Ponta Grossa: Congresso Internacional de Administração.

DGEMN/ Direção Geral dos Edifícios e Monumentos Nacionais (1978). Special protection zone for the walls of the Lamego Castle and Cistern. http://www.monumentos.gov.pt/Site/APP_PagesUser/SIPA.aspx?id=5000

Duarte, E. (2016). *A Valorização do Património Arqueológico na Qualificação da Experiência Turística: O caso de Bracara Augusta (Braga).* Master's dissertation, Instituto Politécnico de Castelo Branco, Castelo Branco.

Estatísticas, gráficos e indicadores de Municípios em Portugal. (2022). PORDATA. https://www.pordata.pt/Municipios.

García-Delgado, F. J., Martínez-Puche, A., & Lois-González, R. C. (2020). Heritage, tourism and local development in peripheral rural spaces: Mértola (Baixo Alentejo, Portugal). *Sustainability, 12*(21), 9157. https://doi.org/10.3390/su12219157.

Griffin, T., Giberson, J., Lee, S. H. M., Guttentag, D., Kandaurova, M., Sergueeva, K., & Dimanche, F. (2017). Virtual reality and implications for destination marketing. In *International Conference of the Travel and Tourism Research Association June 20–22.* Quebec City, Canada.

Guerra, J. P., Pinto, M. M., & Beato, C. (2015). Virtual reality-shows a new vision for tourism and heritage. *European Scientific Journal, ESJ, 11*(9), 49–54.

Ham, S. (2016). *Interpretation: Making a difference on purpose.* Fulcrum publishing.

Hart-Robertson, M. (2015). Heritage interpretation, place branding and experiential marketing in the destination management of geotourism sites. *Translation Spaces, 4,* 289–309. https://doi.org/10.1075/ts.4.2.06rob.

Hill, M. M., & Hill, A. (2012). *Preparação final do questionário. Investigação por questionário* (2º). Edições Sílabo.

Liu, Y. (2020). Evaluating visitor experience of digital interpretation and presentation technologics at cultural heritage sites: A case study of the old town, Zuoying. *Built Heritage, 4*(1), 1–15. https://doi.org/10.1186/s43238-020-00016-4.

Lynch, M. F., Duinker, P. N., Sheehan, L. R., & Chute, J. E. (2011). The demand for Mi'kmaw cultural tourism: Tourist perspectives. *Tourism Management, 32*(5), 977–986. https://doi.org/10.1016/j.tourman.2010.08.009.

Mantzou, P., Bitsikas, X., & Floros, A. (2023). Enriching cultural heritage through the integration of art and digital technologies. *Social Sciences, 12*(11), 594.

Miranda, J. M. (2001). *Guía práctica para la interpretación del patrimonio: el arte de acercar el legado natural y cultural al público visitante* (2ª ed.). Junta de Andalucía, Consejería de Cultura.

Monterey, G. D. (s.d.). *Terras ao Léu-Lamego.* Tip. Sociedade de Papelaria, Lda.

Murta, S. M., & Albano, C. (2005). Interpretação, preservação e turismo: uma introdução. In S. M. Murta, & C. Albano (Org.) (Ed.), *Interpretar o patrimônio: Um exercício do olhar* (pp. 9–12). Belo Horizonte: UFMG.

Murta, S. M., & Goodey, B. (2005). Interpretação do patrimônio para visitantes: Um quadro conceitual. In S. M. Murta, C. Albano (Org.) (Ed.), *Interpretar o patrimônio: Um exercício do olhar* (pp. 13–46). Belo Horizonte: Editora UFMG.

Pinto, M. (2022). Lamego-cidade histórica e monumental. In *Cidades Portuguesas.* Lamego, https://cidadesportuguesas.com/lamego/.

Rahaman, H., & Kiang, T. B. (2017). Digital heritage interpretation: Learning from the realm of real-world. *Journal of Interpretation Research, 22*(2), 53–64. https://doi.org/10.1177/1092587217022002.

Richards, G. (2016). El turismo y la ciudad: hacia nuevos modelos? *Revista CIDOB d'Afairs Internacionals, nº 113,* 71–87. https://doi.org/10.1177/1092587217022002.

Salgueiro, T., 2006. Oportunidades e transformação na cidade centro. *Finisterra, XLI*(81), 9–32. https://doi.org/10.18055/Finis1460.

Sharpley, R. (2003). Tourism, modernisation and development on the island of Cyprus: Challenges and policy responses. *Journal of Sustainable Tourism, 11*(2 & 3), 246–265. https://doi.org/10.1080/09669580308667205.

Sol, H., De Brito, M. P., Coelho, J. P., Figueira, L. M., Pratt, C., & Lopes, E. R. (2017). Rooting a new event in its place: The case of Festa Templária, Tomar, Portugal. *International Journal of Event and Festival Management, 8*(3), 324–345. https://doi.org/10.1108/IJEFM-10-2016-0068.

Tilden, F. (2009). *Interpreting our heritage.* University of of North Carolina Press.

Vieira, I., Gama, R., Fernandes, D., Oliveira, I., & Mota, M. (2022). Tourism information platform for low-density territories: The Douro region. *International Journal of Tourism Policy, 12*(2), 223–241. https://doi.org/10.1504/IJTP.2022.122237.

Wiltshier, P., & Clarke, A. (2012). Tourism to religious sites, case studies from Hungary and England: Exploring paradoxical views on tourism, commodification and cost-benefits. *International Journal of Tourism Policy, 4*(2), 132–145. https://doi.org/10.1504/IJTP.2012.049000.

Yin, R. K. (2010). *Estudo de caso: Planejamento e métodos* (4th ed.). Bookman.

Yu, C. P., Chancellor, H. C., & Cole, S. (2011). Examining the effects of tourism impacts on resident quality of life: Evidence from rural midwestern communities in USA. Paper presented at the TOSOK International Tourism Conference, in July in Korea.

Smart Tourism Destinations: To Be or not to Be? The Case of Romanian Rural World Heritage Sites

Cornelia Pop and Liliana Guran

Abstract The aim of the present research is to investigate the current situation of Romanian rural World Heritage Sites (WHSs) as potential smart tourism destinations. The trend toward developing smart destinations even within rural areas, though slower than in the case of cities and towns, is gathering momentum. Previous studies have shown the standing of identified Romanian WHSs within the broader settings of Romanian rural tourism development. However, since 2017 Romania's WHS list was extended by one natural site, which includes multiple rural localities, and a new cultural site in a rural area nominated in 2021 (Rosia Montana, a rural locality). The research question for the present preliminary study was: are the Romanian rural WHS localities smart destinations? The paper adopts an exploratory multiple case study methodology where several bounded case studies are examined using multiple data collection methods. The originality of this paper consists of: (a) investigating the aspects of smart tourism in relation to rural WHSs in Romania; (b) the inclusion of all the current localities associated with WHSs, while most of the previous studies have the tendency to concentrate either on one rural destination or a geographical region. The answer to the research question is **no**. The investigated rural WHS destinations cannot ignore the unstoppable trend of smart tourism. Though late in introducing the digital technologies might prove a blessing in disguise since it allows the rural WHS destinations to learn from other destination's experiences and to customize their future smart offer to their specific standings.

Keywords Rural tourism · Smart tourism · Creative tourism development · World heritage sites · Romania

---

C. Pop (✉) · L. Guran
Faculty of Business, Babes-Bolyai University, Cluj-Napoca, Romania
e-mail: cornelia.pop@ubbcluj.ro

A. L. Negruşa et al. (eds.), *Sustainable and Resilient Businesses in the Global Economy*, Springer Proceedings in Business and Economics,
https://doi.org/10.1007/978-3-031-75883-6_7

Introduction

The past 10–15 years observed a surge in the development of a variety of information communication technologies (ICTs) that impacted all aspects of life. The fragmented tourism industry benefited from this development to the current level of tourism sector being dependent on ICTs (Ammirato et al., 2018; Benckendorff et al., 2014). ICTs became intertwined in the connection between various tourism and related services providers, inserted in the management of tourism products, influenced and impacted tourist experience(s) at a destination, therefore including the smart phenomenon into tourism sector (Xiang et al., 2015; Gajdosík, 2018).

Though academic literature considers smart tourism concept to derive from smart city literature (Buhalis & Amaranggana, 2014; Mora et al., 2017) the affinity of tourism to the implementation of ICTs allowed smart tourism to emerge and thrive, becoming the current stage in tourism development (Gajdosík, 2018). Some authors consider smart tourism as unstoppable given the continuous and permanent implementation of ICTs in the individuals' habits and quotidian life, therefore having a lasting influence on (potential) tourists also (Ballina, 2022b) who might become 'smart' by default, as highlighted by Weaver and Moyle (2019).

Smart tourism applied to tourism destinations implies an increased reliance of these destinations on the implementation of various ICTs infrastructures within (almost all) their aspects and features, from destination management to the business environment within, to tourist experiences (Ammirato et al., 2018; Gelter et al., 2021; Gretzel et al., 2015). Therefore, smart tourism is transforming tourism destinations (Weaver & Moyle, 2019) by being one of the most important drivers of their development (Buhalis & Amaranggana, 2014; Gelter et al., 2021). By becoming smart, tourist destinations can increase their visibility, enhance their differentiating features, and become more competitive (Buhalis & Amaranggana, 2014; Jovicic, 2017).

World Heritage Sites (WHSs) have already the notoriety related to their nomination and WHS label which has almost the power of a brand (Hassan & Rahman, 2015; Van der Zee et al., 2024). If WHSs are situated in rural areas, they are expected to enhance tourism development (Iatu et al., 2018; Iorio & Corsale, 2010). When combined with smart tourism, the pulling power of WHS destinations should increase several-fold, though (to the best of authors' knowledge) currently no academic research investigated this connection. The present study tries to cover a small portion of this research gap by investigating the Romanian WHSs situated in rural areas and their status as smart tourism destinations (STDs). This preliminary study is structured as follows: it reviews the literature related to STDs and rural areas, presents some selected information regarding Romanian WHSs in rural areas, discusses the existence of ICTs applications and the online presence of these destinations, and concludes.

Literature Review

The studies on STDs are numerous and investigate various cities and countries (Ballina, 2022a). However, the main trend of the research studies show a focus on the use of ICTs applications in relation to STDs (Hughes & Moscardo, 2019). This main trend is caused chiefly by the ICTs being embedded within a tourism destination environment, supporting tourists in all the travel/trip stages: before the trip, during the trip, and post-trip. Nonetheless, Buhalis and Amaranggana (2014) and Ammirato et al. (2018) consider that one of the most important transformations is the provision of destination-wide access to real-time information. However, this transformation is (probably) the most difficult to achieve given the needed investments in correlation with an intimate understanding of the destination's tourist profile(s). Therefore, several studies indicate that STDs use ICTs mainly to change their relationship with the (interested) tourists (Femenia-Serra et al., 2019), to present themselves for achieving an increased value for their attributes in the tourism market (Jovicic, 2017). This attitude is triggered chiefly by the fact that most of the time the tourists have limited knowledge on the destination they intend to visit, and ICTs applications could provide valuable insights of the respective destination (Buhalis & Amaranggana, 2014). The fact that within STDs the ICTs are implementing mainly applications related to tourist attractions and the available public services is pointed out by Ballina (2022a), which further shows that the needed presence of smart business within STDs is often overlooked.

All of the above promoted, at least implicitly, the idea that the widespread use of ICTs alone could make it possible for a tourism destination to be considered 'smart' (Baggio et al., 2020). Though, as shown by Boes et al. (2016), while the presence of ICTs infrastructures and applications (called hard smartness) is a must for an STD, the complementary presence of soft smartness (with 4 components: leadership, innovation, human capital, and social capital) should be carefully considered. Constructing an STDs is a complex and integrative process that requires patience, strategic planning and management and continuous adaptation to changes (Boes et al., 2016; Gajdosík, 2018; Gretzel, 2018). Some studies advance the idea that STDs should move beyond the (current) traditional approach which is focused on ICTs and envisage a post-smart development (Coca-Stefaniak & Seisdedos, 2020). This idea is based on the fact that the invasive interference of ICTs is not always in concordance with the tourists' appetite, willingness and/or capacity to use the offered applications/information/knowledge, as underlined by Weaver and Moyle (2019). A truly smart tourism destination should allow the tourists at least a partial digital disengagement during their trip in order to enjoy the physical attractions and engage with the residents since the increased reliance on ICTs and smart devices might discourage direct interactions with the locals and diminish the tourist satisfaction when the smart facilities are not ready available on-site (Pribadi et al., 2021; Weaver & Moyle, 2019).

Given the proliferation of ICTs determined STD to become the most used strategy in tourism development during the last decade, followed and applied by mature and emerging destinations (Pranita, 2023). However, the implementation and application of STD outside the urban areas is difficult, as highlighted by Gretzel (2018) and Ballina (2022b). There is a need for STD to expand in regional and rural areas (Coca-Stefaniak & Seisdedos, 2020) but the STD concept continues to suffer from conceptual imprecision (Gelter et al., 2021) increasing the difficulties in applying it outside urban destinations. There is no one-size-fits-all solution for STDs, a customized approach is recommended (Ballina, 2022b). As Gretzel (2018) points out, it is essential to avoid the direct export of the urban STD operating principles toward regional and/or rural areas. The specificities of each new rural area and the unique experiences they can provide must be the path to follow, in a trial to create smart village destinations (Ballina, 2022b).

The scarcity of academic literature regarding smart tourism, STD, and smart village related research is pointed out by Ballina (2022a, b) and Rahoveanu et al. (2022). Further, Ballina (2022a) shows that rural destinations have made little progress in adopting smart tourism mainly due to delays in implementing digital technologies. This is in line with the findings of Salemink et al. (2017) who pointed out important differences in data infrastructure quality between urban and rural areas, and also the lower levels of education and (digital) skills in rural areas with a negative impact on the adoption and use of ICTs. Furthermore, it should not be overlooked that in rural areas, often deeply anchored in traditions, the ICTs must be well understood and supported by the local communities in order to be implemented and used to create an STD (Garau, 2015).

For Romania, the academic literature is also concentrated on the main municipalities like Brasov, Cluj-Napoca (Briciu et al., 2020; Coros et al., 2023) similar to international literature (Ballina, 2022a). The Romanian literature concerning smart tourism and rural areas is scarce. However, two recent studies tried to investigate if and how smart tourism can be applied in several Maramures villages (Ciolac et al., 2022) and in 3 villages in Marginimea Sibiului (Moise et al., 2023) both with inconsistent results.

As mentioned in the introduction, the status of WHSs as STDs is quasi-not investigated, some seldom studies refer either to the use of some smart applications for solving destination problems (Du Cros & Kong, 2020) or the implementation of some smart solutions within the chosen destinations (Hatami & Koramaz, 2023).

Therefore, the originality of this paper consists of: (a) to the best of the authors' knowledge, no academic study investigated the aspects of smart tourism about rural WHSs in Romania; (b) the study includes all the current localities associated with WHSs, while most of the previous studies tend to concentrate either on one rural destination (Iorio & Corsale, 2010) or a geographical region (Lupu et al., 2019).

Data and Research Methodology

The 80 rural localities included in the present study (see Annex 1) were selected based on previous studies of Pop and Georgescu (2019) and Pop and Balint (2020). They were grouped based on the two main categories (cultural and natural) of WHSs and sub-categories (see Table 1). All the 80 localities are communes. A commune is the smallest administrative unit in Romania and can include from one to several villages. In Annex 1, where the WHS is located in a component village of the respective commune, the village is mentioned in brackets.

The data for tourist accommodations, tourist arrivals, and tourist activity for the period 2001–2022 were extracted from NIS (National Institute of Statistics) via Tempo Online (http://statistici.insse.ro:8077/tempo-online/). Based on this information, averages and absolute variations were calculated. Since not every commune reported data for 2001, the relative variation (in %) could not be calculated for all communes. Therefore, only absolute variations could be reported.

The research question for the present preliminary study was: are the Romanian rural WHS localities smart destinations?

To answer this question the authors tried to identify the existing applications from two sources: the communes' websites and Google Play. The search was focused chiefly on apps for assisting tourists, as the review of literature suggested their dominance. Further, the related tourism information was searched on the commune's websites. Facebook pages and Instagram pages were also searched. In the end, the Google search results using "visit (locality name), (county)" were reported.

Similar to Gajdosík (2018) the paper adopts an exploratory multiple case study methodology where several bounded case studies are examined using multiple data collection methods, as described above.

Selected Information Regarding Romania WHSs in Rural Areas

Romania hosts 9 WHSs, 7 being included in the cultural category and 2 being in the natural category (https://www.worldheritagesite.org/country/Romania). Within the present study only 5 cultural WHSs are included since the remaining 2 are in urban areas (Monastery of Horezu in the town of Horezu and Sighisoara medieval city center in the municipality of Sighisoara).

Details regarding the year of designation are included in Table 1. Table 1 also includes selected data regarding the tourist activity by categories and sub-categories of WHSs. In the end, Table 1 also presents the evolution for all rural WHSs localities, the rural localities, and all the localities in Romania. However, the absolute variation for these 3 levels is not highly relevant given the scale of numerical values. The relative variation between 2022 and 2001 shows the following:

Table 1 Tourism activity (average data for 2001–2022)

Category, sub-category of WHS, year of designation	No. of localities (communes)	Lodgings/ absolute variation between 2022 and 2001	Arrivals/ absolute variation between 2022 and 2001	Occupancy rate (%)/ absolute variation between 2022 and 2001 (%)	Length of stay (days)/ absolute variation between 2022 and 2001
Cultural	23	90 (+187)	25,697 (+55,570)	6.73 (+10.53)	1.16 (+1.20)
Fortified churches, 1993	7	13 (+19)	3,348 (+9,204)	6.48 (+15.47)	0.91 (+1.36)
Painted churches, 1993	5	48 (+89)	16,566 (+31,287)	13.23 (+13.80)	1.87 (+1.24)
Dacian fortresses, 1999	4	6 (+21)	1,281 (+1,471)	6.55 (+3.34)	1.07 (−0.47)
Wooden churches, 1999	6	22 (+57)	4,483 (+13,196)	6.56 (+10.71)	1.77 (+2.05)
Mining cultural landscape, 2021	1	1 (+1)	19 (+412)	0.83 (+18.28%)	0.19 (+4.21)
Natural	**57**	**276 (+447)**	**45,064 (+64,585)**	**10.06 (+9.94)**	**1.12 (+0.68)**
Danube Delta, 1991	19	132 (+344)	32,490 (+42,795)	14.60 (+19.58)	1.17 (+1.75)
Primeval beech forests, 2017*	38	74 (+103)	12,298 (+21,790)	5.52 (+5.13)	1.06 (+0.15)
Total WHSs	**80**	**297 (+634)**	**70,485 (+120,155)**	**8.13 (+10.11)**	**1.17 (+0.83)**
Total rural	*2,862*	*2,676 (+4,798)*	*1,294,649 (+1,997,154)*	*19.91 (−0.82)*	*2.60 (−1.58)*
Total Romania (without Bucharest)	3,180	***6,084 (+8,719)***	***6,933,160 (+6,549,695)***	***25.77 (−0.036)***	***2.59 (−1.08)***

*Note** While the first designations of primeval beech forests across Europe were made in 2007, as https://www.worldheritagesite.org/country/Romania mentions, the designations for the primeval forests for Romania came only in 2017 as shown byPop and Balint (2020)

Source Authors' calculations based on NIS via Tempo online data

(a) lodgings within rural WHSs localities increased by 541.88%, for all rural localities they increased by 410.44%, and at the national level the increase was of 272.38%;
(b) tourist arrivals for WHSs localities grew by 624.05%, for all rural localities increased by 325.39%, and at the national level the growth was of 148.73%;
(c) the occupancy rate for WHSs localities increased by 303.60%, for all rural localities registered a decrease of −3.96%, and at the national level the decrease was even higher, of −12.04%;
(d) the length of stay for WHS localities grew by 108.05%, for all rural localities decreased by −44.01%, and at the national level, the decrease was of −34.62%.

From data in Table 1 and the relative variations presented above, the rural localities overall registered positive developments between 2001 and 2022, growing faster from lodgings and tourist arrivals viewpoints and registering progress in occupancy rate and length of stay. However, one must stress that the level of occupancy rate and length of stay remains very low for the WHSs localities, despite the recorded growth.

While Table 1 presents various aspects of tourist activity, it does not show the evolution of the number of rural localities and rural localities hosting WHSs that offered accommodations and reported tourist arrivals. Table 2 presents this information. Table 2 shows that, on average for 2001–2022, about 89% of all the communes reported tourist arrivals, while only about 86% of WHSs localities reported tourist arrivals, highlighting the existence of communes that do not attract tourists despite offering accommodation.

Findings and Discussions

Apps (Applications)

None of the 80 rural localities under scrutiny provides links to apps that can assist tourists in reaching their respective destinations and/or accessing more information regarding the local tourist attractions, leisure activities, and available accommodations. Details regarding various available apps, mainly for tourist attractions, are presented in Table 3.

Apart from the app 'Visit Delta' (available in Google Play), no other collective app (for tourism) dedicated to either of the other WHSs could be found. Also, there were no search results for collective apps regarding tourism for the counties of Alba, Constanta, Mehedinţi, Sibiu, Suceava, Tulcea, and Valcea that might include the rural localities hosting WHSs.

Various apps were identified for transport alternatives to the 80 localities under scrutiny, the most frequent appearance being that of Moovit.

Table 2 Number of communes reporting lodgings and tourist arrivals

Year	National level		WHSs localities	
	No. of communes reporting lodgings	No. of communes reporting tourist arrivals	No. of communes reporting lodgings	No. of communes reporting tourist arrivals
2001	354	286	27	17
2002	377	296	26	16
2003	375	305	29	19
2004	417	343	32	24
2005	441	384	32	31
2006	455	381	33	32
2007	461	407	38	35
2008	469	420	37	34
2009	528	465	38	34
2010	531	470	30	30
2011	535	512	33	32
2012	601	557	39	34
2013	618	594	40	33
2014	631	598	41	36
2015	674	643	48	40
2016	699	694	51	43
2017	797	765	54	45
2018	855	778	58	52
2019	862	785	57	53
2020	885	759	59	53
2021	982	836	64	58
2022	1,019	858	65	60

Source Authors' calculations based on NIS via Tempo online data

With only two dedicated/personalized apps and no links to other apps, one can safely say that smart tourism related to Romanian rural localities hosting WHSs is in an embryonic form. One might argue that the presence of collective apps might represent progress, but for those interested in WHSs browsing through various destinations can represent an inconvenience.

Websites and tourism dedicated pages

Of the 80 communes under investigation, only one (Zavoi, Caras Severin county) does not have its own official website. Some information about this commune can be obtained via a collective domain called 'ghidul primariilor'. Within these websites the presence of a page dedicated to tourism/tourism attractions was investigated.

Table 3 Apps for visiting the selected destinations

Type of app	Locality/localities	Type of WHS	Site
Dedicated/ personalized app 'Cetatea Prejmer' in Google play	Prejmer, Brasov county	Cultural	Fortified churches
Dedicated/ personalized app https://cetateacalnic.ro/en/#app Available via https://cetateacalnic.ro/en/	Calnic, Alba county	Cultural	Fortified churches
Collective app 'Brasov Tourism' app (official) in Google Play	For all the 4 localities situated in Brasov county	Cultural andṇ Natural	Fortified churches/ Primeval beech forests
Collective app 'Visit Caras Severin' app (official) in Google Play	For all the 17 localities situated in Caras Severin county	Natural	Primeval beech forests
Collective app 'Visit Gorj' app in Google Play	Pades, Gorj county	Natural	Primeval beech forests
Collective app 'Visit Harghita' app (official) in Google Play	Darjiu, Harghita county	Cultural	Fortified churches
Collective app 'Discover Hundeoara' app (official) in Google Play	For all the 4 localities situated in Hundoara county	Cultural and Natural	Dacian fortresses/ Primeval beech forests
Collective app 'Visit Maramures' app in Google Play	For all the 10 localities situated in Maramures county	Cultural and Natural	Wooden churches and Primeval beech forests
Collective app 'Visit Mures—Travel Guide' app in Google Play	Saschiz, Mures county	Cultural	Fortified churches
Collective app 'Visit Delta' app in Google Play	For all the 19 localities related to Danube Delta	Natural	Danube Delta
Collective app 'Romania Salbatica' app in Google play	For the localities related to the natural WHSs	Natural	Danube Delta/ Primeval beech forests

Source Authors' compilations based on Google searches, last accessed April 15, 2024

The results are presented in Fig. 1. While 49% of communes present at least a dedicated tourism page (sometimes several pages have tourism content like: lodgings, tourism attractions, local customs and traditions, local crafts) the percentage of pages containing good and interesting information are relatively few. Most of the tourism pages have sketchy presentations, exhibit photos without explanations or excessively present local Orthodox monasteries despite being linked to natural WHSs. Furthermore, only 15 communes (19%) mention that some of their tourist attractions are WHSs, as follow: 4 of the communes with fortified churches (Bunesti, Prejmer, Saschiz, Biertan); 3 of the communes with painted churches (Arbore, Manastirea Humorului, Vatra Moldovitei); 3 of the communes with Dacian fortresses (Bosorod, Orastioara de Sus, Sasciori); 2 communes with wooden churches (Barsana, Sisesti); 2 communes related to Danube Delta (Corbu, Jurilovca), and 1 commune related to primeval beech forests (Sinca).

The results suggest that the communes related to cultural WHSs (12 out of 23) are more aware of the pulling-power of their tourism attraction under the UNESCO umbrella, while most of the communes related to natural WHSs (17 out of 19 for Danube Delta; 37 out of 38 for primeval beech forests) ignore this pulling-power. For the localities related to primeval beech forests the ignorance might be excused

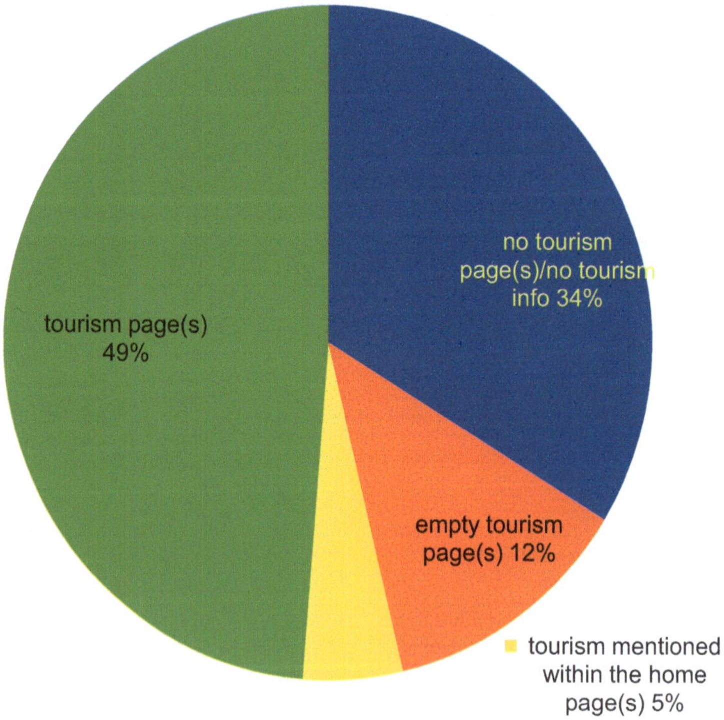

Fig. 1 Tourism pages within communes' websites. *Source* Authors' compilation based on respective communes' websites

since the designation of the WHSs was made only in 2017 and there is no official list of the localities connected with this UNESCO site with multiple locations. However, the absence of mentioning Danube Delta status as WHS (inscribed as UNESCO site since 1991) is difficult to explain. One possible cause for this absence might be that the people living within and near to Danube Delta consider the site so well-known that it needs no further mentioning. Table 1 seems to justify this situation since Danube Delta registers the highest number of arrivals.

It must be noted that in the cases of 6 communes, distinctive dedicated to tourism websites were identified, including key words like 'visit' (the locality) or the name of the WHS or infotourism. These 6 localities are Calnic, Bunesti (Viscri), and Prejmer hosting fortified churches, Budesti—hosting a wooden church, Rosia Montana with its mining cultural landscape, and Zavoi related to natural primeval beech forests. However, the difficulty in finding these alternative websites (there is no link to these websites from the main pages of the respective communes in 5 cases, while for Zavoi there is no official website) does not improve the online presence of these localities.

Facebook, Instagram and X(twiter)

All the 80 communes under investigation have at least one Facebook page. Though, the information provided by these Facebook pages is difficult to investigate since some of the pages are official pages, managed by the commune, while others are unofficial pages (some without any posting). Furthermore, some pages are open to the public, others are profiles, and several are groups, mostly private. Hence, the only information somehow similar is represented by the number of followers/friends/members. Neither investigated Facebook page has more than 10,000 followers. Only 2 commune pages (Prejmer—fortified church, Jurilovca—Danube Delta) have more than 5 thousand followers/friends/members, 5.4 k and respectively 9.6 k.

They are followed in descending order by 4 commune pages having between 4.0 and 4.9 thousand followers/friends/members; 8 commune pages with 3.0–3.9 thousand followers/friends/members; 13 commune pages with 2.0–2.9 thousand followers/friends/members; 28 communes with 1.0–1.9 thousand followers/friends/members; and 25 communes with less than 1 thousand followers/friends/members (data collected as April 15, 2024). The information above shows, despite the mixed type of pages, a low online profile for the investigated communes.

Similarly with the case of websites, for 8 localities alternative Facebook pages were found, dedicated to tourism. In 4 cases (Sarichioi and Jurilovca—Danube Delta, Sisesti and Ieud—wooden churches) the respective Facebook page contains 'Visit (the locality)', with 2.2 k followers, 8.6 k followers, 1.1 k followers, and 2.0 k followers, as of Apr.15, 2024. For other 2 localities, the Facebook page is dedicated to the local tourism information centre: Zavoi—related to primeval beech forests, 275 followers and Nufaru—Danube Delta, 643 followers. The remaining 2 localities are related to Facebook pages dedicated to primeval beech forests: Stulpicani to Slatioara Forest (63 followers) and Salatrucel to the National (Natural) Park Cozia-Salatrucel (1.6 k followers). Probably other localities under scrutiny might have similar alternative pages but they do not appear when searching using 'visit (the locality)' on the first page. The results for these 8 alternative pages are comparable with the results for the Facebook pages presented in the previous paragraph.

The situation is no better when Instagram pages were considered. For 12 communes there were no Instagram pages (two communes related to cultural WHSs: Vatra Moldovitei—painted church and Bosorod—Dacian fortress; two communes related to Danube Delta: Sacele and C.A. Rosetti; the remaining 8 communes are related to primeval beech forests). Of the remaining 68 Instagram pages, only 3 are dedicated to tourism pages (for Valea Viilor and Biertan, both hosting fortified churches, and Perisani, related to primeval beech forests). The Instagram page for Toplet deserves a mention since it tries to present the beauty of the place, though being managed by an private person. The remaining 64 Instagram pages are rather collective pages provided by Instagram based on tags or hashtags, including a variety of photos, only few presenting the related WHSs. One must have already knowledge about the respective WHSs to identify them in photos/posts.

The presence on X(twitter) is almost nonexistent. Only two communes have an X(twitter) account dedicated to WHS/tourism: Prejmer—fortified church and Rosia Montana—mining cultural landscape. The county of Mehedinti has an X(twitter) account where the localities related to WHS—primeval beech forests (4 communes) benefit from various information and short presentation films. The remaining 74 communes are absent from X(twitter).

Some details regarding WHSs destinations in relation with their online presence

Annex 1 presents details regarding the 80 localities investigated in this research. The data show that there seems to be no relation between (the modest) online presence of some localities and their tourist activity. For example, Prejmer, with the fortified church, might be considered the locality having a higher visibility online (dedicated app in Google Play, alternative website, X(twitter account), mentioning of WHS on the dedicated tourism page within the official page, over 120,000 results on Google search), though it averaged less than 1,000 tourists between 2001 and 2022, has a low occupancy rate and a modest (1.55 days) length of stay. Within the fortified churches category, Bunesti (Viscri) seems to attract more tourists and have a better occupancy rate and length of stay, despite a slight decrease. These results come without a dedicated app and an X(twitter) account. Nonetheless, Viscri is known to host also a house owned by King Charles the 3rd of Great Britain and he is known for his periodic visits to the place, visits mentioned by the local and international media.

Within the category of fortified churches two other localities draw the attention: Calnic and Valea Viilor, both not offering (according to NIS records) any tourist accommodation between 2001 and 2022. They are the only 2 communes from the category hosting cultural WHSs in this situation. While in both cases day trips might bring tourists this category of tourists is not reported. Further investigations are needed to understand this state of affairs.

In the case of painted churches (of Moldavia) the most visited destination is Sucevita despite the very modest presence online. However, the repute of Sucevita due to the painted monastery date since long, though the length of stay is under

2 days. One must mention that the modest average figures for Arbore are due to its starting to report tourist activity only in 2021.

The localities hosting the Dacian fortresses have modest tourist activity while they are almost not visible online. Banita, despite offering accommodation, does not report any tourist arrivals; this might be related to being almost absent online. Bosorod has a very low tourist activity despite offering accommodations since 2015 and trying to include WHSs information on its official website. Orastioara de Sus, which hosts 3 of the fortresses, experiences a slower growth in tourist arrivals compared with Sasciori; Orastioara de Sus also exhibits a decrease in occupancy rate and length of stay. Though, both Orastioara de Sus and Sasciori have almost a similar presence online.

The localities hosting the wooden churches seem to have been chosen by tourists based on other sources than online presence. While in the case of Sisesti (with 2 WHSs) it seems to be a relation between its better presence online (compared with other wooden churches destinations), Desesti, ranking second in tourist arrivals, is contradicting the case of Sisesti. These contradictions exist for the remaining 4 localities related to wooden churches.

In the case of Rosia Montana, one must wait since this locality started to report tourist activity only in 2022, while the designation of the WHS came in 2021.

The case of Danube Delta, the oldest designated WHS in Romania, in 1991, is also full of contradictions. While Jurilovca seems to be a popular online presence, it has a lower tourist activity compared with Murighiol which has a lower profile online. The examples can continue taking the 17 remaining localities case by case, though the limited space of this paper lets the data speak for themselves.

For the localities related to primeval beech forests the situation is the most confusing. They have the lowest profile online and vary from localities with no accommodations (8) to localities with no arrivals (2) to 4 localities that attract 1,500 or more tourists per year (on average), though without even a modest presence online. For the last 4 localities, the link with primeval beech forests seems irrelevant since they have reported tourist activity since 2001, long before the designation of the natural WHSs in 2017. Further in-depth analyses are required for a better understanding of the situation.

As the data in Annex 1 show, the feedback on the Romanian WHSs is, in general, average. The best feedback was received by the painted church destinations, slightly over 4 (out of 5), but the rating is decreasing. With a modest online presence, one can only presume that the ratings are offered mostly by those interested in religious tourism. The wooden churches and fortified churches have close ratings of around 3.6, respectively 3.5, while the fortified church's locations have a better online presence. Danube Delta localities have a rating of 3.3, a relatively modest one since this destination had more than 3 decades to develop and improve. Dacian fortresses improved in the space of 1 year from 2.9 to almost 3.0 rating. The primeval beech forests-related localities have an under-average, though slightly improving rating from 2.62 to 2.67 which can be explained by the relatively recent designation of this WHS in Romania, since 2017. The lowest rating, as of Apr. 15, 2024, is for Rosia

Montana, the most recently designated in 2021. The ratings provided for Romanian WHSs are in line with the combined information presented in Annex 1.

Conclusions

Based on the data presented above, the answer to the research question: *are the Romanian rural WHS localities smart destinations?* is **no**. Even their online presence is modest and this is reflected in tourist activity (low), the rankings available on the WHS website (https://www.worldheritagesite.org/country/Romania), and the low number of results generated by Google search (Annex 1).

This situation is not unique, the research of Rudwiarti et al. (2021) for rural destinations in Indonesia and of Zhao and Zhang (2021) for rural destinations in China shows that there is no integration of villages' tourism potential and smart tourism and a poor online presence.

The situation in Romania can be explained through various causes. One is related to the ICTs infrastructures spread at the rural level. The European Union (EU) report on the Digital Economy and Society Index for Romania (2022) shows that the country ranks (overall) in the 27th place, with a score of 30.6 compared with 52.3 at EU level. For 3 of the index components (human capital, integration of digital technology, and digital public services) Romania ranks last (27th), though it is worth noting that, within the human capital dimension, the percentage of individuals with at least basic digital content creation skills is a decent 41% versus the 66% for EU. According to the same report, Romania ranks 15th in the case of connectivity, though with a score of 55.2, below the EU 59.9 score. The data are confirmed by Moise et al. (2023) findings that in the villages from Marginimea Sibiului about 78% of the locals use the Internet and about 74% own at least one smart device. How and for what the ICTs are used, is completely another aspect. This information confirms the findings of Boes et al. (2016) showing that the presence of digital technology is a must but not sufficient for an STD, with the leadership and human dimension proving to be also important.

While a preliminary study, the present research did not investigate the human capital dimension within Romanian rural WHS destinations, though the findings of Ilie et al. (2022) consider the conservative attitudes of the rural population and the limited knowledge regarding the complex use of digital technologies to be important barriers to the implementation of smart tourism in rural destinations. Furthermore, the management of the investigated rural WHSs is poor. The destination management organizations (DMOs) are quasi-absent. Of the current 19 DMOs authorized by tourism authorities in Romania (https://turism.gov.ro/web/autorizare-turism/) only 3 can be linked to the WHS rural localities: DMO Sucevita-Putna, DMO Bucovina, and DMO Maramures. An official website is missing for each of the named DMOs and therefore they cannot be associated with WHS destination management. The study of Chasovschi (2019) pointed out the poor profile of DMOs in Romania and seems that nothing improved during the past 5 years.

Moreover, the concept of community-based tourism, probably a better option for further developments within the investigated rural WHSs destinations is barely understood and scarcely applied in Romania, as underlined by Havadi Nagy and Espinosa Segui (2020).

The absence of an integrated vision at the national level regarding the WHSs further generates deficient coordination and support at the local level, which is best reflected by the lack of awareness of local authorities (and communities) of the importance and attractiveness of the hosted WHSs, as suggested by Pop and Georgescu (2019).

Due to the pervasive smart technologies, one can argue that all tourists are smart (using smart devices and applications) by default (Weaver & Moyle, 2019). Therefore, rural destinations should implement smart tourism components. Being late in applying smart tourism, Romanian rural WHS destinations can learn from other experiences and customize the way ICTs will be introduced in the destination, considering that the integrated applications that assist tourists in preparing the trip and on-site should be developed first, as suggested by Ballina (2022b).

Annex 1 Detailed information for localities hosting WHSs (averages for 2001–2022 for lodgings, arrivals, occupancy rate, and length of stay)

Commune and county	Lodgings/ absolute variation between 2022 and 2001	Arrivals/ absolute variation between 2022 and 2001	Occupancy rate (%)/absolute variation between 2022 and 2001	Length of stay (days)/absolute variation between 2022 and 2001	Rating according to https://www.worldheritagesite.org/		Other tourism attractions	Dedicated apps/ Instagram pages/ X-twitter	Google results using 'visit (locality), (county)' as of Apr.15, 2024
					Feb. 19, 2023	Apr. 15, 2024			
Fortified churches—cultural WHS									
Calnic, Alba	0/0	0/0	0/0	0/0	3.54 (50 votes)	3.55 (59 votes)	Sebes-Apold vineyard	Dedicated app	50,600
Valea Viilor, Sibiu	0/0	0/0	0/0	0/0			Tarnave vineyard	Instagram page dedicated to the church	35,600
Bunesti (Viscri)*, Brasov	6 (−1)	1,368 (+5,170)	14.42 (+28.31)	2.14 (-0.71)					126,000 (for Viscri)
Prejmer, Brasov	3 (+7)	915 (+2,472)	9.12 (+12.36)	1.55 (+1.59)				Dedicated app Dedicated X/twitter account	127,000
Darjiu, Harghita	1 (+1)	40 (+143)	4.09 (+9.45)	0.75 (+1.93)					17,300
Saschiz, Mures	2 (+7)	558 (+902)	9.32 (+15.80)	0.93 (+1.95)			Tarnave vineyard		46,300
Biertan, Sibiu	1 (+5)	467 (+517)	8.46 (+11.42)	1.01 (+2.02)			Tarnave vineyard	Instagram page dedicated to tourism	116,000

(continued)

Annex 1 (continued)

Commune and county	Lodgings/absolute variation between 2022 and 2001	Arrivals/absolute variation between 2022 and 2001	Occupancy rate (%)/absolute variation between 2022 and 2001	Length of stay (days)/absolute variation between 2022 and 2001	Rating according to https://www.worldheritagesite.org/		Other tourism attractions	Dedicated apps/Instagram pages/X-twitter	Google results using 'visit (locality). (county)' as of Apr.15, 2024
					Feb. 19, 2023	Apr. 15, 2024			
Painted churches—cultural WHS									
Arbore, Suceava	1 (+3)	46 (+623)	1.53 (+16.36)	0.32 (+2.75)	4.09 (37 votes)	4.02 (45 votes)			301,000
Manastirea Humorului, Suceava	14 (+31)	4,191 (+12,264)	18.70 (+7.99)	2.45 (0.02)					53,900
Patrauti, Suceava	1 (+2)	252 (+172)	13.48 (+6.91)	2.67 (+1.86)					37,900
Sucevita, Suceava	26 (+39)	10,276 (+12,330)	15.17 (+11.11)	1.85 (+0.91)			Resort of national interest		97,000
Vatra Moldovitei, Suceava	6 (+14)	1,801 (+5,898)	17.28 (+26.65)	2.08 (+0.65)			Resort of local interest		43,000
Dacian fortresses—cultural WHS									
Sasciori (Capalna), Alba	1 (+3)	427 (+1,056)	11.93 (−15.04)	1.99 (+2.40)	2.88 (35 votes)	2.99 (42 votes)			29,800 (for Capalna)
Banita, Hunedoara	1 (+2)	0/0	0/0	0/0					24,000

(continued)

Annex 1 (continued)

Commune and county	Lodgings/ absolute variation between 2022 and 2001	Arrivals/ absolute variation between 2022 and 2001	Occupancy rate (%)/absolute variation between 2022 and 2001	Length of stay (days)/absolute variation between 2022 and 2001	Rating according to https://www.worldheritagesite.org/		Other tourism attractions	Dedicated apps/ Instagram pages/ X-twitter	Google results using 'visit (locality), (county)' as of Apr.15, 2024
					Feb. 19, 2023	Apr. 15, 2024			
Bosorod (Luncani)*, Hunedoara	1 (+3)	3 (0)	0.91 (0)	0.15 (0)					8,460 (for Luncani)
Orastioara de Sus (Costesti, Gradistea de Sus)*, Hunedoara	4 (+13)	851 (+415)	13.34 (−1.67)	2.15 (−4.29)					74,000 (for Costesti) 26,500 (for Gradistea de Sus)
Wooden churches—cultural WHS									
Barsana, Maramures	5 (+13)	829 (+1,636)	6.00 (+6.38)	1.69 (+1.84)	3.65 (32 votes)	3.61 (37 votes)	Resort of local interest		143,000
Budesti, Maramures	2 (+8)	401 (1,768)	5.24 (+9.74)	1.43 (+2.21)					32,000
Desesti, Maramures	4 (+8)	1,010 (+3,439)	7.59 (+17.15)	2.01 (+2.33)					36,800
Ieud, Maramures	2 (+6)	253 (+1,280)	5.83 (+8.26)	1.81 (+2.03)					62,900
Poienile Izei, Maramures	8 (+17)	559 (+2,055)	3.90 (+9.62)	2.00 (+2.13)					38,100

(continued)

Annex 1 (continued)

Commune and county	Lodgings/absolute variation between 2022 and 2001	Arrivals/absolute variation between 2022 and 2001	Occupancy rate (%)/absolute variation between 2022 and 2001	Length of stay (days)/absolute variation between 2022 and 2001	Rating according to https://www.worldheritagesite.org/		Other tourism attractions	Dedicated apps/Instagram pages/X-twitter	Google results using 'visit (locality), (county)' as of Apr.15, 2024
					Feb. 19, 2023	Apr. 15, 2024			
Sisesti (Plopis, Surdesti)*, Maramures	2 (+5)	1,431 (3,018)	10.79 (+13.10)	1.66 (+1.74)					20,900 (for Plopis) 34,500 (for Surdesti)
Mining cultural landscape—in danger									
Rosia Montana, Alba	1 (+1)	19 (+412)	0.83 (+18.28%)	0.19 (−4.21)	2.70 (16 votes)	2.65 (24 votes)		Dedicated X/twitter account	105,000
Danube Delta—natural WHS									
Istria, Constanta	0/0	0/0	0/0	0/0	3.31 (29 votes)	3.33 (37 votes)	Istria-Babadag vineyard		167,000
Mihai Viteazu, Constanta	0/0	0/0	0/0	0/0			Istria-Babadag vineyard		159,000
Sacele, Constanta	0/0	0/0	0/0	0/0					126,000
Corbu, Constanta	2 (+12)	320 (+1,776)	40.52 (−40.59)	3.02 (+2.98)					153,000
Baia, Tulcea	1 (+5)	155 (1,689)	2.95 (−18.30)	0.42 (+1.58)			Istria-Babadag vineyard		261,000
Bestepe, Tulcea	1 (+5)	557 (1,703)	11.78 (+20.42)	0.87 (+2.31)					22,300

(continued)

Annex 1 (continued)

Commune and county	Lodgings/absolute variation between 2022 and 2001	Arrivals/absolute variation between 2022 and 2001	Occupancy rate (%)/absolute variation between 2022 and 2001	Length of stay (days)/absolute variation between 2022 and 2001	Rating according to https://www.worldheritagesite.org/		Other tourism attractions	Dedicated apps/Instagram pages/X-twitter	Google results using 'visit (locality), (county)' as of Apr.15, 2024
					Feb. 19, 2023	Apr. 15, 2024			
C.A.Rosetti, Tulcea	1 (+7)	127 (477)	5.44 (+21.17)	1.09 (+3.87)					27,500
Ceamurlia de Jos, Tulcea	1 (+2)	87 (289)	7.00 (+31.77)	0.42 (+1.89)			Istria-Babadag vineyard		22,200
Chilia Veche, Tulcea	2 (+10)	155 (363)	8.79 (+23.26)	0.77 (+3.39)					44,500
Crisan, Tulcea	21 (+19)	3,564 (6,735)	23.87 (−5.31)	2.85 (−2.55)					85,100
Jurilovca, Tulcea	24 (+64)	4,919 (+2,502)	44.33 (+50.17)	2.04 (+2.07)			Istria-Babadag vineyard		30,800
Mahmudia, Tulcea	8 (+30)	2,139 (+1,046)	26.91 (+13.13)	1.74 (+2.76)					76,400
Maliuc, Tulcea	6 (+12)	918 (+1,053)	18.33 (+27.08)	2.45 (+4.10)				Empty Instagram page	61,000
Murighiol, Tulcea	44 (+125)	12,519 (+13,881)	30.18 (+5.17)	1.95 (+0.77)			Sarica-Niculitel vineyard		40,100
Nufaru, Tulcea	3 (+8)	594 (+326)	13.31 (+16.37)	1.15 (+2.60)			Sarica-Niculitel vineyard		32,700
Pardina, Tulcea	1 (+1)	10 (+26)	1.47 (+8.61)	0.35 (+2.38)					21,400

(continued)

Annex 1 (continued)

Commune and county	Lodgings/ absolute variation between 2022 and 2001	Arrivals/ absolute variation between 2022 and 2001	Occupancy rate (%)/absolute variation between 2022 and 2001	Length of stay (days)/absolute variation between 2022 and 2001	Rating according to https://www.worldheritagesite.org/		Other tourism attractions	Dedicated apps/ Instagram pages/ X-twitter	Google results using 'visit (locality), (county)' as of Apr.15, 2024
					Feb. 19, 2023	Apr. 15, 2024			
Sarichioi, Tulcea	1 (+7)	324 (+1,797)	6.68 (+40.88)	0.54 (+2.73)			Istria-Babadag vineyard		49,900
Sf.Gheorghe, Tulcea	15 (+33)	6,058 (+8,845)	33.37 (+45.97)	2.23 (−0.74)					142,000
Valea Nucarilor, Tulcea	2 (+4)	44 (287)	2.43 (+14.38)	0.40 (+1.84)			Istria-Babadag vineyard		25,000
Primeval beech forests—natural WHS									
Ciuhici, Caras-Severin	0/0	0/0	0/0	0/0	2.62 (91 votes)	2.67 (100 votes)			14,100
Cornereva, Caras-Severin	0/0	0/0	0/0	0/0					28,700
Lapusnicu Mare, Caras-Severin	0/0	0/0	0/0	0/0					19,600
Naidas, Caras-Severin	0/0	0/0	0/0	0/0					15,900
Sopotu Nou, Caras-Severin	1 (+1)	2 (+49)	0.15 (+3.22)	0.08 (+1.82)					2,800

(continued)

Annex 1 (continued)

Commune and county	Lodgings/ absolute variation between 2022 and 2001	Arrivals/ absolute variation between 2022 and 2001	Occupancy rate (%)/absolute variation between 2022 and 2001	Length of stay (days)/absolute variation between 2022 and 2001	Rating according to https://www.worldheritagesite.org/		Other tourism attractions	Dedicated apps/ Instagram pages/ X-twitter	Google results using 'visit (locality), (county)' as of Apr.15, 2024
					Feb. 19, 2023	Apr. 15, 2024			
Toplet, Caras-Severin	0/0	0/0	0/0	0/0				Instagram page that seems dedicated to tourism manage by a private person	41,400
Podeni, Mehedinti	0/0	0/0	0/0	0/0					11,400
Stulpicani, Suceava	1 (+1)	19 (+427)	1.27 (+27.94)	0.10 (+2.29)					41,300
Berislavesti, Valcea	0/0	0/0	0/0	0/0					25,200
Muereasca, Valcea	0/0	0/0	0/0	0/0					24,600
Sinca, Brasov	4 (+7)	870 (+3,631)	5.82 (+0.99)	2.13 (−7.59)					33,700
Sinca Noua, Brasov	1 (+1)	92 (+367)	9.71 (+36.36)	1.78 (+2.12)					27,500
Bozovici, Caras-Severin	2 (+1)	320 (−66)	10.01 (−0.74)	2.42 (−2.64)					39,700
Carbunari, Caras-Severin	1 (+1)	4 (+49)	0.52 (+4.83)	0.32 (+2.88)					6,340

(continued)

Annex 1 (continued)

Commune and county	Lodgings/absolute variation between 2022 and 2001	Arrivals/absolute variation between 2022 and 2001	Occupancy rate (%)/absolute variation between 2022 and 2001	Length of stay (days)/absolute variation between 2022 and 2001	Rating according to https://www.worldheritagesite.org/		Other tourism attractions	Dedicated apps/Instagram pages/X-twitter	Google results using 'visit (locality), (county)' as of Apr.15, 2024
					Feb. 19, 2023	Apr. 15, 2024			
Ciclova Romana, Caras-Severin	1 (+2)	44 (+587)	3.24 (+28.06)	0.26 (+2.45)					17,300
Dalboset, Caras-Severin	1 (+1)	18 (0)	3.56 (0)	1.04 (0)					15,700
Mehadia, Caras-Severin	6 (+9)	2,341 (+591)	16.34 (−13.55)	2.48 (−0.30)			Resort of local interest		35,200
Prigor, Caras-Severin	1 (+2)	113 (+357)	5.52 (+14.07)	0.92 (−1.87)					33,300
Racasdia, Caras-Severin	1 (+1)	0/0	0/0	0/0					11,600
Sasca Montana, Caras-Severin	4 (+13)	1,909 (+4,819)	16.00 (+12.98)	2.36 (−4.71)					19,400
Socol, Caras-Severin	1 (+1)	29 (0)	5.92 (0)	1.78 (0)			Independent vine-growing center Moldova Noua		23,900
Teregova, Caras-Severin	2 (+1)	247 (−288)	21.67 (−69.02)	4.58 (−8.00)					19,700
Zavoi, Caras-Severin	11 (+12)	1,497 (+870)	9.78 (−7.41)	2.48 (−3.03)					30,700

(continued)

Annex 1 (continued)

Commune and county	Lodgings/ absolute variation between 2022 and 2001	Arrivals/ absolute variation between 2022 and 2001	Occupancy rate (%)/absolute variation between 2022 and 2001	Length of stay (days)/absolute variation between 2022 and 2001	Rating according to https://www.worldheritagesite.org/		Other tourism attractions	Dedicated apps/ Instagram pages/ X-twitter	Google results using 'visit (locality), (county)' as of Apr.15, 2024
					Feb. 19, 2023	Apr. 15, 2024			
Pades, Gorj	2 (−1)	120 (+591)	6.00 (+15.21)	1.46 (+1.88)					4,570
Rau de Mori, Hunedoara	8 (+21)	1,695 (2,985)	22.89 (2.85)	2.73 (−4.01)					61,400
Baiut, Maramures	1 (+2)	49 (+297)	2.01 (+3.49)	0.79 (+1.54)					24,500
Botiza, Maramures	15 (0)	637 (+822)	3.97 (+0.64)	1.97 (−0.65)			Resort of local interest		38,200
Grosii Tiblesului, Maramures	1 (+2)	3 (+44)	0.96 (+14.19)	0.21 (+2.00)					4,010
Lapus, Maramures	1 (−2)	0/0	0/0	0/0					9,940
Balta, Mehedinti	1 (+4)	54 (+1,056)	1.26 (+21.99)	0.17 (+2.03)					43,600
Isverna, Mehedinti	1 (+5)	248 (+1,080)	5.25 (+13.50)	0.99 (+3.01)					13,600
Obarsia Closani, Mehedinti	1 (+1)	20 (+84)	4.38 (+24.38)	0.49 (+3.74)					16,600

(continued)

Annex 1 (continued)

Commune and county	Lodgings/absolute variation between 2022 and 2001	Arrivals/absolute variation between 2022 and 2001	Occupancy rate (%)/absolute variation between 2022 and 2001	Length of stay (days)/absolute variation between 2022 and 2001	Rating according to https://www.worldheritagesite.org/		Other tourism attractions	Dedicated apps/Instagram pages/X-twitter	Google results using 'visit (locality) (county)' as of Apr.15, 2024
					Feb. 19, 2023	Apr. 15, 2024			
Crucea, Suceava	1 (+4)	405 (+1,296)	14.07 (+28.12)	1.21 (+2.31)					36,400
Barbatesti, Valcea	1 (+3)	68 (+153)	6.23 (+2.60)	1.40 (+1.49)					24,300
Costesti, Valcea	3 (+5)	522 (+1,027)	18.32 (+4.88)	2.66 (+1.80)					31,300
Perisani, Valcea	1 (+2)	31 (+351)	2.41 (+22.78)	0.42 (+2.59)				Instagram page dedicated to tourism	18,200
Racovita, Valcea	1 (+3)	286 (+379)	6.60 (+2.07)	1.27 (+1.04)					20,800
Salatrucel, Valcea	1 (0)	655 (+232)	5.90 (+0.42)	1.64 (−0.28)					20,500

*Note** The localities in brackets are the villages where the respective WHSs are situated. These villages are under the administration of the specified commune
Sources authors' compilations based on NIS via Tempo online data, https://www.worldheritagesite.org/, Order 1205/2018 of Ministry of Agriculture regarding wine-growing areas, https://turism.gov.ro/web/autorizare-turism/, and online searches

References

Ammirato, S., et al. (2018). Smart tourism destinations: Can the destination management organizations exploit benefits of the ICTs? In *Evidences from a multiple case study*. Cardiff: Springer.

Baggio, R., Micera, R., & Del Chiappa, G. (2020). Smart tourism destinations: A critical reflection. *Journal of Hospitality and Tourism Technology, 11*(3), 407–423.

Ballina, F. (2022a). Smart business: The element of delay in the future of smart tourism. *Journal of Tourism Futures, 8*(1), 37–54.

Ballina, F. J. (2022b). Smart concept in rural tourism: A comparison between two phases (2016–2019). *Revista de Economia e Sociologia Rural, 60*(1).

Benckendorff, P., Sheldon, P., & & Fesenmaier, D. (2014). *Tourism information technology* (2nd ed.). Oxford: CABI.

Boes, K., Buhalis, D., & Inversini, A. (2016). Smart tourism destinations: Ecosystems for tourism destination competitiveness. *International Journal of Tourism Cities, 2*(2), 108–124.

Briciu, A., Briciu, V. A., & Kavoura, A. (2020). Evaluating how 'smart' Brasov, Romania can be virtually via a mobile application for cultural tourism. *Sustainability, 12*(13).

Buhalis, D., & Amaranggana, A. (2014). Smart tourism destinations. In Z. T. I. Xiang (Ed.), *Information and communication technologies in tourism* (pp. 553–564). Springer.

Chasovschi, C. (2019). The evolution of destination management organizations (DMOs) in Romania. *The USV Annals of Economics and Public Administration, 19*(2(30)), 15–24.

Ciolac, R., et al. (2022). Smart tourist village-an entrepreneurial necessity for Maramures rural area. *Sustainability, 14*, 8914.

Coca-Stefaniak, J., & Seisdedos, G. (2020). Smart urban tourism destinations at a crossroads–being 'smart' and urban are no longer enough. In: A. M. Morrison & J. Coca-Stefaniak (Eds.), *The routledge handbook of tourism cities* (pp. 45–59). London: Routled.

Coros, M., Popescu, D., Pop, I., & Gheorghe, G. (2023). Smart city, smart tourism. In A. C. M. Negrușa (Ed.), *Remodelling businesses for sustainable development, ICMTBHT Springer proceedings in business and economics* (pp. 167–188). Springer.

Du Cros, H., & Kong, W. (2020). Congestion, popular world heritage tourist attractions and tourism stakeholder responses in Macao. *International Journal of Tourism Cities, 6*(4), 929–951.

European Union/EU. Digital economy and society index (DESI), Romania. https://digital-strategy.ec.europa.eu/en/policies/desi-romania. (Online)

Femenia-Serra, F., Perles-Ribes, J., & Ivars-Baidal, J. (2019). Smart destinations and tech-savvy millennial tourists: Hype versus reality. *Tourism Review, 74*(1), 63–81.

Gajdosík, T. (2018). Smart tourism: Concepts and insights from Central Europe. *Czech Journal of Tourism, 7*(1), 25–44.

Garau, C. (2015). Perspectives on cultural and sustainable rural tourism in a smart region: The case study of Marmilla in Sardinia (Italy). *Sustainability, 7*, 6412–6434.

Gelter, J., Lexhagen, M., & Fuchs, M. (2021). A meta-narrative analysis of smart tourism destinations: Implications for tourism destination management. *Current Issues in Tourism, 24*(20), 2860–2874.

Gretzel, U. (2018). From smart destinations to smart tourism regions. *Investigaciones Regionales-Journal of Regional Research, 42*, 171–184.

Gretzel, U., Sigala, M., Xiang, Z., & Koo, C. (2015). Smart tourism: Foundations and developments. *Electron Markets, 25*, 179–188.

Hassan, A., & Rahman, M. (2015). World Heritage site as a label in branding a place. *Journal of Cultural Heritage Management and Sustainable Development, 5*(3), 210–223.

Hatami, M., & Koramaz, T. (2023). Smart solutions for heritage sites: Florence and Yazd. *Preservation, Digital Technology & Culture, 4*(143–155), 52.

Havadi Nagy, K., & Espinosa Segui, A. (2020). Experiences of community-based tourism in Romania: Chances and challenges. *Journal of Tourism Analysis: Revista De Analisis Turistico, 27*(2), 143–163.

Hughes, K., & Moscardo, G. (2019). ICT and the future of tourism management. *Journal of Tourism Futures, 5*(3), 228–240.
Iatu, C., Ibanescu, B., Stoleriu, O., & Munteanu, A. (2018). The WHS designation–a factor of sustainable tourism growth for Romanian rural areas? *Sustainability, 6*(626), 10.
Ilie (Marin), N., Oprea, I., Turcea, V., & Parnus (Rusu), A. (2022). New dimensions of rural communities' development in Romania–smart village concept. *Scientific Papers. Series Management, Economic Engineering in Agriculture and Rural Development, 22*(2), 425–436.
Iorio, M., & Corsale, A. (2010). Rural tourism and livelihood strategies in Romania. *Journal of Rural Studies, 26*(2), 152–162.
Jovicic, J. (2017). From the traditional understanding of tourism destination to the smart tourism destinations. *Current Issues in Tourism, 3*(276–282), 22.
Lupu, C., Brochado, A., & Stoleriu, O. (2019). Visitor experiences at UNESCO monasteries in Northeast Romania. *Journal of Heritage Tourism, 14*(2), 150–165.
Moise, G., et al. (2023). Can we talk about smart tourist villages in Marginimea Sibiului, Romania? *Sustainability, 15*(7475).
Mora, L., Bolici, R., & Deakin, M. (2017). The first two decades of smart-city research: A bibliometric analysis. *Journal of Urban Technology, 24*(1), 3–27.
Pop, C., & Balint, C. (2020). Romanian newly-nominated WHS and current tourism development in the associated rural localities. In: P. Nistoreanu (Ed.), *New Trends and Opportunities for Central and Eastern European Tourism* (pp. 223–245). s.l.: IGI Global.
Pop, C., & Georgescu, A. (2019). Romanian rural World Heritage sites and tourism development. *Entrepreneurial Business and Economics Review, 7*(1), 135–158.
Pranita, D. (2023). Post-smart tourism destination: Have we been wise enough? In D. V. F., et al. (Eds.), *Proceedings of the International Conference on Vocational Education Applied Science and Technology (ICVEAST 2023), Advances in Social Science.*(pp. 2352–5398). s.l.: Atlantis Press.
Pribadi, T., Tahir, R., & Yuliawati, A. (2021). The challenges in developing smart tourism: A literature review. *InfoTekJar: Jurnal Nasional Informatika dan Teknologi Jaringan, 5*(2), 23–27.
Rahoveanu, M., et al. (2022). Perspectives on smart villages from a bibliometric approach. *Sustainability, 14*(10723).
Rudwiarti, L., et al. (n.d.). Smart tourism village, opportunity, and challenge in the disruptive era. *IOP Conf. Series: Earth and Environmental Science, 780*(012018), 2021.
Salemink, K., Strijker, D., & Bosworth, G. (2017). Rural development in the digital age: A systematic literature review on unequal ICT availability, adoption, and use in rural areas. *Journal of Rural Studies, 54*, 360–371.
Van der Zee, E., Camatti, N., Bertocchi, D., & Shomali, K. (2024). UNESCO World Heritage Site label and sustainable tourism in Europe: A user-generated content analysis. *Regional Studies* 1–16.
Weaver, D. B., & Moyle, B. D. (2019). 'Tourist stupidity' as a basic characteristic of 'smart tourism': Challenges for destination planning and management. *Tourism Recreation Research, 44*(3), 387–391.
Xiang, Z., Tussyadiah, I., & Buhalis, D. (2015). Smart destinations: Foundations, analytics, and applications. *Journal of Destination Marketing and Management, 4*(3), 143–144.
Zhao, L., & Zhang, D. (2021). Research on the path of revitalizing rural tourism from the perspective of smart tourism. *E3S Web of Conferences, 251*, 03037.

An Analysis of the Impact of Climate Change on Tourism Destination. Policies and Strategies for the Mediterranean

Mariaclaudia Cusumano

Abstract We are assisting in rapidly intensifying the impact of Climate Change (CC) globally on tourism destinations due to a pervasive anthropogenic change on a planetary scale of abiotic nature. The Mediterranean is an interesting area of interest since it is a Climate hotspot. Indeed, the Mediterranean will be particularly affected by Climate Change, primarily by the increase in temperature, secondly of other consequences in terms of environmental impact, such as: (a) Lack of water, (b) Rising of the sea level, (c) Water tropicalization and acidification, (d) Extreme events, (e) Forest fires. Due to the geographical location and morphology, the Mediterranean area is experiencing a higher risk of Global Warming and Climate Change than other European regions. The visible consequences of the Environmental crisis on Travelling causing us to reflect on our consumption-driven lifestyle (Rantala et al. in Annals of Tourism Research, 105, Article 103733, 2024). This work, through a literature review, aims to analyze the impact of Climate Change on tourism to better reflect on tourism in the Anthropocene and contribute to a rethink and radically co-transform tourism. Our approach will investigate the bidirectional relationship in the Mediterranean. In the context, it is appropriate to quote the paradox of Sustainable Development and tourism since this always requires the exploitation of natural resources as the cultural identity. Moreover, the intensification of extreme events as other phenomena brings us to reinterpret Tourism in the light of the Anthropocene (Crutzen & Stoermer in The Anthropocene. Yale University Press, pp. 479–490, 2000). Climate Change (CC) is a prominent challenge for tourism (Bizzarri & Petrana in Rivista di Scienze del Turismo-Ambiente Cultura Diritto Economia 8(1–2), 1, 2018). The phenomenon transcends traditional boundaries of policymakers (Becken et al., Journal of Sustainable Tourism 2810, 1–22, 2020). This issue requires approaches that go beyond conventional institutional responsibilities by implementing cross-sectoral policies both for mitigation and adaptation to favor Sustainable Tourism. This work aims to raise awareness of this prominent issue that is already affecting the Mediterranean. The paper also aims to identify the impact of Climate Change on tourism and provide solutions for public governance in ensuring

M. Cusumano (✉)
University of Milano-Bicocca, Milan, Italy
e-mail: mariaclaudia.cusumano@unimib.it

A. L. Negrușa et al. (eds.), *Sustainable and Resilient Businesses in the Global Economy*, Springer Proceedings in Business and Economics,
https://doi.org/10.1007/978-3-031-75883-6_8

the sustainability of Tourism, both for residents and tourists. The contribution is structured through the following research questions:

(i) How do Climate change, Anthropocene, and tourism affect each other?
(ii) How does Tourism impact Climate Change?
(iii) Which are the main factors of CC that impact on Tourism Destination?
(iv) Which solutions have been identified to face climate change in the Mediterranean area?

The main finding of this research shows that policymakers are currently underestimating this relationship. There are no serious strategies or policies taken. Moreover, tourism can play a crucial role in driving the Ecological transition, considering the pivotal economy of tourism at the global level, including the transportation system.

Keywords Climate change · Coastal tourism · Extreme weather events · Adaptation · Resilience planning

Introduction

2022 was the warmest year in the last 100.000 years.[1] We are assisting in rapidly intensifying the impact of Climate Change globally. The pervasive anthropogenic change on a planetary scale is affecting on tourism destinations (Chakraborty, 2022). The Mediterranean area, for its geographical collocation and its morphology, is experiencing a higher risk of Global Warming and Climate Change (CC). Indeed, the Mediterranean area is known as a hotspot (Darmaraki et al., 2019; Angra et al., 2022), in other words, this area is sensitive to Global Warming due to its proximity on the equator, both air and water have prominently risen in temperature in recent years.

As many researchers report, the Climate Crisis represents one of the most significant challenges of the 21st century (Change IPCC, 2023; Becken et al., 2020; Angelini et al., 2015,), due to its negative consequences (Bizzarri & Pedrana, 2018).

This challenge necessitates a firm commitment from the scientific community to develop innovative solutions. These solutions include, for instance, the development of renewable energy sources and implementing carbon capture and storage technologies. The visible consequences of the Environmental crisis on Travelling causing us to reflect on our consumption-driven lifestyle (Rantala et al., 2024).

Such measures could be implemented across all sectors of our daily lives, including the tourism industry. However, the Climate Crisis challenge requires more than individual efforts. It demands proactive and bold political action. Policymakers must implement cross-sectoral and multi-level policies to address the Climate Crisis, recognizing the weight of their responsibility in this global issue.

[1] https://climate.copernicus.eu/copernicus-2023-hottest-year-record.

The tourism industry, a global economic powerhouse, is currently grappling with the pressures of Climate Change, a pervasive anthropogenic impact. It is urgent to face tourism in the Anthropocene (Chakraborty, 2022). Immediate action is needed to address this issue.

Paradoxically, this industry, which is heavily impacted by climate change, also contributes significantly to it (Bizzarri & Petrana, 2018). The relationship between climate change and tourism is quite intricate. Climate change is already impacting investment, planning, and operations in the tourist industry (Song, 2023). The anthropogenic causes of Climate Change, which are tangible in phenomena such as deforestation, burning of fossil fuels, and land use change, are well-established in the scientific community, but understanding how implementing a change is still a complex issue, even though the intricate relationship between Climate Change and tourism underscores the need for immediate action.

The tourism industry is acutely vulnerable to Climate Change. It has been noted that there are some vulnerable places in the Anthropocene (Chakraborty, 2022), and its effects are likely to disrupt destination management, tourist flows, and even lead to the loss of tourism assets. Climate Change poses direct and indirect threats to tourism, including extreme events, safety concerns, water shortages (Gössling et al., 2012), and biodiversity loss, among others (UNWTO, 2019). The literature underscores the vulnerability of many areas worldwide, each with its specific threats (Scott et al., 2015). The Mediterranean is an area very sensitive to CC; this would impact tourism. (Bizzarri & Petrana, 2018).

Researchers have studied the nexus and interaction between Climate Change and Tourism and strategies to approach the decarbonization transition (Song, 2023). Since 2012, the number of publications has increased, but after 2016, the number of publications tripled and still arises (Dube, 2024). The nexus was a subject of interest for geographers, economist, and sociologists. Researchers applied interdisciplinary methodology in many studies to investigate this thematic area (Becken et al., 2020).

The implications of Climate Change in tourism are already tangible in many areas of the world, some of which are considered particularly vulnerable. Figure 1 shown the Global Climate Risk for tourism in 181 countries. As is possible to note, the Mediterranean area and the Global South have higher risks than the North, as Global Warming is the main threat to our Planet (Scott & Gössling, 2018). If the *business as usual* (BAU) continues to grow, we will encounter the scenario shown in Fig. 3. Consequently, many areas of the World will not be viable or visitable, so policymakers must take urgent action to mitigate Climate Change also for the tourist sector by supporting the Global Transition to a low-carbon economy, (and) strengthening resilience at a local level against climate risks (Scott & Gössling, 2018).

On the one hand, tourism impacts on Climate Change as it contributes to Environmental degradation through air pollution (Zhang, 2022), and CO_2 emission. Soil degradation in tourism is directly responsible for causing Climate Change. On the other hand, climate change impacts territories and tourism destinations. Tourism and climate change are bidirectional relationships in which both phenomena are, at the same time, cause and effect (Scott & Gössling, 2018).

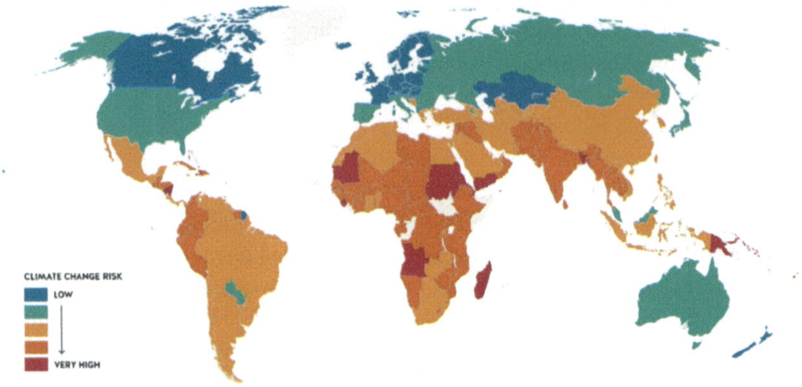

Fig.1 Global climate risk for tourism, *Source* (Scott & Gössling, 2018)

Currently, tourism is one of the fastest-growing economic sectors worldwide. Projections for the future attest that by 2030, the number of tourists will reach 1.8 billion international arrivals, which accounts for an increase of 3% (UNWTO, 2019). This expansion will increase carbon emissions if the *business as usual* (BAU) or *tourism as usual* (TAU) continues to grow by considering only the economic growth without considering the environment, local community, and tourists' well-being impact.

Climate Change will affect tourism supply in the Mediterranean area by reducing and changing tourism seasonality.

The main objective of this work is to investigate the effects of climate change on tourism in the Mediterranean area, which is a vulnerable area of the climate change risk industry (Scott & Gössling, 2018). This work aims to identify the impact of Climate Change on tourism to support public governance in ensuring the sustainability of Tourism, both for residents and tourists. Expertise and involvement are crucial in this endeavor.

Mediterranean area emerges as an interesting case study since there is a High concentration of Cultural Heritage as a Coastal tourism destination, especially for Italy (Colloca & Lipari, 2022). The Mediterranean is a climate hotspot of Biodiversity that will be particularly affected by temperature. This phenomenon will have significant consequences in terms of impact both on the environment and in the tourism sector: (a) lack of water (Gössling et al., 2012)., (b) rising of the sea level, (c) water tropicalization and acidification, (d) extreme events, (e) forest fires, (f) the increase of the temperature (Bizzarri & Pedrana, 2018; Angra et al., 2022).

The relationship between climate change and tourism is under investigation and is limited. The two issues are mainly studied as separate identities. In the light of the intensity of the impact of Climate Change, it is necessary to debate the challenges and opportunities in the tourism sector.

The study aims to illuminate the relationship between tourism and Climate Change to raise awareness and action in vulnerable territories to prevent risks and disasters. This work aims to raise attention to the impact of Climate Change, mainly in tourism, which is a sector that is exceptionally responsible and vulnerable to CC. This responsibility also reverses socio-economical aspects, as Tourism in many areas of the World has become an essential slice of the economy depending on their community.

The contribution is structured through the following research questions:

(i) How do climate change, Anthropocene, and tourism affect each other?
(ii) How does Tourism impact Climate Change (CC)?
(iii) Which are the main factors of CC that impact on Tourism Destination
(iv) Which solutions are identified to face issues in which we are occurring in the Mediterranean area?

The study has been divided into the following sections: an introduction on CC, Sect. 2 presenting the methodology, Sect. 3 a Literature Review on the impact of Tourism on the environment, Sect. 4 a Literature Review on Climate Change in Tourism, and Sect. 5 examining strategies to support the transition, mainly in the Mediterranean area, Sect. 6 conclusions and considerations.

Methodology

Through a literary review, this work follows a theoretical socio-economic as environmental analysis of the main concepts and factors that characterize the bidimensional relationship between tourism and Climate Change, investigating how the two factors influence each other in the light of the Anthropocene. Secondly, assessing the implementation of the policy environment on tourism will help identify policy gaps and strategies for the tourism sector.

This paper aims to analyze and unify two aspects of the same medal, as a need to face one of the most prominent sectors at the global level. The literary review will comprehensively analyze interdisciplinary studies that would put in light on the correlation between the two fields as exploring strategies for climate adaptation and resilience planning to mitigate the adverse effects of coastal tourism.

Tourism on Climate Change, A Focus on Environmental Aspects

Studies have focused on the impact of Tourism on specific aspects of the CC, mainly focusing on natural resources (Amelung & Moreno, 2009), losing biodiversity (Morrison, 2022) and the impact of tourism on water shortage (Gössling et al., 2012), energy consumption as an intensive industry, CO_2 Emissions, and rising sea levels. Global Warming could impact tourism by reducing the climate's comfort.

Below are analyzed only some of those most relevant for the Mediterranean. In Italy, Global Warming is affecting the Water Cycle, with lesser rains but more intense. Italy has the highest level of hydric stress in Europe. Moreover, as climate hotspots have a higher risk of climate risk and dryness, the risk of dryness increases to 80% in Southern Italy. In addition, we must mention that Italy has an old water network infrastructure responsible for dispersing 42% of water for household use. This aspect of decarbonization is relevant as water scarcity has already impacted hydrothermal production, which has been reduced by 37% (Ronchi et al., 2023).

Currently, the Tourism industry is heavily reliant on fossil fuel energy, a situation that urgently needs to change (Spector & Higham, 2019). With tourism contributing to 8% of CO_2 emissions, a significant factor in climate change, the need for a shift to sustainable energy sources is more pressing than ever. The transportation system, which forms the backbone of international and domestic tourism, is the largest contributor to this issue (UNWTO, 2019).

For instance, Cruise tourism is one of the significant contributors to the economic sector as in Tourism. Cruises consume energy for various purposes, including transport, accommodation, entertainment, and more, only for 2018 this sector accounted for $150 billion (Lloret et al., 2021).

However, this tourism model impacts cities and communities more in terms of environmental factors than in economic and employment factors. Once Cruise ships reach the port, they release waste, air pollution (Zhang, 2022), and sea pollution into the city (Cusumano, 2022). Day visitors contribute to the local economy for a pinch but do not take advantage of accommodation or local restaurants. On the other hand, cities affected by this phenomenon testify invasion of by daily visitors, who contribute mainly to the crowds.

At the same time, ships Cruises are known as heavy carbon emitters. According to the report of Friends of Earth, cruises produce 421.43 kg of CO_2 per day. Cruise Ships are expected to double, reaching 20 billion trips (UNWTO, 2019). Cruise companies are developing Sustainable strategies to continue to attract their consumers. However, they remain one of the most pollutant sectors that look at BAU but use Greenwashing strategies instead of reducing carbon fuel emissions.

Another significant issue that demands our collective attention is the impact of the aviation system on the environment. In 2016, CO_2 emissions from transport, including both passenger and freight transport, reached 7,230 million tonnes globally, representing 23% of all man-made CO_2 emissions. By 2030, CO_2 emissions for transportation purposes are estimated to grow by 21% compared to 2016 (UNWTO, 2019). This underscores the urgent need for collective action. Each of us, as part of the global community, has the power to make a difference and reduce these alarming figures.

The aviation system is the most significant contributor to emissions. A large share of emissions is emitted at flight altitude. Airplane emissions depend on the distance traveled, and short travel produces more CO_2 in Km. Aviation produces about 2% of all CO_2 emissions worldwide. Figure 2 shows the emission of international aviation by region. Europe shows the highest rate of emissions, while air in Africa and the Middle East is low, which reflects the global consumer pattern.

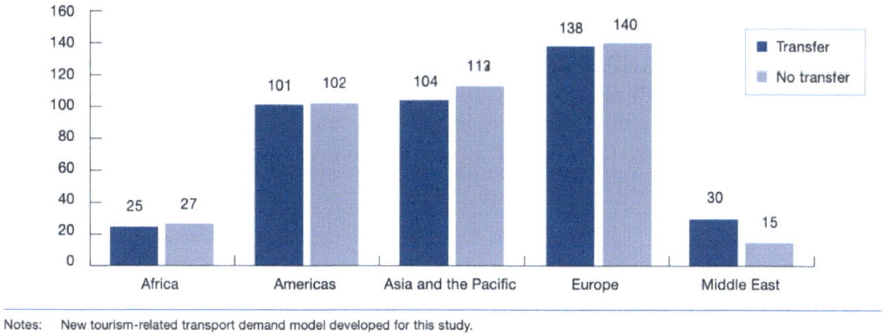

Fig. 2 Emissions of International aviation by region according to transfer and no-tranfer (2016) (Mt of CO_2)

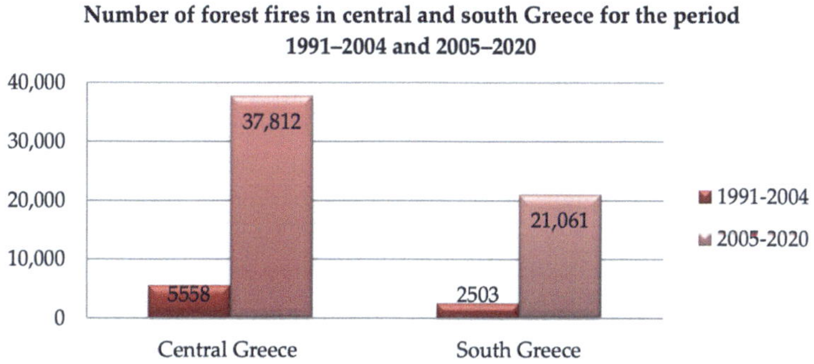

Fig. 3 Number of forest fires in central and south Greece for the period Number of forest fires in central and south Greece for the period 1991–2004 and 2005–2020. *Source* Angra et al. (2022)

The aviation system contributes to anthropogenic Climate Change, accelerating the environmental crisis. If the current trend of tourism (TAU) continues to grow as expected, the impact of Tourism on Climate Change will continue to arise. In terms of economic investment to support decarbonization in the Tourism sector, studies have found that investing in the tourism sector will have a prominent impact on the mitigation of the vital role in the global economy compared to other sectors (Scott et al., 2015).

Researchers are paying more attention to adaptation (Gössling et al., 2023) rather than mitigation to identify measures to prevent the potential effect of Climate Change on tourism by developing resilience (Becken et al., 2020). However, in recent decades, there has been increased awareness and knowledge of Environmental issues as the attention to the anthropic impact on Tourism, with the consequence that new forms of tourism have been raised, such as Sustainable Tourism, Slow tourism, and Sports tourism (Lo Verde, 2023) as Rural tourism, Community-based tourism, Nature tourism (Samora-Arvela et al., 2023). According to Fletcher, Ecotourism responds better to the challenge of Anthropocene tourism by fixing the "disaster capitalism"

(2018). Ecotourism is as a key means to address growing environmental and social problems in pursuit of sustainable development. Other nature-based approach are focusing on the authenticity in tourism in landscape (Bueddefeld & Erickson, 2022).

Tourism's environmental impact is increasing, attesting to annual growth rates of over 5% (UNWTO, 2018). This phenomenon is directly associated with growing CO_2 emissions.

The Impact of Climate Change on Tourism Destination

The impact of climate change is spread out on different scales, both territorial and regional scales, and in many economic sectors (Bizzarri & Petrana, 2018).

According to the IPCC, if the current trend of global emissions continues to grow as expected, the surface temperatures will increase by an estimated 2.6–4.8 °C. The global alert is not exclusively Global Warming but also affects the switch in climate instability, which will cause severe, pervasive, and irreversible impacts during this decade (Angelini et al., 2015). Large coastal and other areas would be transformed and undermined as the ecosystem altered (Scott et al., 2015).

The Mediterranean region hosts exceptional biological diversity and socio-cultural richness. In the latest assessment of the IPCC, it emerges that in the area, there is a significant risk of big Climate Change due to the scenario of temperature which has a range of temperature increasing 0.9–5.6 °C, which is one of the most vulnerable are the area due to the proximity to equator, this will be impacting sea Levels, draught, coastal flood (IPCC, 2023). Moreover, Climate Change will impact social and human vulnerability in terms of health, such as heat waves, which are particularly devastating for elderly people and chronically ill, such as obese or cardiopathic. The phenomenon may result in premature death. Few studies have been focused on Human Health (Elsayed, 2023) and social aspects, such as the perception of risk and safety of the territories in Tourism destinations. Travel decisions are driven by individual selfishness and any threat to freedom (Hindley & Font, 2014).

Forest fire. Global Warming facilitates forest fires, mainly during heat waves, which have become more frequent in recent years. Forest fires occur especially when there is high temperature, which causes permanent damage to nature by favoring further disasters such as floods during the winter due to land instability. Moreover, Forest fires are responsible for releasing air pollution into the atmosphere. Forest fires reduce the destination's attractiveness (Angra et al., 2022). These phenomena represent a serious threat to tourism, as they can reduce the attractiveness of tourist destinations. The burned areas and environmental deterioration are perceived as health risks, uncomfortable, and unsafe (Arabadzhyan et al., 2021). Fire events deteriorate the destination's image of the environment and landscape by reducing natural, cultural, and touristic value attributed by potential tourists as public perception.

Mediterranean is particularly vulnerable to CC in its effects, such as forest fires, the complicity of drouths, significant increases in air temperature, particularly in summers, and heat waves, which create conditions for Natural disasters such as Forest fires. Recently, in the Mediterranean area, the frequency and severity of drought have increased significantly due to the temperature rise (Angra et al., 2022). Indeed, this phenomenon is particularly relevant in summer and tourism seasonality. As this practice has been carried out for a long history, conversely to extreme floods or storms, forest fires are not so associated in public opinion with Global Warming but more a criminal attitude, even though evidence (Angra et al., 2022) shows as increased when came across the heat of waves, windy condition, and other sdverse weather conditions due to complex phenomena. This phenomenon changes the state of places, which, for instance, is expected to rise to 40 days, and Global Warming has increased. As shown in Fig. 3, the number of forest fires from 2005 to 2020 is more than quadrupled in Greece compared to 1991–2024. This situation contributes to rising flood event risks in Greecean territories.

The Portuguese government estimated the impact of forest fires on Tourists. It was estimated that burned areas in 2030 will reach € 17–24 million for domestic and € 18–38 million for inbound tourism due to the loss of attractivity of the natural environment, as tourist places are losing comfort (Arabadzhyan et al., 2021).

Italy is one of the European countries with the highest expenditures due to Climate Change, with more than 90 billion of euros for extreme events. Other countries have registered loose for less than 3 billion of euros (Ronchi et al., 2023) (Fig. 4).

The recent assessment of the IPCC on the Mediterranean region says that "(this) is the world's leading tourist destination and one of its busiest shipping routes. Climate change strongly interacts with other environmental problems in the Mediterranean Basin, resulting from urbanization, land use change, overfishing, pollution, biodiversity loss, and degradation of land and marine ecosystems (Ali et al., 2022). It has been observed that a growing number of impacts across the entire basin are now being attributed to climate change, mainly to climate hazards. In economic terms,

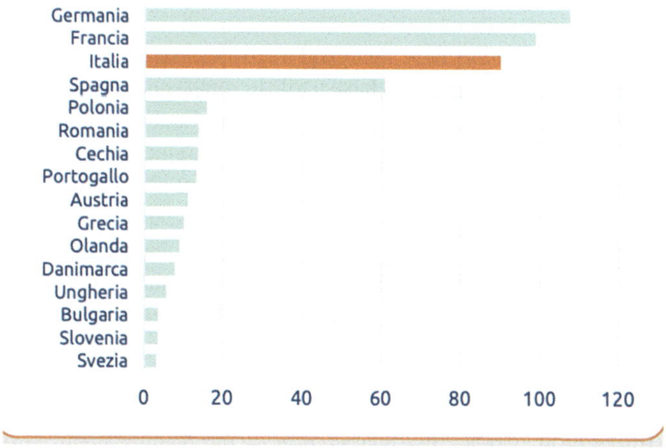

Fig. 4 Economic damage due to exstreme events 1980–2020 in euro, source (Ronchi et al., 2023)

this change will have a different impact: In low-income countries of the Basin, such as Morocco, it is estimated a reduction will be a reduction of 3% of GDP; in MENA countries, it will be 10–13%, in Southern Europe, approximately 2% GDP (Ali et al., 2022).

Coastal Tourism, a significant component of the global tourism sector, is a key vulnerability factor for many tourist destinations. The literature on Tourism and Climate Change primarily focuses on coastal or marine tourism, often referred to as the 3S (Sea, sun and sand) (Scott et al., 2015). Coastal areas, including islands, are fragile ecosystems and the most susceptible to the impact of Climate Change, experiencing specific Climate Change hazards (Arabadzhyan et al., 2021). Understanding the impacts of tourism flow is crucial, as it can bring about significant changes in tourism destinations. Not only does it affect the environment, but it also influences the local community's habits and attitudes. Environments that experience repeated environmental disasters may lose their appeal. This could lead to a decrease in tourism arrivals due to factors such as the loss of species, the increase of exotic invasive species, or the degradation of the landscape (Arabadzhyan et al., 2021). For example, coastal areas in which flora or fauna are disappearing or favoring the entrance of new species due to Climate change or pollution can affect local activities. For instance, Callinectes sapidus, a "blue crab" in the Adriatic Sea (in Italy), has been threatening fishing activities. If the trend continues in the future, this phenomenon shortly will affect Tourism by reducing the activity of the Adriatic Sea due to the "blue crab," which is invading the lagoon ecosystem of the Adriatic Sea and the Mediterranean Sea due to the Sea Tropicalization, in other words, the water heating is creating a favorable habitat for the blue crab and other aliens species, every nine days one new alien's species if found in the Mediterranean Sea.

Tourists are drawn to the coast's natural beauty, but Climate Change is also impacting biodiversity. These physical changes have socio-economic implications for the tourism industry, with a loss of biodiversity potentially leading to a reduction in ecosystem goods and services, and the economic benefits of tourism (Burke et al., 2008).

Only a few research studies have analyzed the impact of ocean acidification and Tropicalization on tourist destinations, have a role in some sports activities in tourism, such as sunbathing, snorkeling, diving, and glass-bottom boating (Arabadzhyan et al., 2021), while many studies focus on arise sea level and extreme events. Further studies should investigate the impact of water heating on tourism as it has been revealed to be an additional threat in terms of reducing attractivity. Despite the importance and many benefits of the coral reef in Oceania and the habitat change, this will represent an obstacle to tourism in Australia. The coral reef is pivotal in the blue economy as this natural beauty is a tourism destination for worldwide visitors who want to snorkel, dive, and sport fish. Tobago and St. Lucia islands comprised about 46 percent and 47% of their respective GDPs 2005. Not all visitors come to these lovely islands because of coral reefs, but for many, it is an essential component of the islands' attraction (Burke et al., 2008); the degradation of the Natural environment in the future can impact tourism. Not all visitors come to these lovely islands because of coral reefs, but for many, they are an important component of the islands' attraction

(Burke et al., 2008). The degradation of the Natural environment in the future can impact tourism.

European Alps, the effect of climate change on the uniqueness of the glacial landscape, "its role on motivation is crucial, given the potential loss of attractiveness of the sites and the development of last-chance tourism" (Salim et al., 2022), the climate changes that are occurring will conduct in losing attractivity of glacier tourism sites. In glacier tourism, visitors are mainly attracted by nature and see the glaciers, or for sport, skiing-based tourism (Steiger & Abegg, 2018). In this form of tourism, the dependence on natural resources is more vital than in other forms. In the future, to quote the best scenario, "A warming of 1 °C would lead to a reduction of snow reliable ski areas to (91%) significantly smaller ski areas at the rim of the Alps are affected by climate change (Steiger & Abegg, 2018).

Bangladesh is particularly vulnerable due to its low-lying coastal areas and high population density. The tourism sector in Bangladesh is particularly relevant in terms of economic development, which is mainly attractive as a Coastal tourism destination, but this geographical configuration makes Bangladesh particularly vulnerable. The main threat consisted of rising sea levels and the risk of inundation, extreme weather events such as cyclones and storm surges, and coastal erosion. Global Warming puts tourist facilities and infrastructure such as Restaurants, Hotels, and Resorts at risk as those facilities are exposed to the risk of damage and destruction due to sea-level rise. Moreover, coastal erosion plays a role, as it diminishes the width and quality of beaches, a major tourist attraction (Louis et al., 2024).

Fiji is an acclimate-tourism hotspot. Therefore, Climate Change may affect the comfort of tourists and slowly cause a decrease in the attractivity of tourism and flows. This prominent vulnerability is due to the risk of sea level rise leading to coastal erosion and coastal inundation, warming sea surface temperatures leading to coral bleaching, increases in frequency as the intensity of tropical cyclones, floods, scarcity of water resources, EI Nino- related drought and increased incidence of diseases such as dengue fever and diarrhea (Government of Fiji, 2005).

South Africa is characterized by never-ending subtropical summer. The KwaZulu-Natal (KZN) bay encompasses 600 km of the Indian Ocean coast. It attracts every year nearly eight million tourists, mainly beach lovers and surfers, 850 000 foreigners visited KZN in 2018, resulting in the tourism industry has developed in the area as the most significant sector of the area, which accounts for roughly 20 billion to the province's GDP (Mgadle, 2022). The dependence on tourism in this province, as the primary source of growth, makes this area particularly sensitive to Climate Change. The area is vulnerable to natural hazards, storms, flooding, and sea-level rise, factors that may alter the competitiveness of the area's attractiveness. Global Warming in South Africa is expected to rise to 3 °C; this will cause drastic changes in precipitation patterns, increased malaria incidences, increased aridity, incidences of drought, and a severe decrease in biodiversity. The primary threat is decreasing precipitation, which will reduce tourists' satisfaction by causing an impact. Those changes will affect the tourism industry and GDP. As previously said, KwaZulu-Natal province mainly depends on tourism. This situation is similar to that of other countries with similar climates and tourist characteristics, such as Tanzania, Zambia, Botswana, and Zimbabwe (Mgadle, 2022).

The latest aspect that here we would focus on is the consequence of the increasing temperature and *heat of waves* that may cause a limitation in footwalk and walkability, but also accommodations have started installing cooling systems to face the heat of waves (Bizzarri & Pedrana, 2018). Studies have focused on the impact of Climate Change on Heath (Filippelli et al., 2020), by focusing on an increase of heat-related injuries and deaths, respiratory problems associated with increased ozone and allergens in the air, and changing disease profiles as the range of ticks increases. Mosquito-related diseases evolve and temperature impacts on Health (Filippelli et al., 2020), but only a few studies have been dedicated to Climate Change, tourism, and Health to the direct effect (Day et al., 2021).

Main Finding as Tourism Policy to Struggle Climate Change

This research advances the nexus among tourism- and climate change as in the Anthropocene. Countries, mainly in Europe, also thanks to the new initiative of the European Commission for the sustainable development of *the Blue economy*, which incorporates coastal tourism, maritime safety and security, and preserving ecosystems and biodiversity, have started initiatives to create (a) safer and more secure maritime space, (b) an innovative and resilient Blue economy, (c) better governance of the sea is a strategy to fight Climate Change, restore biodiversity, and use marine resources responsibly (European Commission, 2017). Unfortunately, as happened for Scandinavian countries, the national climate strategies do not pay enough attention to tourism adaptation needs nor present adaptation actions for tourism actors to adapt and mitigate Climate Change (Landauer et al., 2018). The lack of tourism-specific climate change knowledge **hinders** climate action across the business, policy, planning, and investor domains (Scott et al., 2023). The need for anticipatory adaptation planning emerges (Becken et al., 2020). Mitigation policies should have an integrated approach, embedding all sectors of the economy, including tourism, as linked to each other to address the challenges of Climate Change. Politics should support strategies and resilience-building measures to support the decarbonization of tourism. Acting directly in tourism to reduce the source of CO_2 emissions at the local level to prevent causes means transforming accommodations such us Hotels B&B into an efficient building with zero emissions through the following actions:

1. heat insulation,
2. geothermal cooling system for air conditioning,
3. installation of renewable energy sources such as solar panels or wind,
4. innovating technologies in more efficient systems are continuously replaced.

Those strategies would be necessary for building up a zero-emission community by providing tools to contrast potentially, but high-level, climate disasters. Managers' attitudes and beliefs are not just necessary. They are essential in planning how to minimize or mitigate the negative impacts of Climate Change on the tourist accommodation industry. Their attitudes play a crucial role in implementing measures to minimize the impact of how tourism contributes to Global Warming through greenhouse gas emissions. Significant changes must be made by acting on managers' attitudes and beliefs to understand the risks and opportunities for changing the *business as usual* toward the Ecological Transition. This empowerment and responsibility can lead to a more proactive approach in the face of Climate Change.

Rantala et al., by following the degrowth approach, propose human–nature connectedness and interdependency in proximate settings to address a sustainable existence as a nature-based tourism lifestyle by involving local workers in the transformation process (2024).

In this sense, it is necessary to raise awareness of the impact of Climate Change in the community and among tourism stakeholders to identify risks and weaknesses and develop strategies for addressing them.

It is crucial to build adaptation systems for CC. These systems, such as planning in touristic environments, swift alert and evacuation systems, should be robust and adaptable to change. They should be able to face uncertainty and adapt to changes in case of natural disasters like hurricanes or tsunamis. The capacity to adapt corresponds to the capacity to develop new adaptive pathways. This emphasis on the urgency and necessity of these measures can invoke a sense of responsibility and prompt action, including the risk of Climate Change in the agenda of policymakers for fostering climate resilience through the development of adaptive management plans, zoning regulations, and building codes that account for climate-related risks (Louis et al., 2024).

The Fijian tourism strategy recommends "Enforce building codes that promote climate resilient infrastructure" as promoting renewable energy utilization in the tourism industry.

A transformative approach is needed in tourism policy and practice to face the Anthropocene. This approach implies that tourism needs to be measured in specific relation to the boundaries and limits *vis-a-vis* the Earth and humanity at the global scale" and "reshaping of the Earth for human purposes and to climate change" (Gren & Huijbens, 2014).

Other strategies, main for Coastal tourism, which mostly affects Sicily, consist of diversifying Tourism Products by deseasonalizing tourism and furnishing a wide offer of tourism supply, such as cultural tourism, eco-tourism, Eno gastronomy, or other sustainable tourism practices, in supporting economies and communities at the local level.

Implementing monitoring and assessment at the local level, both for Climate and tourism, to improve decision-making would help evaluate phenomena and develop strategies.

The relationship between Climate Change and Tourism is being investigated and implemented in the Mediterranean, both in academia and in terms of political attention. The limitations of this study are due to the need for more specific papers available in the Mediterranean area. Despite this prominent lack, we have found other studies with specific characteristics related to climate and type of tourism. These studies have allowed us to identify some strategies to address Climate Change in Tourism in the Mediterranean. Future studies could focus on developing other aspects of Climate Change in tourism, such as its impact on local communities and cultural heritage, which should be covered here to raise awareness of this prominent issue.

Results, Discussion and Recommendations

Climate Change is an issue that transcends traditional boundaries of policymaking, on the contrary, it requires approaches that go beyond conventional institutional responsibilities (Becken et al., 2020) by implementing cross-sectoral policies. A necessary effort is needed to "integrate the humanities and social sciences more fully into transdisciplinary environmental change research programmes" (Gren & Huijbens, 2014).

Moreover, it is necessary to quickly proceed with the adoption of a strategy for the abatement of CO_2, specifically for Tourism, as a strategic sector by investing in energy-efficient systems, in the expansion of renewable energy power, and reducing fossil fuel. Decarbonizing Tourism means decarbonizing global economies and staying within Planetary limits and boundaries. Moreover, Scott reveals that the cost of implementing carbonization can be easily supported with the taxes and the fees that companies already pay without any significant increase (Scott et al., 2015).

Transforming the tourism sector into a decarbonized sector means facing the global challenge of Climate Change (Scott & Gössling, 2018) in Tourism and advancing long-term competitiveness by enforcing private and public investment. This investments requires a strong commitment and integration between the public and private sectors, enhancing destination competitiveness and resilience. (Landauer et al., 2018). Climate Change is a prominent challenge, including for the tourism sector. If the BAU as TAU continues to grow, tourist seasons and attractions will be seriously threatened by Climate Change in the next decades. The magnitude of future climate change risks will be determined by today's choice (Scott & Gössling, 2018). Since business as usual model is no longer possible, new strategies of cooperation among the private and public sectors for the tourism sector should be implemented. Moreover, it is necessary to decouple the tourism sector from the use of natural resources, including greenhouse gas emissions; this would ensure the sector's activity (UNWTO, 2019).

The tourism sector has a responsibility to face Climate Change through appropriate prevention strategies, ensure the well-being of tourists and continue ensuring that tourism destinations, including coastal tourism, have a prominent heritage to preserve for future generations.

There is a need to act promptly to mitigate territory and tourism industry. Every economic investment will cause lesser than covering the cost of potentially disaster (Galeotti & Roson, 2012). Identifying the nexus between Tourism and Climate Change was only the first step of the research. The vulnerability assessment of coastal tourism in the Mediterranean has specific vulnerabilities related to the geographical configuration that require specific action to prevent risk and disasters.

The findings emphasize the urgent need for adaptation and resilience strategies to cope with Climate Change, which will ensure tourism for future generations in the Mediterranean area.

There is a need to understand further how it has become necessary to research and develop a national strategy for Climate Change adaptation to address tourism flow and ensure the sustainability of tourism for future generations. The future of tourism depends on the implementation of strategic policy that will be implemented today (Scott & Gössling, 2018). Further research should focus on the economic Health Impact of Climate Change as the Catastrophe of the Anthropocene on tourism to track trends, gaps, and future direction. Other studies should be developed to support and facilitate local policymakers in implementing tourism policies to adapt to Climate Change.

References

Ali, E., Cramer, W., Carnicer, J., Georgopoulou, E., Hilmi, N. J. M., Le Cozannet, G., et al. (2022). Cross-chapter paper 4: Mediterranean region. In H. -O. Pörtner, D. C. Roberts, M. Tignor, E. S. Poloczanska, K. Mintenbeck, A. Alegría, M. Craig, S. Langsdorf, S. Löschke, V. Möller, A. Okem, & B. Rama (Eds.), *Climate change 2022: Impacts, adaptation and vulnerability. Contribution of working group II to the sixth assessment report of the intergovernmental panel on climate change* (pp. 2233–2272). Cambridge: Cambridge University Press. https://doi.org/10.1017/9781009325844.021.

Amelung, B., & Moreno, A. (2009). Impacts of climate change in tourism in Europe. In L. Szabo, & J. Ciscar Martinez (Eds.), *PESETA-tourism study* (EUR 24114 EN) (JRC55392). Publications Office of the European Union.

Angelini, A., Farioli, F., Mattioli, G. F., & Scalia, M. (2015). Le due crisi: Crisi del capitalismo e crisi ambientale. Una Soluzione Sostenibile. *Culture Della Sostenibilità, 8*(16), 96–114.

Angra, D., & Sapountzaki, K. (2022). Climate change affecting forest fire and flood risk—facts, predictions, and perceptions in Central and South Greece. *Sustainability., 14*, 13395. https://doi.org/10.3390/su142013395.

Arabadzhyan, A., Figini, P., García, C., González, M. M., Lam-González, Y. E., & León, C. J. (2021). Climate change, coastal tourism, and impact chains–a literature review. *Current Issues in Tourism, 24*(16), 2233–2268. https://doi.org/10.1080/13683500.2020.1825351.

Becken, S., Whittlesea, E., Loehr, J., & Scott, D. (2020). Tourism and climate change: Evaluating the extent of policy integration. *Journal of Sustainable Tourism, 2810*, 1–22. https://doi.org/10.1080/09669582.2020.1745217.

Bizzarri, C., & Pedrana, M. (2018). Gli impatti dei cambiamenti climatici sul turismo. Un'analisi delle politiche di intervento. *Rivista di Scienze del Turismo-Ambiente Cultura Diritto Economia, 8*(1–2), 1. https://doi.org/10.7358/rst-2017-01-bipe.

Bueddefeld, J. N., Erickson, B. (2022). Wild bears, real bears and zoo bears: Authenticity and nature in Anthropocene tourism. *Tourist Studies, 22*(4), 373–392. https://doi.org/10.1177/14687976221131894. Epub 2022. PMID: 36440384; PMCID: PMC9684654.

Burke, L., Greenhalgh, S., Prager, D., & Cooper, E. (2008). *Coastal capital: The economic contribution of Tobago's coral reefs*. World Resources Institute.

Chakraborty, A. (2022). Geodiversity and tourism sustainability in the anthropocene. *Tourism and Hospitality, 3*(2), 496–508. https://doi.org/10.3390/tourhosp3020031.

Change IPCC. (2023). *Climate change 2022–impacts, adaptation and vulnerability: Working group ii contribution to the sixth assessment report of the intergovernmental panel on climate change* (1st ed.). Cambridge University Press. https://doi.org/10.1017/9781009325844.

Colloca C., & Lipari L. (2022). Attractiveness of coastal towns in the South of Italy. International tourist flows and consumption in Naples, Bari, Cagliari, Messina and Reggio Calabria. *Fuori Luogo Journal of Sociology of Territory, Tourism, Technology, 13*(3), 157–169. https://doi.org/10.6093/2723-9608/8985.

European Commission. (2017). *Western Mediterranean: Actions for the sustainable development of the blue economy.* Brussels, https://ec.europa.eu/commission/presscorner/detail/en/IP_17_924.

Crutzen, P. J., & Stoermer, E. F. (2000). The Anthropocene. In *The 'Anthropocene'* (pp. 479–490). Yale University Press. https://doi.org/10.12987/9780300188479-041.

Cusumano, M. (2022). The UNESCO serial property of Arab-Norman Palermo. An assessment from a sustainable development perspective. *Fuori Luogo. Rivista di Sociologia del Territorio, Turismo, Tecnologia, 13*(3), 171–179. https://doi.org/10.6093/2723-9608/9222.

Darmaraki, S., Somot, S., Sevault, F., Nabat, P., Narvaez, W. D. C., Cavicchia, L., Djurdjevic, V., Li, L., Sannino, G., & Sein, D. V. (2019). *Future evolution of marine heatwaves in the Mediterranean sea.* Springer.

Day, J. Chin, N., Sydnor, S., Widhalm, M., Shah, K., & Dorworth, L. (2021). Implications of climate change for tourism and outdoor recreation: An Indiana, USA, case study. *Climatic Change*, 169. https://doi.org/10.1007/s10584-021-03284-w.

Dube, K. (2024). Evolving narratives in tourism and climate change research: Trends, gaps, and future directions. *Atmosphere, 15*(4), 455. https://doi.org/10.3390/atmos15040455.

Elsayed, M. R. A. E. (2023). The impact of climate change on international tourism: Evidence from Egypt. *International Journal of Energy Economics and Policy, 13*(2), 379–390. https://doi.org/10.32479/ijeep.14045.

Filippelli, G., Freeman, J. Gibson, P., Jay, S., Moreno M. M., Ogashawara, I., Rosenthal, F., Wang, Y., & Wells, E. (2020). Climate change impacts on human health at an actionable scale: A state-level assessment of Indiana, USA. *Climatic Change*, 163. https://doi.org/10.1007/s10584-020-02710-9.

Fletcher, R. (2018). Ecotourism after nature: Anthropocene tourism as a new capitalist "fix." *Journal of Sustainable Tourism, 27*, 1–14. https://doi.org/10.1080/09669582.2018.1471084.

Galeotti, M., & Roson, R. (2012). Economic impacts of cli- mate change in Italy and the Mediterranean: Updating the evidence. *Journal of Sustainable Development, 5*(5), 27–41.

Gössling, S., Peeters, P., Hall, C. M., Ceron, J.-P., Dubois, G., Lehmann, L. V., & Scott, D. (2012). Tourism and water use: Supply, demand and security–an international review. *Tourism Management, 33*, 1–15. https://doi.org/10.1016/j.tourman.2011.03.015.

Gössling G., Balas, M., Mayer, M., Sun, Y. Y. (2023). A review of tourism and climate change mitigation: The scales, scopes, stakeholders and strategies of carbon management. *Tourism Management, 95*, 2023. 104681. ISSN 0261-5177.

Government of Fiji. (2005). *Fiji's first national communication under the framework convention on climate change.* Suva, Fiji.

Gren, M., & Huijbens, E. H. (2014). Tourism and the anthropocene. *Scandinavian Journal of Hospitality and Tourism, 14*(1), 6–22. https://doi.org/10.1080/15022250.2014.886100.

Hindley, A., & Font, X. (2014). Ethics and influence sin tourist perceptions of climate change. *Current Issues in Tourism, 20*(16), 1–17.

IPCC Climate Change. (2023). Synthesis report. Contribution of working groups I, II and III to the sixth assessment report of the intergovernmental panel on climate change. In H. Lee, & J. Romero (Eds.), *IPCC* (pp. 35–115). Geneva, Switzerland. https://doi.org/10.59327/IPCC/AR6-9789291691647.

Landauer, M., Goodsite, M. E., & Juhola, S. (2018). Nordic national climate adaptation and tourism strategies–(how) are they interlinked? *Scandinavian Journal of Hospitality and Tourism, 18*(sup1), S75–S86. https://doi.org/10.1080/15022250.2017.1340540.

Lloret J., Carreño A., Carić H., San, J., & Fleming L. E. (2021). Environmental and human health impacts of cruise tourism: A review. *Marine Pollution Bulletin, 173*(Part A), 112979. ISSN: 0025-326X. https://doi.org/10.1016/j.marpolbul.2021.112979.

Lo Verde F. M., Corvo P., & Pepe V. (2023). *Sociologia dei consumi turistici.* Pearsons, MyLab.

Louis, F., Lucas, D., & Elisha, B. (2024). *Climate change and coastal tourism: Impacts, adaptation, and resilience planning in Bangladesh.* Tourism and Hospitality.

Mgadle, A. (2022). An examination of climate change impacts on coastal tourism in Durban. Dissertation thesis. https://doi.org/10.13140/RG.2.2.32200.60166.

Morrison, A. M. (2022). Editorial: Land issues and their impact on tourism development. *Land, 11*, 658. https://doi.org/10.3390/land11050658.

Rantala, O., Höckert, E., Anttila, S., Ranta, S., & Valtonen, A. (2024). Proximity and tourism in the anthropocene. *Annals of Tourism Research, 105*, Article 103733. https://doi.org/10.1016/j.annals.2024.103733.

Ronchi, E., Barbabella, A., Montanini, C., & Ferruccio, V. (2023). *Troppa o troppo poca: l'acqua in Italia in un clima che cambia Special Report.* Retrieved April 30, 2024, from https://italyforclimate.org/wp-content/uploads/Acqua-e-clima-Special-Report-2023-Italy-for-Climate.pdf.

Salim, E., Mayer, M., Sacher, P., Ravanel, L. (2022). Visitors'motivations to engage in glacier tourismin the European Alps: Comparison of six sites in France, Switzerland, and Austria. *Journal Sustainable Tourism*, 1–21. https://doi.org/10.1080/09669582.2022.2044833.

Samora-Arvela, A., Vaz, E., Ferreira, J., & Panagopoulos, T. (2023). Tourism, climate change and well-being: The products' diversity as an opportunity. In *Geography of happiness.* Contributions to Regional Science. https://doi.org/10.1007/978-3-031-19871-7_7.

Scott, D., & Gössling, S. (2018). *Tourism and climate change mitigation: Embracing the Paris agreement–pathways to decarbonisation.* European Travel Commission, Brussels.

Scott, D., Gössling, S. M., Hall, C., & Peeters, P. (2015). Can tourism be part of the decarbonized global economy? The costs and risks of alternate carbon reduction policy pathways. *Journal of Sustainable Tourism.* https://doi.org/10.1080/09669582.2015.1107080.

Scott, D., Hall, C. M., Rushton, B., & Gössling, S. (2023). A review of the IPCC sixth assessment and implications for tourism development and sectoral climate action. *Journal of Sustainable Tourism*, 1–18. https://doi.org/10.1080/09669582.2023.2195597.

Song, W. (2023). Climate change and tourism sustainability on Jeju Island landscape. *Sustainability, 15*, 88. https://doi.org/10.3390/su15010088.

Spectora, S., & Highamb, J. E. S. (2019). Space tourism in the Anthropocene. *Annals of Tourism Research, 79*, 102772. https://doi.org/10.1016/j.annals.2019.102772.

Steiger, R., & Abegg, B. (2018). Ski Areas' competitiveness in the light of climate change: Comparative analysis in the Eastern Alps. In: D. Müller, & M. Więckowski (Eds.), *Tourism in Transitions.* Geographies of Tourism and Global Change. Cham: Springer. https://doi.org/10.1007/978-3-319-64325-0_11.

UNWTO. (2019). Transport-related CO_2 emissions of the tourism sector—modelling results. In *World tourism organization and international transport forum.* Madrid: UNWTO. https://doi.org/10.18111/9789284416660.

Zhang, L. Y. (2022). Exploring the effects of tourism development on air pollution: Evidence from the panel smooth transition regression model. *International Journal of Environmental Research and Public Health, 19*(14), 8442. https://doi.org/10.3390/ijerph19148442.

Sequential Move Model for Quantity and Quality Competition in the Restaurant Industry

Flávio Ferreira and Guilherme Koerich

Abstract This paper examines a competition in both quality and quantity between two asymmetric restaurants. The model that we consider is a Stackelberg duopoly on quantities (sequential decision-making), after a simultaneous choice of quality. Our findings indicate that the restaurant with the best reputation provides a greater variety of higher-quality food compared to its competitor. We also examine how variations in restaurants' reputations impact market equilibrium outcomes. We demonstrate that an increase in the reputation gap between restaurants boosts the profits of both establishments, enhances consumer surplus, and improves social welfare. Additionally, in comparison to the scenario under the Cournot model (simultaneous decision-making), small restaurant's profits are lower under Stackelberg model, while the opposite holds for the large restaurant's profits and for social welfare.

Keywords Game theory · Stackelberg model · Food industry · Reputation

Introduction

This study offers a nuanced exploration into the intricacies of the interplay between culinary quality and the number of meals offered within the competitive landscape of the restaurant industry. As consumer preferences evolve, the demand for diverse dining experiences presents restaurants with the challenge of balancing exceptional

F. Ferreira (✉)
UNIAG, CiTUR, ESHT, Polytechnic of Porto, rua D. Sancho I, n.º 981, 4480-876 Vila do Conde, Portugal
e-mail: fff@esht.ipp.pt

G. Koerich
SIGMO, PGC, Instituto Federal de Educação, Ciência e Tecnologia de Santa Catarina, Universidade Federal de Santa Catarina, Rua 14 de Julho, 150, Florianópolis 88075-010, Brazil

A. L. Negrușa et al. (eds.), *Sustainable and Resilient Businesses in the Global Economy*, Springer Proceedings in Business and Economics,
https://doi.org/10.1007/978-3-031-75883-6_9

quality with the provision of an appropriate number of meals. Employing a comprehensive research methodology, this investigation integrates a theoretical market analysis to unravel the strategic decision-making processes behind quality and meal quantity setting.

In this study, we examine a market consisting of only two restaurants that compete based on their decisions regarding quantity and quality. Key concepts of duopoly (a market with only two firms) are related to strategic behaviour, and the analysis of strategic behaviour is in the core of Game Theory. Game Theory examines mathematical models that represent strategic interactions between rational decision-makers.

By synthesizing these analyses, the study aims to identify optimal strategies that align quality and meal quantity, providing restaurateurs with insights into creating comprehensive dining experiences that cater to the diverse needs of contemporary consumers.

This integrated approach not only enhances practical strategies for restaurant management but also contributes to academic understanding by shedding light on the nuanced relationship between quality and the number of meals in the competitive landscape. The findings serve as a valuable resource for industry practitioners, guiding them in making informed decisions that strike a harmonious balance between culinary excellence and meeting the diverse meal preferences of their clientele. Ultimately, this research aims to establish a framework for achieving a successful integration of quality and meal quantity in restaurant competition, fostering dining experiences that are both memorable and accommodating.

Food quality has received particular attention worldwide. So, food industry managers must provide great importance to quality. This paper aims to explore competition in terms of both quantity and quality within the food industry. We employ a three-stage dynamic Stackelberg model, a widely used method in the literature on industrial organization. Firstly, both restaurants simultaneously choose their quality levels. Subsequently, they must determine quantities. For these latest strategic variables, we use the model proposed by Stackelberg (1934): One restaurant behaves as a follower, determining its output based on the fixed output of the other restaurant; Meanwhile, the second restaurant acts as a leader, anticipating the follower's response and choosing its output accordingly. In earlier literature, the concept of endogenous leader–follower dynamics has frequently been embedded within the framework of a timing game engaged in by oligopolistic firms. Hamilton and Slutsky (1990) created an 'extended game' framework wherein each company determines the timing of production. Several research papers have discussed the effects of broadly defined informational heterogeneity between firms (Mailath, 1993; Norman, 1997). Conversely, When oligopolists share initial uncertainty about market demand, an early producer can capitalize on less information to secure a strategic advantage through commitment. In contrast, a later producer, armed with more information, lacks this initial advantage of commitment. Firms will be able to control not only the quantity and price variables but also the quality of their products. Hence, industrial managers should prioritize considerations of price, quantity, and quality as the primary factors (Lambertini & Tampieri, 2017). Shen et al. (2021) examined the decisions made by food enterprises regarding yield and quality under conditions with and without resource constraints. Utilized Cournot models to analyse the equilibrium yield and quality, demonstrating that enterprises experiencing strong product substitutability tend to enhance both the yield and quality of their products.

We also analyse how restaurant reputation affects equilibrium outputs. Moreover, we also conduct a comparison of our results regarding the Stackelberg competition in quantity (sequential decisions), with those obtained by Chen et al. (2018) in the context of Cournot competition (simultaneous decisions). Stackelberg model is widely used to analyse other market structures (Ferreira & Bode, 2017; Ferreira et al., 2021).

Nie and Chen (2014) analysed a model of oligopoly within the food industry, where the firms offer identical products and compete by setting quantities simultaneously. Their findings revealed that as the number of firms increases, both the output per firm and the price of food decrease. Additionally, the profits for each firm diminish with an increase in the number of firms, while social welfare improves as more firms enter the market.

Yang and Nie (2016) examined the asymmetrical competition in the restaurant industry, focusing on product substitutability through the formulation of a two-step game model. The equilibrium outcomes were determined under both Cournot and Stackelberg models.

Chen and Nie (2016) examined a mixed duopoly competition within the retaurant industry, consisting of one profit-driven firm and another firm that incorporates corporate social responsibility (CSR) into its goals. In their theoretical construction, the firms first select the quality for their products, followed by determining the quantities. At every phase of the game, the firms decide simultaneously. The study revealed that CSR enhances both the quality and quantity of food produced by the CSR-oriented firm, while it decreases both for the profit-maximizing firm.

Ferreira et al. (2021) investigated a food market featuring two restaurants with differing reputations competing first in quality and then in quantity. In the quality stage, the restaurant with the lower reputation takes on a leadership role. In the quantity stage, both restaurants make their decisions simultaneously. Their findings revealed that the leadership position of the smaller restaurant can be advantageous, as its profit surpasses that of its competitor when the reputation difference between them is minor. Additionally, they concluded that a disparity in reputation increases total quality investments and total outputs at equilibrium.

This paper is organized as follows: The model is established in the next section. After that, we describe the model, analyse it, and present the results. Furthermore, and we also provide a comparison between sequential and simultaneous decisions. The paper concludes in the final section.

Research Purpose and the Methodological Approach

We will utilize a game theoretical model integrated into the decision-making process for our research. This model depicts a scenario where various actors pursue their self-interest to maximize profit, with the payoff contingent upon the collective actions of all players. Game theory, a mathematical framework, serves as the analytical tool to dissect strategic interactions among diverse individuals or entities.

To manage the research's complexity, we limit our focus to just two restaurants in the study—a small restaurant R_1 and a large restaurant R_2—in the market. Both restaurants compete by varying their quality and quantity choices.

We adopt Chen et al. (2018)'s market structure, with the modification that restaurants make their decisions sequentially, instead of simultaneously. The representative consumer maximizes

$U(q_i, x_i) - p_1q_1 - p_2q_2,$

where q_i is the quantity offered by restaurant R_i, x_i is the quality associated with restaurant R_i and p_i is its price, with $i = 1, 2$. The function U, which depends on both quantity and quality, is defined as follows:

$U(q_i, x_i) = (\alpha + \beta x_1)q_1 + (\alpha + (\beta + \tau)x_2)q_2 - \frac{1}{2}(q_1^2 + q_2^2) - \gamma q_1q_2,$

where x_i represents the quality attributed to restaurant R_i; α is a positive constant, β is the fundamental reputation; τ represents the disparity in reputation between the two restaurants, and γ quantifies the extent to which products from the two restaurants can be substituted for each other, with $i = 1, 2$. To simplify, we assume the reputation rating of the small restaurant equal to 5/2, and we fix $\gamma = 1/2$. Therefore, our attention is directed towards the disparity in reputations. U can be restated as

$U(q_i, x_i) = \left(\alpha + \frac{5}{2}x_1\right)q_1 + \left(\alpha + \left(\frac{5}{2} + \tau\right)x_2\right)q_2 - \frac{1}{2}\left(q_1^2 + q_2^2\right) - \frac{1}{2}q_1q_2.$

Therefore, the inverse demand functions for the two restaurants are expressed as

$p_1 = \alpha + \frac{5}{2}x_1 - q_1 - \frac{1}{2}q_2,$

$p_2 = \alpha + \left(\frac{5}{2} + \tau\right)x_2 - q_2 - \frac{1}{2}q_1.$

We hypothesize that the overall costs incurred by small and large restaurants depend on both quantity and quality, defined by

$C_i = \frac{1}{2}q_i^2 + \frac{1}{2}x_i^2 + q_ix_i,$

$i = 1, 2$. The first term represents the investment costs in quantity, the second term signifies the investment costs in quality, and the final term denotes the interaction between quantity and quality investments. So, the profit π_1 of the small restaurant R_1 can be determined by

$$\pi_1 = \left(\alpha + \frac{5}{2}x_1 - q_1 - \frac{1}{2}q_2\right)q_1 - \frac{1}{2}q_1^2 - \frac{1}{2}x_1^2 - q_1x_1$$

and the profit π_2 of the large restaurant R_2 is defined as

$\pi_2 = \left(\alpha + \left(\frac{5}{2} + \tau\right)x_2 - q_2 - \frac{1}{2}q_1\right)q_2 - \frac{1}{2}q_2^2 - \frac{1}{2}x_2^2 - q_2x_2.$

Social welfare W is determined by

$W = CS + \pi_1 + \pi_2,$

where

$$CS = \frac{1}{2}(q_1^2 + q_2^2) + \frac{1}{2}q_1 q_2$$

is the consumer surplus.

We adopt a three-stage game as follows:

- In the initial stage, both restaurants select their quality standards, x_1 and x_2, at the same time;
- In the second stage, the large restaurant R_2 chooses its optimal quantity q_2;
- In the third stage, the small restaurant R_1 chooses its optimal quantity q_1.

Results

Here, we will examine the model described earlier. To determine the subgame perfect equilibrium, the game will be solved by backwards induction. Beginning from the final stage, the small restaurant R_1 addresses the optimization problem $\max_{q_1} \pi_1$. Finding the solution to the equation $\partial\pi_1/\partial q_1 = 0$ yields

$$q_1 = \frac{2\alpha + 3x_1 - q_2}{6}. \tag{1}$$

Putting this result into π_2 and solving $\partial\pi_2/\partial q_2 = 0$ yields

$$q_2 = \frac{10\alpha - 3x_1 + 6(3 + 2\tau)x_2}{34}. \tag{2}$$

So,

$$q_1 = \frac{58\alpha + 105x_1 - 6(3 + 2\tau)x_2}{204}. \tag{3}$$

Putting these results into π_1 and π_2, and solving, simultaneously,

$$\begin{cases} \dfrac{\partial\pi_1}{\partial x_1} = 0 \\ \dfrac{\partial\pi_2}{\partial x_2} = 0, \end{cases}$$

we get the following quality levels at equilibrium[1,2]:

[1] To ensure $x_i > 0$, we fix $0 < \tau < \left(\sqrt{87} - 9\right)/6$.

[2] In this paper, we denote the equilibrium outputs using the superscript $*$.

$x_1^* = \frac{35\alpha(1-18\tau-6\tau^2)}{28-279\tau-93\tau^2}$ and $x_2^* = \frac{25\alpha(3+2\tau)}{2(28-279\tau-93\tau^2)}$.

Inserting these results into Eqs. (2) and (3) yields the following equilibrium quantities:

$q_1^* = \frac{68\alpha(1-18\tau-6\tau^2)}{3(28-279\tau-93\tau^2)}$ and $q_2^* = \frac{25\alpha}{28-279\tau-93\tau^2}$.

The resulting equilibrium profits are:

$\pi_1^* = \frac{949\alpha^2(1-18\tau-6\tau^2)^2}{6(28-279\tau-93\tau^2)^2}$ and $\pi_2^* = \frac{625\alpha^2(1-18\tau-6\tau^2)^2}{24(28-279\tau-93\tau^2)}$.

It is straightforward to obtain the following result:

Proposition 1. At equilibrium, the large restaurant provides a greater quantity of higher-quality food compared to its competitor. Additionally, the large restaurant generates more profits compared to the small restaurant.

So, this result highlights the competitive advantage of the large restaurant, where it leverages its resources and capabilities to offer a greater quantity and higher quality of food than its rival, satisfying consumer preferences and dominating the market. A possible economic interpretation of the above result is that the restaurant's superior reputation attracts more customers, increasing demand for its offerings. This enables the restaurant to produce and sell greater quantities of food. Higher demand might also allow the restaurant to invest in better ingredients and preparation methods, resulting in higher quality food.

In addition, we get the values of consumer surplus *CS* and social welfare:

$$CS^* = \frac{\alpha^2(15349 - 258264\tau + 1412088\tau^2 + 998784\tau^3 + 166464\tau^4)^2}{18(28 - 279\tau - 93\tau^2)^2}$$

and

$$W^* = \frac{\alpha^2(85909-1510524\tau+9178908\tau^2+6454944\tau^3+1075824\tau^4)^2}{72(28-279\tau-93\tau^2)^2}.$$

Comparative static analysis

In this section, we assess how varying restaurant reputations impact the equilibrium outcomes. Based on the findings outlined earlier,

$$\frac{\partial q_1^*}{\partial \tau} = -\frac{1700\alpha(3+2\tau)}{28 - 279\tau - 93\tau^2} < 0$$

$$\frac{\partial q_2^*}{\partial \tau} = \frac{2325\alpha(3+2\tau)}{(28 - 279\tau - 93\tau^2)^2} > 0$$

$$\frac{\partial (q_1^* + q_2^*)}{\partial \tau} = \frac{625\alpha(3+2\tau)}{(28 - 279\tau - 93\tau^2)^2} > 0$$

$$\frac{\partial x_1^*}{\partial \tau} = -\frac{2625\alpha(3+2\tau)}{(28 - 279\tau - 93\tau^2)^2} < 0$$

$$\frac{\partial x_2^*}{\partial \tau} = \frac{25\alpha\left(893 + 558\tau + 186\tau^2\right)}{\left(28 - 279\tau - 93\tau^2\right)^2} > 0$$

$$\frac{\partial\left(x_1^* + x_2^*\right)}{\partial \tau} = \frac{25\alpha\left(263 + 138\tau + 186\tau^2\right)}{\left(28 - 279\tau - 93\tau^2\right)^2} > 0$$

$$\frac{\partial \pi_1^*}{\partial \tau} = -\frac{23725\alpha^2(3 + 2\tau)\left(1 - 18\tau - 6\tau^2\right)}{\left(28 - 279\tau - 93\tau^2\right)^3} < 0$$

$$\frac{\partial \pi_2^*}{\partial \tau} = \frac{625\alpha^2(3 + 2\tau)\left(161 - 558\tau - 186\tau^2\right)}{4\left(28 - 279\tau - 93\tau^2\right)^3} > 0$$

$$\frac{\partial CS^*}{\partial \tau} = \frac{25\alpha^2\left(8889 + 84874\tau + 78948\tau^2 + 17544\tau^3\right)}{3\left(28 - 279\tau - 93\tau^2\right)^3} > 0$$

$$\frac{\partial W^*}{\partial \tau} = \frac{25\alpha^2\left(37617 + 830272\tau + 805194\tau^2 + 178932\tau^3\right)}{12\left(28 - 279\tau - 93\tau^2\right)^3} > 0$$

Therefore, we one of the primary findings is:

Proposition 2.

(a) The widening gap in reputation between restaurants diminishes the quality and productivity of small restaurants while enhancing those of larger establishments. This overall trend results in an increase in both the overall quality and total output within the market.
(b) The widening gap in reputation between restaurants diminishes profits for smaller establishments while boosting profits for larger ones.
(c) The widening gap in reputation among restaurants enhances both consumer surplus and social welfare.

According to the interpretation of the above statements, the widening reputation gap results in a decline in the small restaurant's quality and output but significantly boosts the large restaurant's performance. The overall effect is beneficial for the market because the total quality of food improves and the aggregate quantity of food increases, driven by the large restaurant's enhanced efficiency and investment in quality. Moreover, the increasing reputation difference shifts customer preference toward the large restaurant, boosting its profits due to higher revenue, economies of scale, and enhanced reinvestment capabilities. Conversely, the small restaurant faces declining profits as it loses customers and revenue, struggles with fixed costs, and possibly lowers its quality in cost-cutting efforts, which further exacerbates its financial difficulties. Furthermore, enhanced reputation of the large restaurant raises consumer surplus through higher satisfaction and availability of quality food. Social welfare increases due to the efficient reallocation of resources and the net gains in consumer and producer surpluses.

Comparisons with simultaneous move model

In this section, we conduct a comparison of the results above with those obtained by Chen et al. (2018) in the context of Cournot competition. At equilibrium, their paper provides us with the following outputs[3]

$x_1^{C*} = \frac{108\alpha(13-216\tau-72\tau^2)}{1079-10368\tau-3456\tau^2}$ and $x_2^{C*} = \frac{468\alpha(3+2\tau)}{1079-10368\tau-3456\tau^2}$,

$q_1^{C*} = \frac{70\alpha(13-216\tau-72\tau^2)}{1079-10368\tau-3456\tau^2}$ and $q_2^{C*} = \frac{910\alpha}{1079-10368\tau-3456\tau^2}$.

Furthermore,

$$\pi_1^{C*} = \frac{1518\alpha^2(13-216\tau-72\tau^2)^2}{(1079-10368\tau-3456\tau^2)^2} \text{ and } \pi_2^{C*} = \frac{1014\alpha^2(253-1286\tau-432\tau^2)^2}{(1079-10368\tau-3456\tau^2)^2},$$

$$CS^* = \frac{7350\alpha^2(169 - 2808\tau + 14616\tau^2 + 10368\tau^3 + 1728\tau^4)^2}{(1079 - 10368\tau - 3456\tau^2)^2}$$

and

$$W^* = \frac{18\alpha^2(97513-1693224\tau+9720648\tau^2+6856704\tau^3+1142784\tau^4)^2}{(1079-10368\tau-3456\tau^2)^2}.$$

From the results above, it is evident that $\pi_1^{C*}<\pi_2^{C*}$. So, as in the sequential move model presented previously, the large restaurant also generates greater profits compared to the small restaurant when both make decisions simultaneously.

Upon comparing the output equilibrium between the sequential move game and the simultaneous move game, the following results emerge:

Proposition 3. Under Cournot competition, the small restaurant achieves a higher equilibrium quality level and quantity compared to Stackelberg competition. Conversely, for large restaurants, the situation is reversed, with Stackelberg competition resulting in higher equilibrium quality level and quantity.

We can interpret this result as follows: In Stackelberg competition, the small restaurant, acting as a follower, must adapt to the leader's output decisions, often resulting in lower equilibrium quality and quantity due to its reduced strategic influence. On the other hand, the large restaurant, as the leader, can set its output first, allowing it to anticipate competitor reactions and optimize its plan, resulting in improved equilibrium quality and quantity.

Proposition 4. The small restaurant achieves higher profits in Cournot competition compared to Stackelberg, whereas the large restaurant experiences higher profits in Stackelberg competition than in Cournot.

This statement suggests that in Stackelberg competition, the small restaurant's profits are lower because it must adapt to the leader's output decisions, reducing its ability to strategically influence the market and optimize its profitability. On the other hand, the large restaurant's profits are higher because it acts as the leader, setting its output first and strategically influencing the market. This allows the large restaurant to optimize its profit by anticipating and shaping the responses of its competitors.

[3] In this paper, the superscript C is employed to denote the Cournot model.

Proposition 5. Under Stackelberg competition, both consumer surplus and social welfare are higher compared to Cournot competition.

Let us give an interpretation of this result. In Stackelberg competition, the leader's ability to anticipate and influence market conditions typically leads to higher consumer surplus and social welfare compared to Cournot competition. This is because the sequential decision-making process allows for better coordination and optimization of output levels, resulting in more efficient resource allocation and greater overall societal benefit.

Conclusions

We examined a competition of both quality and quantity between two restaurants of differing asymmetry. The model we considered is a Stackelberg duopoly on quantities, after a simultaneous choice of quality. The results shows that there are different outputs for each type of restaurant, and the variance in restaurants' reputations significantly influences market equilibrium outcomes, namely on restaurants' profits and social welfare.

We observed that the restaurant with the highest reputation provides greater quantities of food with superior quality compared to its competitor. We also showed that as the reputation gap between restaurants widens, the profits of the larger restaurant increase while those of the smaller restaurant decrease. Additionally, this gap results in heightened consumer surplus and improved social welfare. Moreover, when compared to the scenario in Cournot competition, the small restaurant's profits decline under Stackelberg competition, whereas the profits of the large restaurant increase, as well as for social welfare.

For simplicity of computations, we assumed some parameters as fixed. It was the case of the parameter that measures the extent of product substitutability and the parameter that measures the reputation of the small restaurant. This is, naturally, a limitation of the paper. Analysing how changes in key parameters affect the results is left for future work.

Acknowledgements This work was supported by national funds through FCT/MCTES (PIDDAC): UNIAG, UIDB/04752/2020 (DOI https://doi.org/10.54499/UIDB/04752/2020) and UIDP/04752/2020 (DOI https://doi.org/10.54499/UIDP/04752/2020).

References

Chen, Y.-H., & Nie, P.-Y. (2016). Duopoly competition in food industry considering corporate social responsability. *Quality Assurance and Safety of Crops & Foods, 8*(1), 33–40.

Chen, Y.-H., He, Q., & Paudel, K. P. (2018). Quality competition and reputation of restaurants: The effects of capacity constraints. *Economic Research-Ekonomska Istrazivanja, 31*(1), 102–118.

Ferreira, F., & Bode, O. R. (2017). Licensing by royalties in a Stackelberg model when the follower is the innovator. *AIP Conference Proceedings, 1863*, 050002-1–050002-4.

Ferreira, F. A., Ferreira, F., & Fernandes, P. O. (2021). Price and quality competition in the restaurant industry: Effects of restaurants' reputation. *Journal of Tourism & Development, 37*, 21–29.

Ferreira, F. A., Ferreira, F., & Fernandes, P. O. (2022). Sequential quality competition in the restaurant industry: Effects of restaurants' reputation. In: V. Ş. A. Katsoni (Ed.), *Transcending borders in tourism through innovation and cultural heritage* (pp. 779–789). Springer Proceedings in Business and Economics.

Hamilton, J. H., & Slutsky, S. M. (1990). Endogenous timing in duopoly games: Stackelberg and Cournot equilibria. *Games and Economic Behavior, 2*, 29–46.

Lambertini, L., & Tampieri, A. (2017). Endogenous timing in quality choices and price competition. *Bulletin of Economic Research, 69*(3), 260–270.

Mailath, J. (1993). Endogenous sequencing of firms decision. *Journal of Economic Theory, 59*, 169–182.

Nie, P.-Y., & Chen, Y.-H. (2014). A general equilibrium analysis of food industry considering the food quality. *Agricultural Economics-Czech, 60*(7), 301–308.

Norman, H.-T. (1997). Endogenous Stackelberg equilibria with incomplete information. *Journal of Economics, 66*, 177–187.

Shen, N., Song, Y., Liu, D., & Štreimikienė, D. (2021). Food quality competition among companies and government food safety supervision under asymmetric product substitution. *Amfiteatru Economic Journal, 56*(23), 221–239.

von Stackelberg, H. (1934). *Marketform und Gleichgewicht*. Julius Sringer.

Yang, Y. -C. & Nie, P. -Y. (2016). Asymmetric competition in food industry with product substitutability. *Agricultural Economics—Czech, 62*(7), 324–333.

Factors Shaping Consumers' Experiences of Whiskies Tasting—Case Study Cluj-Napoca

Adina Letiția Negrușa and Oana Ruxandra Bode

Abstract The alcohol beverage industry has been focusing more and more on creating more enjoyable tasting experiences for their customers throughout time. With the resultant expansion of whisky production worldwide and the growing recognition of whisky as a beverage, the practice of whisky tastings at the client level has expanded since attendees are typically more receptive to new and unique tastes and sensations. Whisky's unique sensory qualities have made it one of the world's leading categories of spirit drinks. It is generally acknowledged that taste is the most important factor in determining how consumers choose whisky. Both this beverage and the wine must have a pleasing flavour for the consumer. The present study is mainly focused to determine the factors that play an important role in the motivations driving attendees to whiskies tasting sessions and to identify their overall satisfaction. Motives falling under the push and pull categories have both been studied. V Cramer coefficient statistical analysis was used to investigate the association between the type of consumer and their participation in tasting events. Considering the main goal of the present study, the investigated population consists of domestic consumers who have selected Bardot—Cocktail and Whisky Bar from Cluj-Napoca, Romania, for their whisky tasting experience. These consumers had the opportunity to discover special drinks less known to the public in Romania in one of the restaurants from Cluj-Napoca that make its mark in the hospitality industry. The present paper is one of the first scientific study in Romania that analyzes the factors influencing a whisky tasting experience in Romania.

Keywords Whisky tasting · Cluj-Napoca · Romania · Hospitality industry · Consumer

JEL classification L83 · L80 · M19 · Z32

A. L. Negrușa · O. R. Bode (✉)
Department of Hospitality Services, Faculty of Business, Babeș-Bolyai University, Cluj-Napoca, Romania
e-mail: oana.bode@ubbcluj.ro

A. L. Negrușa et al. (eds.), *Sustainable and Resilient Businesses in the Global Economy*, Springer Proceedings in Business and Economics,
https://doi.org/10.1007/978-3-031-75883-6_10

Introduction

In time, the alcohol beverage industry, which includes the world of beer, wine, cider and whisky, increased its popularity, leading to changing lifestyle patterns for the individuals and higher disposable incomes for the firms. The expanding demand for the consumption of alcoholic beverages all over the world conducted to the growing consumption of whiskey as well, and, in recent years, it has become the best-selling spirit drinks. This beverage is an old traditional drink, existing since 1494. Whisky has become very common among people and is mainly consumed as a recreational or rejuvenating drink. When consumers face the whisky-buying decision for the experience of drinking it, are influenced by different factors, such as the budget, the presentation, the type, the store to shop or the glassware.

The global whiskey market can be broadly divided into Irish, Canadian, American, Scotch and Other whiskeys, but the Scotch whiskey is the main one in terms of market share globally. The global whisky market is estimated to have an exponentially growth in the period of 2024–2032 reaching nearly USD 100.34 billion by 2032 from about USD 69.24 billion in 2023. Therefore, the growth rate of the global whisky market is projected to be approximative 4.20% yearly for the period 2024–2032. The increasing consumption of whisky across different countries is contributing to this market growth.

Although the quantity of whisky consumed is an important metric, knowing the amount consumed per person gives everyone a better idea of who the real whisky fans is in each nation. In 2023, the top five countries in the world for whisky consumption per capita are as follows: France (with an annual per capita consumption of 2.15 L on average), Uruguay (with an annual per capita consumption of 1.77 L on average), United States (with an annual per capita consumption of 1.41 L on average), Australia (with an annual per capita consumption of 1.3 L on average) and Spain (with an annual per capita consumption of 1.22 L on average). Therefore, France and Spain are the top European list countries in terms of whisky consumption per capita.

According to IWSR Country Summary Report 2023 for Romania, in 2022 the spirits drinks own 24.7% of the beverage alcohol market, an increase with almost 5% compared with the previous year. Out of all spirits type (flavoured spirits, rum, vodka, gin and genever), whisky is the leader in terms of value of ecommerce channel in Romania (1.975.000 RON).

Until 2022 the entire quantity of whisky in our country was imported. As specified in this report, for Romania 853,84 000 s 9-L cases in 2022, respectively 701.62 000 s 9-L cases in 2021 were imported. The top five imported whisky included Scotch, US whisky, Canadian whisky, Irish and then Japanese whisky. But October 2022 will remain a historic date for the beverage industry in Romania and beyond; is the date when the first Romanian single malt whiskey was officially launched: Carpathian Single Malt produced by Alexandrion Group the company that in 2022 owned 8,6% of spirits market in Romania.

In terms of consumption, as % share of consumption of whisky by channel in 2022, 45,5% was represented by on-premise whisky consumption and 54,5% by off-premise whisky consumption. The whisky consumption faced an important growth in our country in the last years, increasing from 5,2% in 2018 to 7,3% in 2022. It is forecasted that this increasing trend will continue, reaching 8,8% in 2027. The leading brand of whisky in Romania it is represented by Jack Daniel's US whisky (191,4 000 s 9-L cases in 2022), while the five whisky owners' brands are represented by Pernod Ricard (32,5%), Diageo (29,4%), Brown-Forman (22,5%), William Grant & Sons (3,8%), respectively Beam Suntory (3,3%).

Over time, the global whisky industry, has increasingly focused on figuring out and designing tasting experiences that are more pleasant for their clients. Since consumers at tastings are usually more open to new and novel tastes and experiences, the practice of whisky tastings at the client level has increased with the resulting spread of whisky production across the globe and the increasing reputation of whisky as a beverage. It is recognized as one of the world's leading categories of spirit drinks due to its complex sensory characteristics. Being a complex spirit, whisky is mainly popular and valued among maven consumers. It is generally agreed that what matters the most when it is necessary to identify how consumers select whisky, it is the taste that matters. This beverage, as well as wine, must taste good to the consumer.

In Romania, like in different other countries, whiskey is perceived as a beverage difficult to access and an expensive one. During the last years, some events are organized at the national level to change this perception by demonstrating that whisky can be surprisingly affordable compared to other beverages and to remove mental barriers for consumers, especially young people. The largest event dedicated to whisky culture in Romania is *Whisky Fest Romania*, organized annually in the capital of the country, Bucharest, since 2016. It is the first event dedicated to whisky culture on the Romanian market.

Among other rare events dedicated to whisky culture in Romania we can emphasize the whiskey tasting sessions organized regularly by Bardot bar and restaurant from Cluj-Napoca, having an interesting combination: a British pub, but with a French flavor, and with a twist. It is the place in Cluj-Napoca that mark the hospitality industry and where you can discover the world of whiskies. Bardot is not just a simple bar, but a real institution in the whisky world in Romania, being recognized as the largest seller of Single Malt per portion in the country.

The present study was conducted during *Bardot Whisky Days Festival* in April 12–13, 2024, at BARDOT restaurant from Cluj-Napoca. Even events of this type are rare in Romania and in Cluj-Napoca, consumers participating at the tasting enjoyed among more than 100 assortments. The event gathered nine whisky importers and distributors who have presented the largest whisky producers in the world, but also niche producers. Offering access to whiskey brands never seen before in Romania, and for those new to this world, a memorable initiation.

The research focuses on identifying why people attend whisky tasting events in Cluj-Napoca and how satisfied they are overall with the experience. Additionally, the idea of the present study is that there is no better opportunity to create a loyal customer than when whisky consumers participate at a whisky's tasting festival. There has been no previous research conducted in the area of consumer whisky tasting in Romania, according to the authors knowledge.

The literature review on the whisky industry and whisky tasting events is summarized in the first section of the current study. The study methodology employed is next highlighted, and lastly, the primary research findings and results are revealed.

Brief Literature Review

A food and beverage festival represents the best possibility to reveal off the place where it is organized. Additionally, gift the method of qualitative networking with different businesses, from restaurants and bars to new providers of different other supplementary products. On the other hand, different festival stakeholders have varied perspectives on the significance of a festival and its potential for local development.

An examination of their characteristics reveals that local festivals are typically limited to minority communities and have a regional focus. Vesci and Botti (2019) concluded that these events promote "hit and go" travel, where participants only spend a short time at a place. According to Tanford and Jung (2017), determining the level of satisfaction among attendees is crucial because festival attendance is crucial to the events' success. Regular participants guarantee that the festival will continue to be attended, assuring the event's financial sustainability through their consistent attendance and word-of-mouth promotion.

At the same time, these events are vital to a town's life because they foster a sense of place and community and hence increase social capital (Derrett, 2003). Egresi and Kara (2014) underline that visitors who attend small-scale local festivals spend money, which has an impact on the local economy. This is the case also for the whisky tasting festivals. Furthermore, Folgado-Fernández et al. (2019) underline the significance of small festivals for local economies, pertaining not just to the hospitality and tourism industry, but also to the consumption of different goods.

Food and beverage festivals offer company owners a chance to test and display their products as well as a means of getting first access to the market. These events are above all a place to supply and sell different products and share the supplier's passion for such kind of products (Park & Widyanta, 2022). From the perspective of the participants, people's participation in local life festivals can have a positive impact on their quality of life, especially in the case of residents who are seeing them as a fun way to spend their free time.

Since the alcoholic beverage industry, respectively whisky industry, has faced a significant growth during the last decade, we should consider whisky as a food or beverage product for use in research related to hospitality industry. As it can be found in the literature, studies related to alcoholic beverages tastings has primarily focused on wine (Bruwer et al., 2013; North, 2012; Oberfeld et al., 2009). There is definitely much less scientific study on spirits when compared with wine, as emphasized by Spence, (2011). It is revealed (Sester et al., 2013) that the drink choice, in the case of beer, can change as a function of the environment in which people make their beverage-related decisions. Therefore, further studies are needed and required to evaluate how other drinks, including whisky, are perceived.

Researchers studying consumer motivation have utilized varying methods to address inquiries regarding motivation. People are driven by a variety of reasons to attend tastings sessions; some of these reasons offer motivation. There is a variety of reasons why people attend such kind of events and some of these reasons offer additional advantages beyond the tasting session experience. These advantages could stem from connection, avoidance, achieving dreams, finding oneself, personal development, bonding with family, to taste whisky; to buy whisky; a day out; meet the whisky producers and learn about whisky or doing something different. Additional reasons could be the chance to sample different whiskies, participating in activities, and make purchases. Researchers frequently utilize demographic information, as well as psychographics and lifestyle traits, to investigate motivations.

Different factors can shape individual's preferences at a whisky tasting session. These factors consist of, but are not limited to, expertise of the taster and the order of the options (Mantonakis et al., 2009), the influence of the host (Thaler et al., 2013., Philp & Mantonakis, 2018), their personality, a specific context, social influence from other people participating at the tasting when they emphasize their impressions (Cialdini & Goldstein, 2004), and memory for flavours or whiskies that people have disliked or enjoyed in the past (Tversky & Kahneman, 1974).

Over the coming years, we consider that different research related to hospitality industry must be conducted to emphasize the importance of people's whisky drinking or tasting experiences. So far, it was investigating the role of the shape of the glass in people's perception of whisky and the range of other product extrinsic cues and their influence on the multisensory drinking experience (Spence, 2012, Theatre of whisky website). Nevertheless, the result (Chadwick & Dudley, 1983) suggests that people may not be as good at discriminating between different types of whisky (for example, single malt and blended whisky) as perhaps many of them believe themselves to be. The findings highlighted by Carlos et al. (2013), underline the significant influence that a multisensory atmosphere can have on people's experience and/or enjoyment of a drink (in this case, a glass of whisky).

In the literature, it was highlighted that the sensory characteristics of the whisky supply a great spectrum of aromas, ranging from smoky, fruity or floral (Daute et al., 2021; Lee et al., 2001) (fermented whisky), caramel, toffee-like, vanilla-like (canned) or peach-like, cider-like (Haug et al., 2023). A future research study can be realized to identify if the findings of Sester et al., 2013, are fulfilled in the case of whisky as well: these researchers designed two different environments, one with furniture designed to convey an impression of 'warmth' and the other varying in 'warmth'–one populated with 'cold' furniture.

Research Methodology

The study aims to explore the factors influencing participants' experiences during a whisky tasting event. It examines their motivations for attending, prior experiences with similar events, and whether they are regular whisky consumers. Additionally, the research investigates the key elements that contribute to overall satisfaction at whisky tasting events.

To achieve the main goals of this study a questionnaire made up of 22 questions was used as a method of research. The questionnaire was conducted during *Bardot Whisky Days Festival* in April 2024, at Bardot bar and restaurant from Cluj-Napoca. It is a self-administered questionnaire and consisted of three parts. The first part included questions aiming at gathering information about how the respondents found out about Bardot's whisky tastings events, their reasons for participating in such events, the respondents' orientation to whisky consumption (regular consumer or a novice one), their habits when participating at such events (with their friends or family or alone; if they serve food while tasting whisky or before). The purpose of the second part was to gather information regarding the respondent's consumption behavior (the respondent's most important variable when choosing whiskey: the area from which it originates, the brand, the alcohol level, the flavor; how respondents assess the overall quality of the whiskies tasted, the respondent's degree of satisfaction with their experience at the tasting session). The third part mainly focuses on the demographic's information of respondents.

Diverse types of questions were applied: structured questions (single choice questions, dichotomous questions, interval scale questions, semantic differential questions, matrix questions, ordinal scale questions and Likert type questions), respectively unstructured questions (one attitude question and one open-end question). The questionnaire was designed paying attention to the time required to complete it. This allowed the authors to ensure that the responses were as accurate as possible and could be gathered using an online survey technology that could be self-administered in a fair amount of time.

The questionnaire was conducted on a relatively small sample, adding up to 36 individuals out of 110 who had an interest in completing it. It was presumed that the entire sample size is enough for this research that is purely of an exploratory nature and the respondents were an adequate representation of whisky brand non-loyal or loyal consumers who visited this tasting event. Since the sample members were selected at random, the context was preserved so as not to bias the findings. The sample respondents were of legal drinking age (18 years and older) and consisted of female and female whisky consumers who participated at the Bardot Whisky Days Festival in April 2024, at Bardot bar and restaurant from Cluj-Napoca. All of them were randomly selected after arrival and after they had actually enjoyed their whisky tasting experience for at least two hours. They were kindly asked to complete the questionnaire of approximative 3 min in duration. The sample respondents were assured of the privacy and confidentiality of the answers and of the fact that the results are used only for research purpose.

Results and Discussions

Considering the main objectives of the present research, the study analysis leads to the following findings.

Descriptive statistics

The demographic characteristics highlight that the sample was weighted toward males (69%) relatively young in age, with 69.4% belonging to the millennial age generation (18–34 years old), highly educated (67% have university bachelor's degree or above), and in high income (69% have an annual household income of higher than 4.000 RON). With 47% of whiskey drinkers under the age of 30, and almost half of it over the age of 30. The profession of the respondents who visited the tasting event tended to be an employee (60%) or entrepreneur (26%), while only 14% being students and no respondent was unemployed (Table 1).

Motivations attending to whiskies tasting sessions

The primary objective of the study was to delve into and determine the motivations driving attendees to whiskies tasting sessions. Upon analysis, it becomes evident that respondents were primarily motivated by activities associated with whisky, coupled with the prospect of enjoying themselves. The predominant reason, as indicated in Table 2, was the desire to acquire or discover more information about whisky, closely followed by the opportunity to taste a large variety of whiskies. Additionally, within the top five reasons, factors pertaining to entertainment and social engagement emerged prominently. These aspects seem to be more aligned with experiencing a social gathering rather than focusing solely on the product itself. Enjoying a new experience was another popular reason to attend at the whisky tasting session, indicating that in addition to whisky sampling, an important number of respondents were interest to discover the novelty of this event experience.

Table 1 Demographic features and whisky consumption

Characteristic	Total (%)
Age (years) Youth >30 years	47 53
Male	69
Female	31
Average annual household income Over 4.000 RON <4.000 RON	69.44 30.56
Whisky consumer in general Yes No	66.6 33.4

Source Authors' calculations

Table 2 The importance of motivational factors

Motivational factors	Rank	Push and pull theory	Mean scores	Std. Dev.
Learn more about whisky	1	Push	4.59	0.77
Taste a variety of whiskies flavors	2	Pull	4.44	0.90
New experience	3	Pull	4.22	0.83
Socialize with friends	3	Pull	4.22	0.95
Meet the professionals in whisky tasting	4	Push	4.19	0.98
Experience the atmosphere of tasting events	5	Pull	4.06	1.01
Curiosity	6	Pull	3.88	1.35
Find information about whisky producers	7	Push	3.94	1.28
Find a specific brand of whisky	8	Pull	3.61	1.41
Buy whisky	9	Push	3.06	1.51
Push factors			3.94	
Pull factors			4.09	

Source Authors' calculations

In light of prior research examining attendees' motivations across various event types which include tasting sessions, such as wine or food festival, shows and workshops both pull and push categories of motives have been explored. Aligned with existing literature the pull items entail characteristics or activities associated with whisky tasting, novelty of experience, the atmosphere and socialization opportunities. Push factors, represent internal desires propelling respondents to attend tasting events, such as learning or discovering new information, meeting experts and buying whisky. This nuanced understanding, sheds light on the multifaceted motivations driving individuals to engage in whisky tasting experiences.

Given the observation that there is a notably higher prevalence of pull-type motivations among participants in tasting sessions, it led to the hypothesis that individuals expressing an interest in participating are likely to be regular consumers of whisky.

Consequently, in order to explore the association between the type of consumer and their engagement in tasting events, statistical analysis employed the calculation of V Cramer coefficient, see Table 3. The objective of this approach was to provide insight into how much a consumer's tendency to participate in taste sessions correlates with their typical preferences toward whisky consumption.

The coefficient value of approximately 0.31 obtained suggests a moderate level of association between participation in tasting sessions and the consumer profile. This implies that there is a notable but not overpowering correlation between attending such events and being a whisky consumer. While the coefficient doesn't approach 1, indicative of a strong association, it still signifies a meaningful link between the two variables. From this, we can infer that individuals who participate in these

Table 3 Contingency Table for testing the association between participation in tasting sessions and the consumer profile

Type of whisky consumer/previous tasting experience	First time	Repetitive experience
Consumer (Yes)	10	14
Non-consumer (No)	9	3

Source Authors' calculations

events are generally inclined towards consuming whisky. However, it's crucial to emphasize that this association doesn't establish a causal relationship. Attending a tasting session doesn't necessarily dictate one's status as a whisky consumer, nor does being a whisky consumer guarantee attendance at such events.

Various factors could contribute to this correlation, and there might be unknown variables influencing the association. Despite this correlation, it's notable that non-consumers appear to attend tasting events less frequently after their initial participation compared to consumers. This observation warrants further investigation into potential causative factors.

For event organizers, this finding holds significance as non-consumers represent a substantial segment with potential for market development. Encouraging them to explore different tastes, fostering an interest in whisky, or highlighting other aspects of the tasting experience could attract their participation, thereby expanding the market and enhancing the event's success.

When examining the various motivations behind participation in whisky tasting sessions through the lens of Uysal's theories, it becomes evident that certain factors play pivotal roles in driving individuals' engagement. Foremost among these factors is the desire to expand one's knowledge about whisky, reflecting a curiosity and interest in delving deeper into the intricacies of this beverage. Additionally, the opportunity to enjoy leisure time in a fulfilling manner emerges as another significant motivation, suggesting that participants seek not only education but also relaxation and enjoyment from these sessions. Closely trailing these primary motivations are those linked to socialization and novelty. The appeal of social interaction and the chance to engage with others who share similar interests contribute substantially to the overall allure of whisky tasting events. Similarly, the prospect of experiencing something new, unique, or different adds an element of excitement and anticipation for participants, further enhancing their motivation to attend such gatherings (Table 4).

However, when examining the motivations of whisky consumers versus non-consumers, noteworthy distinctions emerge. For consumers, who most likely already have some familiarity and experience with whisky, socializing and entertainment rank as the most important motivators for participation. This indicates that the social component and enjoyment of these events are major factors in the decision of individuals who have an expertise in whiskey culture to attend.

Conversely, non-consumers are primarily motivated by the desire for escape or novelty and the opportunity to acquire knowledge. For individuals who may be less

Table 4 The order of significance of motivational factor categories

Motivational factors	Uysal and al. theory	Average scores		
		Total	Consumer	Non-consumer
Learning/discovering factors		*4.26*	4.35	4.08
Learn more about whisky	Learning	4.58	4.75	4.25
Find information about whisky producers		3.94	3.96	3.92
Entertainment factors		*4.25*	4.4	3.96
Taste a variety of whiskies flavors	Entertainment	4.44	4.67	4.00
Experience the atmosphere of tasting events		4.06	4.13	3.92
Escape/novelty factors		*4.12*	4.12	4.12
Curiosity	Escape/novelty	4.03	3.96	4.17
New experience		4.22	4.29	4.08
Socialization factors		*4.20*	4.4	3.83
Socialize with friends	Socialization	4.22	4.38	3.92
Meet the professionals in whisky tasting		4.19	4.42	3.75
Economic opportunity factors		*3.33*	3.45	3.08
Buy whisky	Economic opportunity	3.06	3.08	3.00
Find a specific brand of whisky		3.61	3.83	3.17

Source Authors' calculations

familiar with whisky or who have yet to develop a taste for it, the appeal lies in seeking new experiences and broadening their understanding of the spirit. This underscores the diverse array of motivations that can influence individuals' participation in whisky tasting sessions, highlighting the multifaceted nature of this pastime.

The findings make it evident that economic incentives, such as the opportunity to purchase beverages or discover specific whisky brands, hold minimal significance as motivating factors for participation among both groups of participants.

The discrepancies seen between the two participant categories, consumers and non-consumers, made it required to look into the statistical significance of the mean scores linked to each motivation element the study explored in more detail. The mean scores of the "Yes" and "No" groups have been compared using the t-test for each motive in order to assess these differences even more.

It was established that a significance level (α) of 0.05 would be adequate to evaluate the outcomes once considering the following hypothesis:

(H0): There are no significant differences in the mean scores for each motive between individuals who are and those who are not consumers of whisky.

(H1): Proposes that there are significant differences in the mean scores for each motive between the two groups.

Based on the outcomes presented in Table 5, significant differences were observed between the "Yes" and "No" groups for motives such as experiencing the atmosphere of tasting events, learning more about whisky, tasting a variety of whisky flavours, finding a special brand of whisky, and meeting whisky tasting professionals. However, for other motives, the differences were not statistically significant. These findings suggest that certain motivations play a role in the decision to participate in tasting events, depending on whether individuals are whisky consumers or not.

Regarding the motivation to learn more about whisky for whisky consumers attending tasting sessions, it appears to be somewhat less important for those attending this type of event for the first time compared to those with previous experience in this area. This leads us to the idea that as a consumer gains more information by participating in these events, they become more motivated to attend other tasting events to learn more about this alcoholic beverage. Concerning non-consumers, there is no distinction in learning motivation between those who attend for the first time and those who have experienced whisky tasting sessions before.

Attendee's overall satisfaction and perception of tasting session' experience

The degree of participant satisfaction with the experience and their assessment of the contributing elements are definitely important considerations when analysing the factors that impact the tasting experience. With an average score of 4.77 and a standard deviation of only 0.42, the participants' satisfaction with the tasting session is very high and statistically significant. Among the aspects that influence attendees' perceptions of the tasting session experience, the reliability of the information provided during the events is more important than the diversity and quality of the whisky brands offered (see Table 6). This raises the question of how much these

Table 5 Compares motivational factors between whisky consumers versus non-consumers

Motivation factor		p Value
To learn more about whisky	$t(68) = 1.69$	$p = 0.032$ (p value < 0.05)
To learn about whisky producers	$t(68) = 0.54$	$p = 0.591$ (p value > 0.05)
To experience the atmosphere of tasting events	$t(68) = 2.12$	$p = 0.037$ (p value < 0.05)
To taste a variety of whisky flavors	$t(68) = 2.03$	$p = 0.018$ (p value < 0.05)
Out of curiosity	$t(68) = 0.24$	$p = 0.812$ (p value > 0.05)
For a new experience	$t(68) = 0.34$	$p = 0.736$ (p value > 0.05)
To spend time with friends	$t(68) = 1.25$	$p = 0.216$ (p value > 0.05)
To meet whisky tasting professionals	$t(68) = 2.07$	$p = 0.042$ (p value < 0.05)
To purchase different whisky brands	$t(68) = 0.32$	$p = 0.748$ (p value > 0.05)
To find a specific brand of whisky	$t(68) = 2.42$	$p = 0.018$ (p value < 0.05)

Source Authors' calculations

aspects influence overall satisfaction, which is why a Pearson correlation analysis was employed.

The positive correlation coefficient of 0.372 with a significance level of 0.042 indicates that there is a meaningful relationship between the perceived quality of the tasting session's facilities and the overall merit attributed to the event. This finding suggests that attendees who perceive the facilities as remarkable tend to rate the tasting session more favourably. It implies that investments in enhancing the ambiance, comfort, and amenities of the tasting venue could potentially enhance attendees' overall experience and satisfaction, thereby contributing positively to the event's perceived value.

With a correlation coefficient of 0.426 and a significance level of 0.021, the results reveal a significant positive association between the perceived reliability of the information provided during the tasting session and the overall merit of the event. This finding implies that attendees who consider the information provided as trustworthy and accurate are more likely to regard the tasting session as valuable and worthwhile. It underscores the importance of ensuring that the information shared during the event, such as details about whisky production, tasting notes, and brand histories, is credible and engaging. Doing so can enhance attendees' satisfaction and perception of the event's educational and entertainment value.

However, there is no significant correlation between the merit of the tasting session and the variety of whiskies provided ($r = 0.138$, $p = 0.455$), indicating that the variety of whiskies does not significantly influence the perceived merit of the tasting session. This finding indicates that while offering a diverse selection of whiskies may be appreciated by attendees, it may not significantly impact their overall assessment of the event's quality or value. Organizers may need to focus on other aspects, such as the presentation, tasting experience, and educational components, to enhance attendees' perception of the event's merit. In the same manner the quality of the whiskies

Table 6 Analysis of tasting session experience's factors and satisfaction

Perception regarding the aspects which create the tasting session experience	Average score	Std. Dev.	Pearson correlation-Worthfully tasting event and:	
Tasting facilities	4.5	0.73	$r = 0.372$	$p = 0.042$ (significant)
Reliable information provided	4.69	0.70	$r = 0.426$	$p = 0.021$ (significant)
Variety of whiskies	4.61	0.87	$r = 0.138$	$p = 0.455$ (not significant)
Quality of whiskies	4.08	1.15	$r = 0.258$	$p = 0.155$ (not significant)

Source Authors' calculations

provided during the tasting session seams that will not significantly influence the event's quality perception.

Conclusions

According to the author's knowledge, in Romania, no prior research has been done about whisky brand loyalty among consumers. This study contributes to the knowledge base of the tasting sessions for whiskies. These sessions create the premises to educate the consumers' drinking behavior and at the same time represent a distinct event host by hospitality units for their customers. This study provides on one hand a theoretical approach, contributing to the understanding of motivations which determine attendees to participate in general at alcoholic beverages tasting sessions and except for wine tasting sessions the rest has been so far neglected in the literature. Thus, the primary motivation for attendees at whisky tasting sessions is the desire to acquire or discover more information about whisky, closely followed by the opportunity to taste a variety of whiskies. This highlights a strong interest in both the educational and experiential aspects of these events. This result is strongly supported by the fact that perceived reliability of the information provided during the tasting sessions and the quality of the event's facilities significantly influence the overall satisfaction of a tasting event. Entertainment and social engagement factors are also considerable among the top reasons for attending whisky tasting sessions. This suggests that participants value not only the product but also the social atmosphere and the chance to enjoy a novel experience. The study reveals that distinguished differences in motivations exist between whisky consumers and non-consumers. Consumers prioritize socializing and entertainment, reflecting their familiarity with whisky culture. In contrast, non-consumers are driven by the desire for new experiences and learning opportunities, indicating a curiosity to broaden their understanding of whisky.

The fact that resulted a moderate association between participation in tasting sessions and being a whisky consumer, does not imply causation. This means that regular whisky consumers are more likely to attend tasting sessions, but attending these sessions does not necessarily turn one into a regular consumer.

On the other hand, the study provides important insights to the organizers of these type of events, especially for hospitality units—bars, restaurants—which can be used for a better understanding of the attendees' expectances and to generate more efficient promotional activities and tasting experience. Thus, based on results the organizers of whiskies tasting session should prioritize educational components in their events and may involve training staff systematically and collaborating with knowledgeable whisky experts. Given the importance of social and entertainment factors, organizers should create an appealing social atmosphere. This could involve arranging seating to facilitate conversation, planning interactive activities, and incorporating elements of entertainment, such as live music or themed decorations.

In the same direction, integrating digital experiences such as virtual tours of distilleries, online tasting notes, and interactive apps can enhance the educational component. Digital tools can also provide attendees with additional resources and engagement opportunities beyond the physical event. Themed whisky tasting events, such as regional whisky explorations, historical-themed tastings, or food-pairing sessions, can add novelty and variety. These themes can attract different segments of the market and keep the offerings fresh and exciting.

This study has certain limitations even if it makes a significant contribution to the literature's consideration of whisky tastings sessions. The geographic area analysis (only the Cluj-Napoca city) is the initial boundary, which indicates the need of extending the present research in other main Romanian cities where such kind of events are organized. Secondly, the *Bardot Whisky Days Festival* attendees are the only target group on which the study has focused. The results could not fairly represent Cluj-Napoca's overall whisky consumers' level of satisfaction. Therefore, the present study can be extended for all whisky consumers from Cluj-Napoca. Lastly, it would be intriguing if additional research attempted to validate or disprove the current study's conclusions at the national level by replicating the study's approach in other Romanian main cities.

More research is needed to enhance the current findings even more. A future goal of another study is to identify how to attract younger Romanian consumers to become a main segment of customers who understand the philosophy for whisky.

Acknowledgements The authors thank the Bardot—Cocktail and Whisky Bar team for their support in applying the questionnaire.

References

Bruwer, J., Coode, M., Saliba, A., & Herbst, F. (2013). Wine tourism experience effects of the tasting room on consumer brand loyalty. *Tourism Analysis, 18*(4), 399–414.

Carlos, V., Russell, J., Scott, K., & Charles, S. (2013). Assessing the influence of the multisensory environment on the whisky drinking experience. http://www.flavourjournal.com/content/2/1/23.

Chadwick, S., & Dudley, H. (1983). Can malt whiskey be discriminated from blended whisky? The proof. A modification of Sir Ronald Fisher's hypothetical tea tasting experiment. *BMJ, 287*, 1912–1915.

Cialdini, R. B., & Goldstein, N. J. (2004). Social influence: Compliance and conformity. *Annual Review of Psychology, 55*, 591–621.

Daute, M., Jack, F., Baxter, I., Harrison, B., Grigor, J., & Walker, G. (2021). Comparison of three approaches to assess the flavour characteristics of scotch whisky spirit. *Applied Sciences*. https://doi.org/10.3390/app11041410

Derrett, R. (2003). Making sense of how festivals demonstrate a community's sense of place. *Event Management, 8*, 49–58. https://doi.org/10.3727/152599503108751694

Egresi, I., & Kara, F. (2014). Economic and tourism impact of small events: The case of small-scale festivals in Istanbul, Turkey. *Studia UBB Geographia, 1*, 47–64.

Folgado-Fernández, J. A., Di-Clemente, E., & Hernández-Mogollón, J. M. (2019). Food festivals and the development of sustainable destinations. The case of the cheese fair in Trujillo (Spain). *Sustainability, 11*, 2922. https://doi.org/10.3390/su11102922.

Haug, H., Grasskamp, A. T., Singh, S., et al. (2023). Quick insights into whisky—investigating rapid and efficient methods for sensory evaluation and chemical analysis. *Analytical and Bioanalytical Chemistry, 415*, 6091–6106. https://doi.org/10.1007/s00216-023-04883-5

Lee, K.-Y.M., Paterson, A., & Piggot, J. R. (2001). Origins of flavour in whiskies and a revised flavour wheel: A review. *Journal of the Institute of Brewing, 107*, 287–313. https://doi.org/10.1002/j.2050-0416.2001.tb00099.x

Mantonakis, A., Rodero, P., Lesschaeve, I., & Hastie, R. (2009). Order in choice: Effects of serial position on preferences. *Psychological Science, 20*, 1309–1312.

North, A. C. (2012). The effect of background music on the taste of wine. *British journal of Psychology, 103*, 293–301.

Oberfeld, D., Hecht, H., Allendorf, U., & Wickelmaier, F. (2009). Ambient lighting modifies the flavor of wine. *Journal of Sensory Studies, 24*, 797–832.

Park, E., & Widyanta, A. (2022). Food tourism experience and changing destination foodscape: An exploratory study of an emerging food destination. *Tourism Management Perspectives, 42*, 100964. https://doi.org/10.1016/j.tmp.2022.100964

Philp, M., & Mantonakis, A. (2018). Guiding consumers' evaluation strategies and the probability of order-effects-in-choice. *Journal of Business Research.*

Sester, C., Deroy, O., Sutan, A., Galia, F., Desmarchelier, J. F., Valentin, D., & Dacremont, C. (2013). "Having a drink in a bar": An immersive approach to explore the effects of context on beverage choice. *Food Quality and Preference, 28*, 23–31.

Spence, C. (2011). Crystal clear or gobbletigook? *World of Fine Wine, 33*, 96–101.

Spence, C. (2012). On crossmodal correspondences and the future of synaesthetic marketing: Matching music and soundscapes to tastes, flavours, and fragrance. In *Proceedings of Audio Branding Academy* (pp. 41–54).

Tanford, S., & Jung, S. (2017). Festival attributes and perceptions. A meta-analysis of relationships with satisfaction and loyalty. *Tourism Management, 61*, 209–220.

Thaler, R. H., Sunstein, C. R., & Balz, J. P. (2013). Choice architecture. In S. Eldar (Ed.), *The behavioral foundations of public policy* (pp. 428–439). Princeton University Press.

Theatre of whisky. http://tinyurl.com/c2p7epq.

Tversky, A., & Kahneman, D. (1974). Judgment under uncertainty: Heuristics and biases. *Science, 185*, 1124–1131.

Vesci, M., & Botti., A. (2019). Festival quality, theory of planned behavior and revisiting intention: Evidence from local and small Italian culinary festivals. *Journal of Hospitality and Tourism Management, 38*, 5–15.

The Influence of e-Word-of-Mouth on Travel Behavior: Evidence from Greece

Nikolaos Trihas, Eirini Papadaki, Georgios Zozonakis, and Aikaterini Michelaki

Abstract The emergence of Web 2.0 and social media has led to an increase in the number of consumers using them to search for information about products and services and to carry out transactions. The purpose of the paper is to explore the extent to which Greek consumers use online travel review sites (TripAdvisor, Booking.com, social media, etc.) in three phases of their trip: before the trip, during the trip, and after the trip. The objective is to investigate the extent to which the choices and bookings made by Greeks before and during their trip in hotels, restaurants, etc. are influenced by comments, ratings, and reviews posted by other users on the internet. In order to meet this purpose, a survey was conducted via a structured self-completed questionnaire in a sample of 332 people in Greece. The study's results revealed that Greeks are heavy Internet and social media users. The vast majority of them, before buying tourism products and services, seek and read comments, reviews and ratings of other users on the internet, especially in the pre-travel phase. These reviews concern different tourism businesses, mainly hotels, restaurants, airlines, but also tourist destinations and attractions. For these searches they mainly use popular platforms, such as TripAdvisor, Booking.com, and Google. Participants consider that tourism services, due to their intangible nature, involve a high degree of risk, and they do not trust advertisements as they often prove to be misleading. However, most agree that the freedom of speech and expression that exists online can allow the dissemination of unverified, malicious, derous, or distorted information about a tourism business or destination. The results of the research provide useful implications for tourism businesses and organizations on how to manage online reviews in order to enhance their online image and protect their reputation.

Keywords Online reviews · eWoM · Travel behavior · Decision making · Greece

N. Trihas (✉) · E. Papadaki · G. Zozonakis · A. Michelaki
Hellenic Mediterranean University, Heraklion, Greece
e-mail: ntrihas@hmu.gr

A. L. Negrușa et al. (eds.), *Sustainable and Resilient Businesses in the Global Economy*, Springer Proceedings in Business and Economics,
https://doi.org/10.1007/978-3-031-75883-6_11

Introduction

Tourism services are intangible. The customer of a tourist service provider cannot see, feel, hear and generally have any kind of contact with any of the five senses with the service before purchasing and consuming it (Flipo, 1988). The customer cannot, therefore, evaluate the quality of the services before purchasing and consuming them (Jalilvand & Samiei, 2012). It is information that will help them to minimize the gap between their expectations and their actual experience. Word-of-mouth (WOM) is defined as face-to-face discussion about products and services between two or more people, which is not motivated by the companies providing those products or services (Confente, 2015). The power of WOM relies on consumers discussing openly and honestly, without distortion, about a particular product or service, since they have no commercial interest or incentive (Barber & Wallace, 2009). Consumers tend to trust other consumers' ratings and reviews of a product more than the advertisements created by advertising agencies and corporate marketing departments. Advertisements tend to conceal the negative features of a product by highlighting only the positive ones.

Globally, 5.35 billion people use the internet (69.4% of the population), while 5.04 billion are social media users. The average daily time these users spend on internet is 6 h and 40 min, while the main reason for using the internet with 60.9% is finding information (Datareportal, 2024). The COVID-19 pandemic and the accompanying restrictive measures also prompted people to spend more time searching for information on social media channels (Trihas et al., 2023; Tung & My, 2023). The Internet and its applications (social media etc.) now offer users new ways for personal expression and communication, bringing about critical changes in the behavioral patterns of tourists in the way they perceive, buy, evaluate, search and consume products, services and information. The emergence of Web 2.0 has led to an increase in the number of consumers using it to search for information about products and services and to carry out transactions (Casaló et al., 2015; Hernández-Méndez et al., 2015), so that today we are talking about electronic word-of-mouth (eWOM) (Singh & Kathuria, 2019). eWOM is not just a means of interpersonal communication today but is recognized as a powerful marketing tool (Liang et al., 2013). eWOM has distinct advantages that distinguish it from traditional WOM, such as stronger communication capabilities, broader coverage and wider audience (Wu, 2024).

In this context, the aim of the paper is to explore the extent to which Greek travelers use online review websites (i.e., TripAdvisor, Booking.com, Google, social media, etc.) in order to find information to help them make better travel choices. While there is a large body of research available on eWOM in hospitality and tourism, there is a lack of empirical studies focusing on Greece. This paper intends to partly fill this gap and contribute to the existing literature.

Following this introduction, next section elaborates on a theoretical background related with the concept of e-word-of-mouth in the tourism industry and its effects on travel behaviour. The research methodology and results sections then follow and finally this paper is concluded by highlighting the main conclusions, implications and directions for future research.

Literature Review

Electronic word-of-mouth (eWOM) can be defined as an "*informal, non-commercial digital communication (primarily positive or negative statements) about a brand, product, service or company made by actual or former customers*" (Mukhopadhyay et al., 2023). In the tourism context, tourists seek information about hotels, restaurants, destinations and attractions in official destination websites, travel blogs and social media (Hernández-Méndez et al., 2015). People write online travel reviews motivated by the desire to help a travel service provider improve, to help other consumers' decision-making and also because they experience a feeling of satisfaction and enjoyment when they complete a review (Yoo & Gretzel, 2008). Bakshi et al. (2019) suggest that what motivates travelers posting online reviews is altruism, reciprocity, effort expectancy, habit and hedonism. There is a large number of publications on eWOM, while a primary topic in the field's current research is the effects of eWOM and online reviews on consumer behavior, and more specifically tourist behaviour. Singh and Kathuria (2019) examined the adoption of eWOM among consumers interested in travel services and found that online reviews influence travelers' choices, from the choice of the destination they travel to the choice of the hotel they stay in and the restaurants they dine at. Likewise, Ladhari and Michaud (2015) argue that Facebook users' hotel booking intentions are influenced by the comments their 'friends' have posted on the platform. Specifically, positive comments about a particular hotel increase the likelihood of booking at that hotel. Ye et al. (2011) suggest that online reviews have a significant impact on the online sales of hotel rooms. Other authors argue that eWOM and online reviews have a significant influence on tourists' attitudes toward destinations and visit intentions (Albarq, 2014; Doosti et al., 2016; Hung & Khoa, 2022; Jalilvand & Samiei, 2012; Setiawan et al., 2021; Tham et al., 2013; Tung & My, 2023). More specifically, Su et al. (2022) suggest that online review valence and emotional intensity affect travel intentions.

Singh and Kathuria (2019) suggest that consumers are looking for online reviews about destinations, hotels and restaurants, as they believe that this information will reduce the chance of making a bad decision and help them plan their trips in a better way. However, Roy et al. (2024) argue that making a decision based on online reviews is challenging, as customers are exposed to sponsored or fake reviews that confuse them. Liu and Park (2015) suggest that there is a cognitive cost for travelers looking for online reviews, as it is not always easy to judge whether a review is useful or not. It seems that a combination of both the reviewer (identity, expertise, reputation) and review (star ratings, length, readability etc.) characteristics positively affect the perceived usefulness of reviews. Moreover, Filieri and McLeay (2013) found that there are some dimensions of information quality (accuracy, value-added, relevance, timeliness) influencing the extent to which travelers adopt information from online reviews. Quality and quantity of the information provided in travel review websites are also recognized as significant factors that influence information acceptance (Chopra et al., 2022). Kim and Kim (2020) focus on the importance of the perceived authenticity of online reviews as a factor in shaping travelers' behavioral intentions. Ahmad and Sun (2018) recognize that there are two attributes of the reviewer that influence consumer distrust on online hotel reviews, namely fake identity and ulterior motivation. In the same direction, Guzzo et al. (2022) found that a number of factors can make online travel reviews more credible, namely the website brand, the advisor's expertise, the reviews' sidedness, and consistency, as well as consumer experience. In addition to enhancing the credibility of reviews and ratings and thereby reducing travelers' perceived risk, these factors simultaneously increase the perceived usefulness of the information and ultimately enhance travelers' intention to visit a destination (González-Rodríguez et al., 2022). In any case, the disadvantages of online reviews, according to Hernández-Méndez et al. (2015), lead travelers to be influenced in their travel decisions more by the opinions of family and friends (traditional WOM) than by internet user reviews (eWOM). Cox et al. (2009) also suggest that travelers do not consider online reviews as reliable and trustworthy, so they simply use them as an additional source of information and never rely on them alone to make purchasing decisions. Other researchers suggest that contemporary technologies, such as blockchain and artificial intelligence can potentially enhance consumers' trustworthiness of online reviews in the tourism and hospitality context (Rajaguru, 2024).

Research Methodology

As mentioned in the introductory section, the aim of the paper is to explore the extent to which Greek consumers read comments, ratings and reviews posted on online travel review websites (TripAdvisor, Booking.com, Google, social media, etc.) to make better travel decisions. To meet this aim, a structured self-completed online questionnaire was designed which consisted of four sections and twenty-one questions. This was followed by a pilot test on a sample of 20 people. Based on feedback from the pilot survey, minor modifications were made to the content and structure of the questionnaire. Subsequently, the questionnaire was uploaded to the online platform Google Forms. The snowball sampling was applied to send the questionnaire. The survey lasted from February 1st, 2023 to March 31st, 2023. Finally, 331 valid fully completed questionnaires were collected. The statistical analysis of the data was implemented in the Statistical Package for the Social Sciences (SPSS) version 26.0.

Results and Discussions

Table 1 presents the demographic characteristics of the participants. Female respondents dominate with 74.3% versus 25.7% of men. The majority of the participants (46.8%) belongs to the age group 25–34 and are single (57.1%). The educational level of the participants is remarkably high, as 176 of them (53.2%) are graduates of tertiary (higher) education, while 93 of them (28.1%) hold a Master's or Ph.D. degree. 36.9% of the participants reported an income between €5,000 and €10,000, followed by the other income groups.

In the first part of the survey, participants were asked to describe their travel behaviour. Most of them appear to have developed a strong travel activity, as 63.1% of them take on average one to two trips annually, 21.1% of participants take three to four trips, while a remarkable proportion of them (10.9%) take more than four trips per year. Participants who do not travel reach 4.8% of the sample. The reasons for these trips are multiple and mainly concerned vacation and leisure (292 answers), followed by visiting friends and relatives (131 answers), business reasons (71 answers), education (28 answers), shopping (26 answers), healthcare (19 answers) and other reasons (2 answers). 140 participants travel both within Greece and abroad, while 172 of them

Table 1 Demographic profile of respondents

		N	%
Gender	Male	85	25.7
	Female	246	74.3
Age	18–24	28	8.5
	25–34	140	42.3
	35–44	113	34.1
	45–60	45	13.6
	61+	5	1.5
Marital status	Single	189	57.1
	Married	121	36.6
	Divorced/widowed	21	6.3
Education	Primary	4	1.2
	Secondary	58	17.5
	Tertiary (higher)	176	53.2
	Master/Ph.D.	93	28.1
Annual income	Less than 5,000 €	56	16.9
	5,000–10,000 €	122	36.9
	10,001–20,000 €	101	30.5
	More than 20,000 €	52	15.7
Total		331	100

Source The authors

Table 2 Importance of travel information sources

	Mean	Std. Deviation
Previous personal experience	4.26	0.973
Internet/search engines	4.12	1.008
Review websites	3.93	1.121
Suggestions from family and friends	3.69	1.113
Social media	3.31	1.230
Official websites of tourist destinations	3.23	1.201
Travel websites and blogs	2.96	1.186
Travel guidebooks	2.76	1.234
Mass media	2.23	1.101
Promotional leaflets/brochures	2.09	1.059
Tourism exhibitions	2.06	1.109
Travel agencies/tour operators	2.05	1.083

Notes (a) 1 = Strongly Disagree, 5 = Strongly Agree
(b) Mean is the average of participants' answers on this set of values
(c) Standard Deviation (Std. Deviation) is a measure of how dispersed the data is in relation to the mean
Source The authors

travel only within the country and 19 of them travel abroad. Most participants travel individually (84.3%), a few prefer to travel in groups (1.5%), and the rest of them (14.2%) travel both ways. For their accommodation, the majority of the participants prefer hotels (52.6%), while others choose various other types of accommodation, namely Airbnb (15.1%), rental rooms (15.1%), homes of friends/relatives (13.6%), second homes (1.8%), hostels (1.5%) and camping (0.3%).

Regarding the sources where Greeks seek information about destinations (Table 2), attractions and tourism businesses, it seems that previous personal experience is the strongest factor (Mean = 4.26). The internet and search engines follow (Mean = 4.12), with various websites such as review sites (Mean = 3.93), social media (Mean = 3.31) and official websites of tourist destinations (Mean = 3.23) also playing an important role as sources of travel information. In addition, Greek consumers seem to rely heavily on recommendations from family and friends (Mean = 3.69). In contrast, traditional forms of information that were very important in previous years such as travel guide books (Mean = 2.76), the mass media (i.e., TV, radio, press) (Mean = 2.23), brochures (Mean = 2.09), tourism exhibitions (Mean = 2.06) and travel agencies (Mean = 2.05) seem to have lost their value as sources of information.

In the second part of the survey, the questions focused on the use of the internet and social media. The results revealed that participants are heavy internet users, as 97.3% of them use it daily, while 33.8% spend 3–4 h online every day, 22.7% 5–6 h and another 22.7% spend seven or more hours. To access the internet, they use multiple devices, mainly mobile phones (97.2%), followed by laptops (44.1%), desktops (20.8%) and tablets (17.2%). In addition, the vast majority of them (97.3%) have an account on at least one social network. They are active social media users, since 94.7% of them connect to them on a daily basis, while Facebook is the dominant social network (311 participants use it), followed by Facebook Messenger (268 users), Instagram (246 users), YouTube (234 users), TikTok (115 users) and WhatsApp (99 users).

The last part of the survey concerned online reviews and the extent to which they influence the travel decisions of Greeks. First, 92.7% of the participants agree that before buying a tourism product or service they seek comments, ratings and reviews of other users on the internet. Most of these searches are for hotels (87.6%), followed by restaurants (74%), tourist destinations (58.3%), attractions (52%), airlines (25.1%) and other tourism businesses (18%). The most popular websites that participants visit in order to read the reviews are Google (79.2%), Booking.com (74%), TripAdvisor (64.7%), and business pages on social media (33.5%). Subsequently, respondents were asked to what extent they agree with a series of statements about eWOM and the extent to which it influences their travel choices (Table 3).

Participants agree that they always look for information before choosing a tourist destination or a tourist business to be as confident as possible in their choice (Mean = 4.30). This is linked to the nature of the tourism product, which is intangible. As they cannot experience with any of their senses a tourism service before purchasing and consuming it, purchasing it involves a high degree of risk (Mean = 3.56). For example, there were cases where participants had high expectations of a tourism business or destination which were not fulfilled ultimately resulting in disappointment (Mean = 3.60). This is also the reason why Greek travellers seek comments, reviews and advice from other travellers, mainly friends and relatives, which they trust more than advertisements (Mean = 4.09), which they consider tend to hide negative facts and are often misleading (Mean = 3.58). Greek travellers visit travel review websites in all three phases of a trip: before a trip (Mean = 3.99) and during the trip (Mean = 3.75) looking for useful information, but to a lesser extent also after the trip to write comments and reviews that will help other users in their choices (Mean = 3.14). However, participants are not active members of travel review websites (Mean = 2.61), so they do not systematically contribute with new content. Also, despite the great importance and usefulness that reviews and ratings seem to have in influencing travellers' purchasing decisions, participants do not seem to have been largely motivated in the past by hotel or restaurant owners to write a review (Mean = 3.05).

Although participants largely believe that the ratings and reviews available online are valid and reliable (Mean = 3.20), they also recognise some of their disadvantages. First of all, they agree that ratings and reviews available online are too short and usually omit important information (Mean = 3.16), and it is difficult to interact with the users who have written these reviews to request additional information. In addition, they agree that the freedom of speech and expression that exists online can allow the dissemination of unverified, malicious, slanderous or distorted information about a tourism business or destination (Mean = 3.67). They also agree that they are

Table 3 Statements about e-Word-of-Mouth and travel behavior

	Mean	Std. Deviation
I don't trust the advertisements and photos presented by tourism businesses because they tend to hide the negatives and are often misleading	3.58	0.889
As I cannot try a tourist service (e.g. a hotel) before buying and consuming it, buying it involves a high degree of risk	3.56	0.930
There are cases where I had high expectations of tourism businesses before I visited them and ended up being disappointed	3.60	1.044
I always look for as much information as possible before choosing a tourist destination or a tourist business to be as confident as possible in my choice	4.30	0.804
I trust the feedback and advice of my friends and family about tourism products and services more than advertisements	4.09	0.835
When I book my holidays through a travel agency, I feel more secure as this personal contact is very important to me	2.75	1.216
I visit travel websites *before* my trip looking for comments, ratings and reviews from other users	3.99	1.108
I visit travel websites *during* my trip looking for reviews, ratings and reviews from other users	3.75	1.145
I visit travel websites *after the end* of my trip to write comments and reviews to help other users with their choices	3.14	1.339
I am an active member of travel review sites	2.61	1.391
I believe that the ratings and reviews available on the internet are valid and reliable	3.20	0.892
The ratings and reviews available online are very short and usually omit important information. It is difficult to interact with these users and ask for additional information	3.16	0.993
The fact that the ratings and reviews that exist on the internet come from "invisible" and unknown to me people make these reviews less reliable	2.95	1.053
The fact that the ratings and reviews on the internet are often contradictory is confusing to me	3.05	1.078
I trust the recommendations of my friends and family more than the user reviews on the internet	3.78	1.024
I am much more likely to write about a bad experience I had with a tourism company than a positive experience	2.99	1.388
In the past, owners of accommodation and/or restaurants have encouraged me to write a review of their business	3.05	1.407
In the past accommodation and/or restaurant owners have bribed me to write a good review of their business	1.61	1.045
It has happened in the past that I have changed my decision to travel at the last minute because of some comments I read on the internet	2.28	1.283
It has happened that I have chosen a hotel/restaurant because of its good reviews and ended up being disappointed	3.19	1.226

(continued)

Table 3 (continued)

	Mean	Std. Deviation
The freedom of speech and expression that exists on the internet may allow the dissemination of unverified, malicious, defamatory or distorted information about a tourist business or destination	3.67	1.127
Seeing that the management of a hotel responds to the comments written about it on the internet, I get the impression that the hotel really cares about its customers	3.86	1.033
Responses from the management to the reviews they receive usually increase my chances of booking that hotel (compared to a similar hotel that does not respond to reviews)	3.49	1.118
An aggressive response from management to a bad review prevents me from booking this hotel	3.75	1.171
A proper response from management to a bad review improves my impression of a hotel	3.95	1.003
If I see that a tourist business deletes a negative comment I made, I come back with a newer comment and condemn this fact	3.25	1.392

Notes (a) 1 = Strongly Disagree, 5 = Strongly Agree
(b) Mean is the average of participants' answers on this set of values
(c) Standard Deviation (Std. Deviation) is a measure of how dispersed the data is in relation to the mean
Source The authors

confused by the fact that the ratings and reviews available online are often contradictory (Mean = 3.05), while to a lesser extent they agree that because ratings and reviews on the internet come from 'invisible' and unknown people, it makes them less reliable (Mean = 2.95). For example, it has happened that they have chosen a hotel/restaurant because of good reviews and ended up being disappointed (Mean = 3.19). Because of all these disadvantages, they end up trusting more the advice of their friends and relatives than user reviews on the internet (Mean = 3.78).

Furthermore, apart from the comments, reviews and ratings themselves, the way tourism businesses manage them seems to influence the image that Greek consumers form of them. More specifically, seeing that the management of a tourism business responds to the comments written about it on the internet, participants form the impression that this business really cares about its customers (Mean = 3.86). Moreover, the responses of the management of the business to the reviews it receives usually increase the probability of participants choosing this business (compared to a competing business that does not respond to reviews) (Mean = 3.49). The way and the style of responding seems to be important as well. For example, participants largely agree that an aggressive response from the management of a business to a bad review received will deter them from booking with that business (Mean = 3.75). In contrast, a competent response from a business's management to a bad review received improves their impression of that business (Mean = 3.95). Finally, the practice followed by some tourism businesses of deleting comments and reviews that are negative seems to be inappropriate, since many participants agree that if they find that a tourism business deletes a negative comment they made, they come back with a newer comment criticising that fact (Mean = 3.25).

Conclusions

The evolution of technology, the high penetration of the internet and smartphones among the general population and the rapid growth of social media have brought about major changes in the way consumers communicate, search and buy products and services. In the case of tourism products and services, because they are intangible, consumers try to gather as much information as possible before purchasing and consuming services to minimise potential risk. In this context, the purpose of this paper is to explore whether Greek consumers visit travel review websites and to what extent online reviews, comments and ratings influence their travel behaviour and purchase decisions.

The results of the survey revealed that Greeks are experienced travelers since they make several trips per year with different motivations (mainly for holidays, visiting relatives and friends, and business reasons). Most of them travel individually, mainly within Greece, and prefer hotels for their accommodation. This travel profile confirms previous research also focused on Greek travellers (Trihas et al., 2023, 2024). There now seems to be a shift of consumers from traditional means of information such as travel agencies, travel exhibitions, travel guidebooks, brochures, mass media, to modern innovative means such as the internet, search engines, social media, destination websites, as well as review sites. Also, very popular as sources of information remain both the previous personal experience and the suggestions of family and friends (Word-of-Mouth). Albarq (2014) suggests that reduced consumer trust in advertising has led to an increase in the popularity of eWOM as a source of information. The results of the survey also showed that Greeks are heavy users of the internet and social media, where they log on daily and spend several hours. They access the internet via multiple devices, with mobile phones now dominating. They have accounts on multiple social media, the most popular being Facebook, Facebook Messenger, Instagram, YouTube, TikTok and WhatsApp. These data also confirm previous relevant research conducted in the Greek context (Constantoglou & Trihas, 2020).

The vast majority of participants search for relevant information online, on websites such as Google, Booking.com, TripAdvisor and social media, before buying travel products and services, to help them make better purchasing decisions and reduce the risk of making the wrong choice. The same findings were obtained by Hernández-Méndez et al. (2015) and Singh and Kathuria (2019). In the past, several participants of this survey have experienced feelings of dissatisfaction and disappointment as their experiences in tourism businesses and destinations have failed to meet their pre-consumption expectations. This is the reason why these consumers seek the advice of either relatives and friends or other internet users, which they consider more reliable than advertisements that are usually misleading by hiding negative aspects of the product, as they believe that relatives, friends and internet users do not have any financial benefit to lead them to a particular purchase decision. Although they consider online reviews to be for the most part valid and reliable, they recognise, however, that the credibility of eWOM is less than traditional WOM since the freedom that characterizes the internet can allow the dissemination of unverified, malicious, slanderous or distorted information about a tourism business or destination. Liu and Park (2015), however, suggest that there is a number of quantitative and qualitative characteristics about the reviewer and the review that make a review reliable and useful. Nevertheless, Greeks visit travel review sites in all three phases of a trip: before the trip, during the trip and after the trip to write a review themselves, although the majority of them are not active members of such online communities.

Based on the results of the survey, some implications for tourism businesses and destination management organizations emerge. First of all, owners or managers should understand that reviews and ratings are welcome and should not be intimidating. Customers' reviews are a valuable source of information for businesses. Businesses can use them to better understand the market reaction to the products and services they offer. The entrepreneur is therefore aware of what is being said about his business, while at the same time demonstrating to potential customers that he cares about his business and wants to improve the services he offers. Moreover, Ye et al. (2011) suggest that online reviews have a significant impact on online sales. The managers should therefore encourage their customers to take the time to write a review, which, as the survey results showed, businesses are largely not doing, resulting in valuable information being lost. Jalilvand and Samiei (2012) agree that managers should motivate tourists to participate in online travel communities. Subsequently, businesses and organisations should implement a systematic process for managing their online reputation, which will involve the following steps: (a) gathering reviews and ratings, (b) content and sentiment analysis, and (c) where necessary intervening and addressing the reviews. The latter means that businesses should respond to all reviews they receive, whether positive or negative. If the reviews are positive, they are a free advertisement for businesses and increase sales, whereas if they are negative, they can easily destroy the image and reputation of businesses (Chen & Law, 2016). As the results of the survey showed, when Greek consumers see that a business responds to the reviews they receive, they form the impression that the business really cares about its customers and are more likely to book with that business than with another that ignores the reviews and does not respond to them. The style in which they should respond to reviews is also important, as participants revealed that an aggressive or defensive response to a review from a business prevents them from booking with that business, while a proper response to a bad review improves their impression of the business. Of course, managers should not remove reviews (if they have this option of course) but should respond to the criticism with arguments while protecting the brand image.

This research has one basic limitation. More specifically, the data were collected electronically based on the non-probability snowball sampling method. This means that participants were internet users and not chosen but voluntarily participated in the survey (Saunders et al., 2016), therefore the sample was not random due to the self-response nature of the survey. A new study could be conducted to cover a larger and possibly more representative sample of the Greek consumers. Other studies could focus on the supply side and examine how Greek tourism businesses and organisations manage their online reputation. The rapid advancement also of the internet, big data and artificial intelligence creates new challenges in online reputation management that should be explored.

References

Ahmad, W., & Sun, J. (2018). Modeling consumer distrust of online hotel reviews. *International Journal of Hospitality Management, 71*, 77–90.

Albarq, A. N. (2014). Measuring the impacts of online Word-of-Mouth on tourists' attitude and intentions to visit Jordan: An empirical study. *International Business Research, 7*(1), 14–22.

Bakshi, S., Dogra, N., & Gupta, A. (2019). What motivates posting online travel reviews? Integrating gratifications with technological acceptance factors. *Tourism and Hospitality Management, 25*(2), 335–354.

Barber, P., & Wallace, L. (2009). The power of Word-of-Mouth marketing. *American Libraries, 40*(11), 36–39.

Casaló, L. V., Flavián, C., Guinalíu, M., & Ekinci, Y. (2015). Do online hotel rating schemes influence booking behaviors? *International Journal of Hospitality Management, 49*, 28–36.

Chen, Y.-F., & Law, R. (2016). A review of research on electronic Word-of-Mouth in hospitality and tourism management. *International Journal of Hospitality & Tourism Administration, 17*(4), 347–372.

Chopra, I. P., Lim, W. M., & Jain, T. (2022). Electronic word of mouth on social networking sites: What inspires travelers to engage in opinion seeking, opinion passing, and opinion giving? *Tourism Recreation Research*, 1–14.

Confente, I. (2015). Twenty-five years of Word-of-Mouth studies: A critical review of tourism research. *International Journal of Tourism Research, 17*, 613–624.

Constantoglou, M., & Trihas, N. (2020). The influence of social media on the travel behavior of Greek Millennials (Gen Y). *Journal of Tourism and Hospitality Management, 8*(2), 10–18.

Cox, C., Burgess, S., Sellitto, C., & Buultjens, J. (2009). The role of user-generated content in tourists' travel planning behavior. *Journal of Hospitality Marketing & Management, 18*(8), 743–764.

Datareportal. (2024). Digital 2024–global overview report. Retrieved April 10, 2024, from https://datareportal.com/reports/digital-2024-global-overview-report. (Online)

Doosti, S., Jalilvand, M. R., Asadi, A., Pool, J. K., & Adl, P. M. (2016). Analyzing the influence of electronic word of mouth on visit intention: The mediating role of tourists' attitude and city image. *International Journal of Tourism Cities, 2*(2), 137–148.

Filieri, R., & McLeay, F. (2013). E-WOM and accommodation: An analysis of the factors that influence travelers' adoption of information from online reviews. *Journal of Travel Research, 53*(1), 44–57.

Flipo, J. P. (1988). On the intangibility of services. *The Service Industries Journal, 8*(3), 286–293.

González-Rodríguez, M. R., Díaz-Fernández, M. C., Bilgihan, A., Okumus, F., & Shi, F. (2022). The impact of eWOM source credibility on destination visit intention and online involvement: A case of Chinese tourists. *Journal of Hospitality and Tourism Technology, 13*(5), 855–874.

Guzzo, T., Ferri, F., & Grifoni, P. (2022). What factors make online travel reviews credible? The consumers' credibility perception-CONCEPT model? *Societes, 12*, 50.

Hernández-Méndez, J., Muñoz-Leiva, F., & Sánchez-Fernández, J. (2015). The influence of e-Word-of-Mouth on travel decision-making: Consumer profiles. *Current Issues in Tourism, 18*(11), 1001–1021.

Hung, N. P., & Khoa, B. T. (2022). Examining the structural relationships of electronic word of mouth, attitude toward destination, travel intention, tourist satisfaction and loyalty: A meta-analysis. *GeoJournal of Tourism and Geosites, 45*(4), 1650–1660.

Jalilvand, M. R., & Samiei, N. (2012). The impact of electronic word of mouth on a tourist destination choice. Testing the theory of planned behavior (TPB). *Internet Research, 22*(5), 591–612.

Kim, M., & Kim, J. (2020). The influence of authenticity of online reviews on trust formation among travelers. *Journal of Travel Research, 59*(5), 763–776.

Ladhari, R., & Michaud, M. (2015). EWOM effects on hotel booking intentions, attitudes, trust, and website perceptions. *International Journal of Hospitality Management, 46*, 36–46.

Liang, S. W. J., Ekinci, Y., Occhiocupo, N., & Whyatt, G. (2013). Antecedents of travellers' electronic Word-of-Mouth communication. *Journal of Marketing Management, 29*(5–6), 584–606.

Liu, Z., & Park, S. (2015). What makes a useful online review? Implication for travel product websites. *Tourism Management, 47*, 140–151.

Mukhopadhyay, S., Pandey, R., & Rishi, B. (2023). Electronic word of mouth (eWOM) research–a comparative bibliometric analysis and future research insight. *Journal of Hospitality and Tourism Insights, 6*(2), 404–424.

Rajaguru, R. (2024). Effects of contemporary technologies, such as blockchain and artificial intelligence (AI) in enhancing consumers' trustworthiness of online reviews. *Journal of Hospitality Marketing & Management, 33*(2), 251–259.

Roy, G., Datta, B., Mukherjee, S., Eckert, A., & Dixit, S. K. (2024). How online travel reviews sources affect travelers' behavioral intentions? Analysis with source credibility theory. *Tourism Planning & Development, 21*(3), 299–329.

Saunders, M., Lewis, P., & Thornhill, A. (2016). *Research methods for business students* (7th ed.). Pearson Education Limited.

Setiawan, P. Y., Purbadharmaja, I. B. P., Widanta, A. A. B. P., & Hayashi, T. (2021). How electronic word of mouth (e-WOM) triggers intention to visit through destination image, trust and satisfaction: The perception of a potential tourist in Japan and Indonesia. *Online Information Review, 45*(5), 861–878.

Singh, S. K., & Kathuria, L. M. (2019). Factors influencing e-Word-of-Mouth adoption among consumers availing travel services. *International Journal of Tourism Policy, 9*(2), 155–173.

Su, L., Yang, Q., Swanson, S. R., & Chen, N. C. (2022). The impact of online reviews on destination trust and travel intention: The moderating role of online review trustworthiness. *Journal of Vacation Marketing, 28*(4), 406–423.

Tham, A., Croy, G., & Mair, J. (2013). Social media in destination choice: Distinctive electronic word-of-mouth dimensions. *Journal of Travel & Tourism Marketing, 30*, 144–155.

Trihas, N., Vassakis, K., Kopanakis, I., Nikoloudakis, Y., Kefaloukos, I., Pallis, E., & Markakis, E. (2023). The impact of COVID-19 on travel behavior and holiday intentions. Evidence from Greece. In B. Sharp, R. Finkel, & K. Dashper (Eds.), *Transforming leisure in the pandemic: Re-imagining interaction and activity during crisis* (pp. 42–56). Routledge.

Trihas, N., Zozonakis, G., & Kourgiantakis, M. (2024). Contactless tourism in the post-COVID-19 era: How ready are Greek tourists? In: A. L. Negrusa, & M. M. Coros (Eds.), *Sustainable approaches and business challenges in times of crisis* (pp. 103–118). Springer.

Tung, L. T., & My, D. T. H. (2023). Electronic word of mouth, attitude, motivation, and travel intention in the post-COVID-19 pandemic. *Journal of Tourism and Services, 14*(27), 181–196.

Wu, Q.-M. (2024). The influence of online reviews on the purchasing decisions of travel consumers. *Sustainability, 16*, 3213.

Ye, Q., Law, R., Gu, B., & Chen, W. (2011). The influence of user-generated content on traveler behavior: An empirical investigation on the effects of e-word-of-mouth to hotel online bookings. *Computers in Human Behavior, 27*, 634–639.

Yoo, K. H., & Gretzel, U. (2008). What motivates consumers to write online travel reviews? *Information Technology & Tourism, 10*, 283–295.

The Impact of News About the Recession on Consumer Sentiment in the United States

Dejan Romih and Valentin Toader

Abstract News, whether good or bad, can have an impact on consumer sentiment. In this paper, we examine the impact of news about the recession on consumer sentiment in the United States. Using a vector autoregressive model, we find that an unexpected increase in news about the recession has an immediate impact on consumer sentiment in the United States. To reach this conclusion, research hypotheses are developed based on an extensive review of the literature on the relationship between news and the evolution of consumer sentiment and other macroeconomic variables. The methodology is then explained, the results are discussed, and the paper ends with conclusions.

Keywords Consumer sentiment · News · Recession · United States

Introduction

The economic impact of news in the modern interconnected world is profound and multifaceted. The rapid dissemination of both genuine and fake news through the Internet and social networks can significantly distort people's perceptions of economic conditions. Even when economic fundamentals are robust, the spread of negative news about events, such as deadly viruses, racial tensions, or wars, can create a pervasive atmosphere of negativity, influencing consumer sentiment and behaviour. Furthermore, the reporting of economic news has become increasingly negative in recent years, intensifying this phenomenon (Barua, 2024). Regarding the approach of the paper, the literature review captures the impact of newly received or noteworthy information, about recent or important events disseminated through different channels, from formal, such as TV and the press, to informal, such as social

D. Romih (✉)
Faculty of Economics and Business, University of Maribor, Maribor, Slovenia
e-mail: dejan.romih@um.si

V. Toader
Faculty of Business, Babes Bolyai University of Cluj-Napoca, Cluj-Napoca, Romania

A. L. Negrușa et al. (eds.), *Sustainable and Resilient Businesses in the Global Economy*, Springer Proceedings in Business and Economics,
https://doi.org/10.1007/978-3-031-75883-6_12

networks. However, the analysis performed focusses on TV news about the recession, collected from Stanford Cable TV News Analyzer.

Song and Shin (2019) provide further information on the economic impact of news by demonstrating that news articles can serve as effective sources for generating economic indicators, particularly in the context of consumer sentiment indexes. These indexes, derived from news data, provide qualitative information on the current economic situation and can warn of potential turning points in economic activity.

Starting from turning points, the economists use a practical approach to define a period of recession by focusing on two consecutive quarters of decline in real GDP (Claessens & Kose, IMF, n.d.). Since such a narrow approach may have some drawbacks, the NBER provides a more comprehensive assessment that focuses on various economic measures beyond just GDP, including employment, income, sales, and industrial production. Thus, according to NBER (2024), a recession is defined as a substantial downturn in economic activity that affects multiple sectors of the economy that lasts more than a few months.

The purpose of the present study is to evaluate the relationship between news about the recession and consumer sentiment in the United States. Thus, the paper is organised as follows: The next section contains a comprehensive analysis of the literature about the impact of news on consumer sentiment and other macroeconomic variables, then it follows the methodology and the results and discussions. The paper ends with conclusions.

Literature Review

The economic impact of news is substantial and influences consumer sentiment, behaviour, and economic results through various channels. From shaping perceptions of economic conditions to driving consumer sentiment and behaviour, news media play a pivotal role in the modern economy.

The impact of changes in consumer confidence and future economic activity is assessed through two competing channels: the animal spirit channel, which posits that changes in economic activity are driven by fluctuations in consumer beliefs, and the news channel, which suggests that consumer confidence measures contain information about the present and future state of the economy. The findings indicate that consumer confidence serves as a predecessor to future economic conditions, with news shocks playing a significant role in shaping the dynamics of consumption, output, and inflation (Choi et al., 2023).

Negative bias among consumers, in which people react more strongly to bad news than to good news, is another important aspect reflected in the literature. This bias may arise from evolutionary survival instincts and loss-aversion tendencies. Such reactions to negative news can significantly influence consumer behaviour and economic outcomes (Nguyen & Claus, 2013). Furthermore, He (2017) notes that pessimistic news negatively impacts total employment and new hiring, while its effects on lost jobs are mixed. In the same vein, Eggers et al., (2021) emphasise the role of news about economic recession in shaping economic perceptions and behaviour. They find that the announcement of a recession can lead to a significant reduction in consumer confidence, consumption growth, and GDP growth. Economic agents receive a large proportion of their information about the economy from the media, shaping their perceptions and behaviour accordingly. Furthermore, the intensity of news coverage affects the frequency of expectation updates among consumers, with increased coverage during recessions leading to increased expectations update. This idea is also supported by Doms and Morin (2004), who found that media coverage, especially intense during recessions, significantly impacts consumer sentiment, sometimes deviating from economic fundamentals. High news coverage prompts more frequent expectation updates, countering the "stickiness" of expectations. Due to the non-linear nature of consumer sentiment, changes in sentiment are complex and can produce unexpected outcomes, such as a more positive sentiment despite negative economic reporting. In addition, they identify three channels through which the news media influence consumer perceptions of the economy: providing economic data and opinions, signalling economic conditions through the tone and volume of reporting, and affecting the likelihood of consumers updating their expectations.

The Relationship Between News and Consumer Sentiment

Multiple authors (Anastasiou et al., 2022; Gric et al., 2022; Sandoval & Walsh, 2024) emphasise that consumer sentiment significantly influences economic decisions and activities. Sandoval and Walsh (2024) examined the role of consumer sentiment as a leading indicator of economic trends, which often attracts the attention of policymakers, economists, and the media. Their conclusion is that, despite its importance, the predictive reliability of consumer sentiment is not consistent across different countries and economic measures because the consumer sentiment generally reflects the consumers' attitudes and expectations regarding economic conditions, influenced either by economic fundamentals or by independent fluctuations in beliefs. Gric et al. (2022) consider that household sentiment reflects how optimistic or pessimistic people are about their financial situation and the general economy. This sentiment can be broken down into rational components based on macroeconomic data and irrational components, reflecting excessive optimism or pessimism. As a result, Anastasiou et al. (2022) highlight that when it comes to borrowing money, potential borrowers, in addition to considering their financial situation, also evaluate the overall economic outlook to adjust their perceptions and beliefs. Optimistic economic conditions encourage loan applications, while pessimistic ones deter them. This perception affects the availability of credit and economic activity. Under unfavourable economic conditions, discouraged borrowers avoid profitable projects, leading to misallocation of funds and hindering long-term economic growth.

The importance of economic perceptions, whether based on fundamentals or independent beliefs, represents a recurring theme in the literature. Anastasiou et al. (2022) emphasises that expectations about the future evolution of the economy influence consumer sentiment, optimistic expectations increasing the economic sentiment and leading to increased bank credit supply, as banks use sentiment indicators to inform their decisions. This positive effect reduces discouragement among potential lenders. However, higher volatility in Economic Sentiment Indicator (used by Anastasiou et al. (2022)) shocks increases firms' fear of loan rejection, reducing loan applications and bank credit growth during anxious periods. Banks also cut loan supply when sentiment declines, further discouraging potential borrowers. Notably, smaller firms, especially micro-firms, face a higher probability of being discouraged from applying for loans compared to larger firms.

The impact of media and information on consumer sentiment highlights the role of economic reporting and news shocks in shaping sentiment. Song and Shin (2019) found that consumer sentiment is significantly affected by economic news shocks, which influence aggregate economic activity. Since experts have a broader perspective on the economic phenomenon due to data availability, their opinions expressed in the news articles tend to provide a better explanation of households' economic conditions than individual subjective judgments. Furthermore, the volume and tone of economic reporting affect sentiment independently of actual economic activity and professional opinions (Doms & Morin, 2004). Increased reporting on recessions or layoffs generally relates to a decrease in sentiment. Doms and Morin noted that in the 1990s, news coverage was larger than justified by economic fundamentals, resulting in sentiment measures that were lower than they might have been otherwise. Although media coverage can influence sentiment, one of the most important conclusions of their analysis is that these effects are typically short-lived and last only a few months. Their research also highlights the need to understand how frequently people update their expectations, with "sticky expectations" potentially explaining slow responses to economic shocks. During periods of intense economic reporting, sentiment tends to be less sticky, suggesting that media coverage plays a crucial role in how quickly people adjust their economic expectations (Doms & Morin, 2004).

Due to the extensive use of social networks in our lives, it was shown that social networks play a powerful role in shaping consumer sentiment, with non-sponsored influencer content being particularly influential due to its perceived authenticity and broader economic implications (Zhang et al., 2024). The study indicates two main reasons for this impact. First, social media users perceive non-sponsored posts from influencers as more authentic and genuine compared to sponsored posts, which are seen as commercial and less credible. This perception leads to greater trust and engagement with non-sponsored content. Second, influencer sentiment derived from all posts captures a wider array of economic and social factors compared to sentiment linked solely to posts about listed firms. This broader range of content allows influencers to shape consumer sentiment more effectively. Furthermore, research shows that middle-aged and less educated individuals are more affected by social media influencers than younger and more educated people. This difference is likely due to the rise of TikTok, which has drawn many younger users away from Instagram. Younger audiences in the US now spend twice as much time on TikTok as on Instagram. Educational levels also play a role, as higher education is associated with better rational economic decision making, making educated individuals less susceptible to influencer marketing.

Following the aspects discussed previously, the research hypothesis that will be tested in the paper is that an increase in the frequency of word recession in cable TV news will lead to a decline in the consumer sentiment.

The Relationship Between News and Consumption

News, particularly macroeconomic news, plays a critical role in shaping consumer sentiment, which, in turn, affects consumption patterns. The literature focused on the impact of news on consumption analyses the differences in the impact between positive and negative news, the impact on consumption-saving decisions or the impact on consumption of durable and nondurable goods.

Sandoval and Walsh (2024) argue that, although consumer sentiment does impact consumption, the relationship is complex and influenced by factors, such as economic conditions (income, for example), political environment, and the timing of consumer surveys relative to major events, such as elections. Consumer sentiment could follow economic activity rather than drive it. This suggests that sentiment is influenced by broader economic conditions, rather than being a primary driver of consumption changes. Additionally, consumer sentiment can be significantly influenced by political factors, such as changes in government. For example, consumers tend to have more optimistic views about future economic conditions when the political party they support is at power. Sharp shifts in sentiment often occur immediately after presidential elections, resulting in a change in the ruling party. Therefore, understanding these dynamics is crucial to accurately assessing how changes in sentiment translate into changes in consumer behaviour and overall economic activity. Taking these aspects into account, they conclude that spending intentions, influenced by sentiment, are predictive of actual spending behaviour. This implies that changes in sentiment can lead to corresponding changes in consumer spending patterns, an aspect that is supported also by Nguyen and Claus (2013) who demonstrate that declines in consumer sentiment are associated with real household consumption declines in Australia, whereas rises in sentiment do not show a significant relationship with consumption. Spending intentions are found to be more sensitive to long-term expectations about the health of the national economy compared to short-term expectations. This indicates that consumers' views on future economic conditions play a crucial role in shaping their spending decisions. In this vein, Vinokurov et al. (2018) explain that precautionary motives lead households to save more when income is unpredictable, while uncertainty about investment returns can reduce savings. The decision to save or consume hinges on what is more salient: income stability or returns on savings. Structural changes, pension reforms, and political changes cause more savings and less consumption, addressing the threat of losing current levels of well-being. Additionally, shocks to aggregate sentiment significantly impact fluctuations in newly issued consumer loans. Both rational and irrational sentiments influence credit growth, and irrational beliefs positively affect loan growth during economic booms. Positive sentiment shocks lead to faster credit growth, especially under favourable macroeconomic conditions. In conclusion, macroeconomic policies should consider the unconditioned part of the information that affects households' decisions about consumption and savings. Adjusting policies to account for this can help neutralise the impact of economic uncertainty on consumption and savings behaviour.

When it comes to the impact of different types of news on consumption, the literature highlights a similar view: households react more to bad news than good news (Corrado et al., 2022; Nguyen & Claus, 2013; Shapiro et al., 2022; Uhl, 2012). This asymmetric response is attributed to the ambiguity of information and the aversion of households to ambiguity. Male-headed households appear to be more reactive to positive news and increase their consumption significantly upon receiving good news. Corrado et al. (2022) consider that households with aversion to ambiguity reduce spending more with negative information than increase it with positive information. On the contrary, consumers who love ambiguity tend to increase spending more with good news than they reduce it with bad news. The same conclusion is valid also for television news; Uhl (2012) concluded that positive sentiment in TV news correlates with higher private consumption growth. Higher personal income and lower savings rates also contribute to increased consumption, whereas higher consumer prices prompt immediate consumption due to expectations of future price increases. On the contrary, higher interest rates encourage savings over consumption.

Shapiro et al. (2022) and Nguyen and Claus (2013), in addition to emphasising that positive news sentiment shocks increase consumption, add that these effects on consumption and output are lagging and long-lasting, peaking between 12 and 18 months post-shock before gradually weakening. As a result, policy makers, in order to mitigate these effects, must make economic signals clearer and more transparent. Central bank communication, for example, should aim to align expectations more explicitly to reduce the relevance of ambiguous information (Corrado et al., 2022).

Song and Shin (2019) conclude that optimistic economic perceptions increase consumption, with a stronger impact on durable goods compared to nondurable goods. Durable goods, being more expensive and providing long-term services, require confidence in the economic outlook. Nondurable goods, on the other hand, are subject to immediate purchasing decisions without significant time delays. Therefore, consumer sentiment influences durable goods consumption more significantly due to the confidence required for such purchases.

In conclusion, the study hypothesis that will be investigated in this paper is that a decrease in household consumption may represent the result of a greater frequency of the use of the word "recession" in cable TV news.

The Relationship Between News, Consumer Sentiment and Unemployment

Various studies demonstrate that news and consumer sentiment significantly impact unemployment and labour market activity. For example, Song and Shin (2019) highlight the long-term effects of consumer sentiment on labour market participation, while Nguyen and Claus (2013) underscore asymmetric reactions to economic news. He (2017) further emphasises the powerful influence of pessimistic news on hiring, particularly at the beginning of economic downturns.

However, the relationship is multifaceted. On the one hand, there are studies that analyse the impact that changes in employment/unemployment rates have on consumer sentiment, while other studies focus on the impact that different types of news have on labour market prospects. Nguyen and Claus (2013), in their study on Australian consumers' reactions to economic news, found that consumers exhibit a strong reaction to bad news but remain unaffected by good news. More precisely, consumer sentiment tends to decline after an increase in unemployment. On the contrary, changes in sentiment following decreases in unemployment and increases in share prices are not statistically significant. This asymmetric response was consistent between various demographic groups, including age, homeownership status, and gender, indicating that all groups react negatively to increasing unemployment but not positively to falling unemployment. Song and Shin (2019) support this conclusion, pointing out that higher employment levels lead to increased consumer spending and confidence, while rising unemployment has the opposite effect, demonstrating the complex link between economic sentiment and employment/unemployment rates.

On the other hand, shocks in consumer sentiment have long-lasting effects on economic participation and unemployment rates. Song and Shin (2019) emphasise that changes in the consumer sentiment index can have long-lasting effects on the labour market, and adjustments take considerable time. Furthermore, He (2017) investigated the effects of exposure to pessimistic economic news from newspaper articles on labour market activity, particularly new hiring. He argues that pessimistic news can significantly dampen labour market activity, reducing employment by approximately half a million persons per quarter. This effect is most pronounced at the beginning of a recession, where the initial spike in pessimistic news is strong, although total job losses may be mild initially. As the recession progresses, the impact of pessimistic news diminishes as the media becomes saturated with bad news, and the continued rise in unemployment becomes widely recognised.

Thus, in this article, the investigated research hypothesis states that an increase in the presence of the word 'recession' in cable TV news, leading to a decline in consumer sentiment, determines an increase in unemployment rates.

The Relationship Between News, Consumer Sentiment and Inflation

According to economic theory, news can affect inflation through various channels. For example, news about economic policies, supply chain disruptions, or changes in money supply can influence inflation expectations. If people expect higher inflation, they may demand higher wages and companies might set higher prices, creating a self-fulfilling cycle of inflation. Additionally, news that affects consumer confidence or investor behaviour can also impact economic activity, and thus inflation. The overall effect of news on inflation depends on the nature, credibility, and context of the information disseminated.

News about possible economic changes can lead to expectations of higher or lower inflation. For example, if there is news of an increase in government spending, people might expect inflation to increase because more money will be circulating in the economy. Additionally, when individuals or businesses predict an increase in prices, they often incorporate these predictions into wage discussions and price adjustment clauses (like routine rent hikes). This anticipatory action plays a role in shaping inflation for the upcoming period; as wages and prices increase according to these agreements, the initial expectations validate themselves. Furthermore, when these expectations are grounded in recent trends, inflation tends to follow a consistent pattern over time, leading to persistent inflation (Oner, n.d.).

The news can also influence investors. According to Duggan and Curry (2023), if investors expect inflation to increase, they may demand higher interest rates for loans, which can increase the cost of borrowing and lead to higher prices. Investors prefer a situation of price stability (low and stable inflation rates) which is considered 'healthy', with a more stable money value, consistent demand for goods and services, and predictable pricing. However, higher annual inflation rates determine uncertainty and volatility in the stock market, coupled with a decrease in consumer spending. This can lead to reduced economic growth, negatively impacting investors and causing valuation issues that may result in poor stock market performance.

Accordingly, the study hypothesis under investigation in this article asserts that a rise in the word "recession" in cable TV news corresponds with a rise in inflation rates.

The Relationship Between News, Consumer Sentiment and Interest Rate

To achieve their goals (price stability, sustainable growth, etc.), the central banks adjust their monetary policy interest rate based on a comprehensive assessment of various economic indicators and conditions. Since news and media are almost impossible to cover all aspects assessed by the central bank to reflect or provide reliable sources of information, the impact of news and media on central bank decisions is very difficult to be detected appropriately. However, the news has the capacity to influence society's expectations, influence private sector decisions, and indirectly affect the effects of monetary policy.

For example, Nguyen and Claus (2013) concluded that decreases in the monetary policy interest rate are perceived negatively. Typically, mortgage rates align with changes in the monetary policy interest rate, so a decrease in the policy rate, which lowers mortgage payments, would be expected to boost consumer confidence. However, their empirical results show that decreases in the monetary policy interest rate reduce sentiment, making people less optimistic about the future evolution of the economy. This outcome cannot be attributed to a wealth effect, since a decline in the policy rate would reduce investment income, impacting high-income households and homeowners, but no other groups. A plausible explanation is that consumers interpret the easing of monetary policy as a sign of impending economic weakness, thus lowering sentiment. This finding implies the existence of a 'consumer sentiment channel' in the transmission of monetary policy, highlighting its potential importance.

However, in the long run, Shapiro et al. (2022) conclude that shocks caused by positive news sentiment led to increased consumption, output, and the real rate of government funds. These effects are more lasting, peaking between 12 and 18 months after shock before gradually diminishing. This indicates that positive news sentiment has significant and lasting effects on the economy, particularly influencing interest rates and overall economic activity. The increase in the interest rate could be perceived as a precautionary approach of the central bank to avoid overheating the economy due to a too optimistic reaction generated by the shock of positive sentiments.

Thus, the present study aims to explore the research hypothesis that suggests a fall in the monetary policy interest rate is correlated with increased usage of the term "recession" in cable TV news.

The Relationship Between News, Consumer Sentiment and Stock Market

News, especially bad news, has a significant impact on the volatility of the stock market and investor sentiment. The media's role in influencing the perceptions and contributing to market instability is significant, with the potential to drive speculative bubbles. The relationship between news, sentiment, and market dynamics is complex and cyclical, with varying effects depending on overall market conditions and investor psychology.

In general, good news improves market performance, while bad news typically hinders market growth. However, the influence of news on the stock market is asymmetric; negative news has a greater impact on market volatility. Furthermore, the effect of news on market returns can vary based on the overall situation of the market. For example, in a bullish market, investors may react more strongly to good news than in a bearish market, where the same news might have a muted effect. Investor sentiment also plays a crucial role in stock market fluctuations. High sentiment can lead to overvaluation of stocks due to overly optimistic future cash flow estimates, while low sentiment can result in undervaluation (Van de Kauter et al., 2015).

Economic news primarily affects market volatility rather than stock returns. This means that although stock prices may not necessarily move significantly up or down in response to news, the stability of those prices is often affected (Vinokurov et al., 2018). The significant contribution of the media to market noise may be a primary source of market instability. Media stories often enhance information with elements of emotion or anxiety, affecting investors' perceptions. As a result, media coverage can make investors irrational by altering the character of the information presented. This mediatisation of markets can lead to speculative bubbles, where the risks of the stock market are overlooked. Historical examples, such as the Internet and housing bubbles, illustrate how media coverage can fuel speculative excesses. The history of market bubbles almost coincides with the beginning of newspapers, suggesting that media coverage has long been an influential factor in financial markets. Furthermore, high levels of media pessimism are robust predictors of downward pressure on market prices, although this is typically followed by a reversion to fundamentals. Furthermore, extreme values of media pessimism often predict high market trading volumes. Interestingly, low market returns tend to lead to high pessimism in the media, indicating a reciprocal relationship between market performance and media sentiment. (Van de Kauter et al., 2015).

The study hypothesis examined in this work, which builds on the earlier analysis, emphasises that higher stock market volatility is correlated with the term "recession" being used more frequently in cable TV news media.

Research Methodology

In our study, we analysed the relationship between news about the recession and consumer sentiment in the United States. We used monthly data from January 2010 to February 2024 and a vector autoregressive (VAR) model with two lags and seven variables: the first difference of the natural logarithm of the Consumer Price Index (*D.ln_cpi*), the unemployment rate (*unrate*), the first difference of the natural logarithm of Personal Consumption Expenditures (*D.ln_pce*), the screen time of the word 'recession' in the subtitles of cable news (*news*), the first difference of the natural logarithm of the Consumer Sentiment Index (*D.ln_csent*), the first difference of the natural logarithm of the NASDAQ Composite Index (*D.ln_nasdaq*) and the Federal Funds Effective Rate (*fedfunds*). Data on the Consumer Price Index, unemployment rate, and Personal Consumption Expenditures were seasonally adjusted. We also used Cholesky decomposition to orthogonalise the residuals in our VAR model.

To measure the screen time of the word 'recession' in the subtitles of cable news, we used the Stanford Cable TV News Analyzer developed by Hong et al. (2021). Data on other variables were obtained from the Federal Reserve Economic Data from the Federal Reserve Bank of St. Louis.

Results and Discussions

In this section, we present the main results of our study, focusing on the impact of an increase in the orthogonalised shock to *news* on *D.ln_csent*. As we shall see below, this study has important implications for policy makers who need to monitor cable news.

Figure 1 shows that an increase in the orthogonalised shock to the news leads to an increase in the unemployment rate that is statistically significant two months after the shock. This finding is consistent with the results of He (2017), who found that bad news can have a negative impact on the labour market.

We also found that an increase in the orthogonalised shock to the news leads to a decrease in the first difference of the natural logarithm of the Consumer Sentiment Index, the first difference of the natural logarithm of the NASDAQ Composite Index, and the first difference of the natural logarithm of Personal Consumption Expenditures that is statistically significant only at the time of the shock. As discussed in the literature review, the relationship between news and consumer sentiment is complex and is influenced by a number of economic and psychological factors. Despite this complexity, Song and Shin (2019) found that bad news have a negative impact on consumer sentiment, which is consistent with our findings. This applies also to consumption (Nguyen & Claus, 2013; Shapiro et al., 2022; Song & Shin, 2019). As for the impact of bad news, the picture is clearer because we know that stock markets are driven by fear and greed. And that the impact of bad news is usually felt immediately.

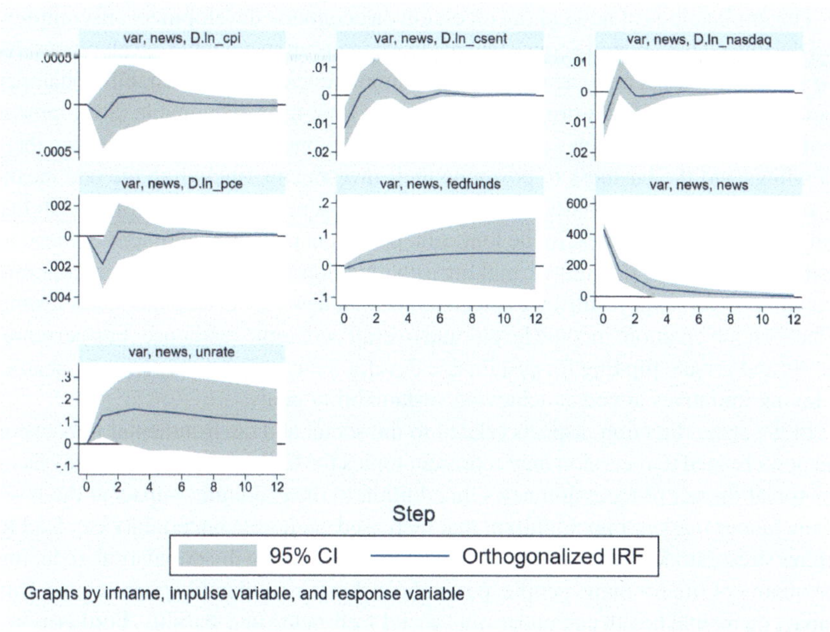

Fig. 1 Results of an increase in the orthogonalised shock to *news* on other variables. *Notes* The horizontal axis of each graph is in units of time, and the vertical axis of each graph is in units of the variable. Each graph shows the impact of an innovation over a 12-month period. *Source* The author's own processing based on data from the Federal Reserve Bank of St. Louis, Hong et al. (2021) and the Stanford Cable TV News Analyzer (https://tvnews.stanford.edu/)

Figure 1 also shows that an increase in the orthogonalised shock to news has no impact on the first difference of the natural logarithm of the Consumer Price Index and the Federal Funds Effective Rate. This means that the unexpected increase in news about the recession has no impact on inflation and monetary policy in the US, which is consistent with the literature.

Conclusions

The results of the VAR analysis show that news about the recession has a negative impact on some economic variables, including consumer sentiment in the United States. This means that the role of the media is important and that the media bears a large part of the responsibility for the state of the economy, especially when it prioritises bad news over good news, fake news over real news, and so on. This means that politics and the media are not completely separate, and policy makers should consider this aspect when they develop macroeconomic policies.

The implications of news about recession on economic development are complex, and the impact is even more consistent when we focus on sustainable development. As our results show, negative news can lead to immediate economic instability, hamper social and environmental progress, and disrupt long-term sustainable development goals. In the short run, news about recession determines households to reduce their spendings and the business to postpone their investments (including the one meant to increase the sustainability of their business), generating short-term instability and declining growth rates. In the long-run, persistent negative sentiment can lead to prolonged economic downturns, making it challenging to sustain long-term economic growth. Based on our results, we can conclude that news about stagnation has lasting effects on job creation, income levels, and overall economic resilience. Furthermore, public and private funding for sustainable development projects may become scarce, delaying initiatives aimed at achieving sustainability goals.

In the same direction, aspects related to the social and environmental impacts of the news related to recession may represent topics for further papers. The analysis of the social impact of recession news, in addition to the economic impact at the level of the labour market, may highlight that increased economic uncertainty can lead to higher stress, anxiety, and other mental health issues among the population, reducing the quality of life for many people, particularly those in vulnerable populations. This impact on mental health can undermine social well-being and stability. Furthermore, if the recession persists, governments facing challenges in allocating resources may decide to cut social safety funds, decisions with significant impacts, especially in the case of low-income and marginalised communities.

In most cases, during economic recessions, businesses and governments prioritise short-term economic survival over long-term environmental goals. As a result, investments in renewable energy, conservation, and other environmental initiatives can be postponed or cancelled. On the other hand, following the decisions of some EU member states to reopen their coal-fired power plants that were closed due to environmental impact with the purpose of reducing the EU energy dependency from Russia, we may conclude that economic pressures could lead to increased exploitation of natural resources as countries and companies seek immediate revenue, potentially causing environmental degradation and undermining sustainability efforts.

To address these challenges, it is essential that policymakers, businesses, and communities focus on building resilience, fostering innovation, and prioritising sustainable practices even in the face of economic adversity. By doing so, it is possible to mitigate the negative impacts of recessions and continue making progress toward a sustainable future.

References

Anastasiou, D., Kallandranis, C., & Drakos, K. (2022). Borrower discouragement prevalence for Eurozone SMEs: Investigating the impact of economic sentiment. *Journal of Economic Behavior and Organization, 194*, 161–171.

Barua, A. (2024). *Consumer sentiment isn't enough to gauge economic outlook.* https://www2.deloitte.com/us/en/insights/economy/spotlight/consumer-sentiment-spending-economic-conditions.html. (Online)

Choi, S., Jeong, J., Park, D., & Yoo, D. (2023). *News or animal spirits? Consumer confidence and economic activity: Redux.* Vienna: SUERF-The European Money and Finance Forum.

Claessens, S., & Kose, M., IMF. (n.d.). *Recession: When bad times prevail.* https://www.imf.org/en/Publications/fandd/issues/Series/Back-to-Basics/Recession. (Online)

Corrado, L., Silgado-Gómez, E., & Yoo, D. W. R. (2022). Ambiguous economic news and heterogeneity: What explains asymmetric consumption responses? *Journal of Macroeconomics,* 72.

Doms, M., & Morin, N. (2004). Consumer sentiment, the economy, and the news media. In *FRB of San Francisco working paper* (October).

Duggan, W., & Curry, B. (2023). *How does inflation affect stocks?* Retrieved June 14, 2024, from https://www.forbes.com/advisor/investing/inflation-and-stock-market/. (Online)

Eggers, A., Ellison, M., & Lee, S. (2021). The economic impact of recession announcements. *Journal of Monetary Economics, 120*, 40–52.

Gric, Z., Ehrenbergerova, D., & Hodula, M. (2022). The power of sentiment: Irrational beliefs of households and consumer loan dynamics. *Journal of Financial Stability,* 59.

He, Z. (2017). The contractionary effect of bad economic news. *Journal of Money, Credit and Banking, 49*(6), 1339–1384.

Hong, J., et al. (2021). *Analysis of faces in a decade of US cable TV news* (pp. 3011–3021). ACM.

National Bureau of Economic Research. (2024). *Business cycle dating.* https://www.nber.org/research/business-cycle-dating. (Online)

Nguyen, V., & Claus, E. (2013). Good news, bad news, consumer sentiment and consumption behavior. *Journal of Economic Psychology, 39*, 426–438.

Oner, C. (n.d.) *Inflation: Prices on the rise.* Retrieved June 12, 2024, from https://www.imf.org/en/Publications/fandd/issues/Series/Back-to-Basics/Inflation. (Online)

Sandoval, H., & Walsh, A. (2024). Sentiments and spending intentions: Evidence from Florida. *Economic Inquiry,* 1–28.

Shapiro, A., Sudhof, M., & Wilson, D. (2022). Measuring news sentiment. *Journal of Econometrics, 228*, 221–243.

Song, M., & Shin, K. (2019). Forecasting economic indicators using a consumer sentiment index: Survey-based versus text-based data. *Journal of Forecasting, 38*, 504–518.

Uhl, M. (2012). And action: TV sentiment and the US consumer. *Applied Economics Letters, 19*(11), 1029–1034.

Van de Kauter, M., Breesch, D., & Hoste, V. (2015). Fine-grained analysis of explicit and implicit sentiment in financial news articles. *Expert Systems with Applications, 42*, 4999–5010.

Vinokurov, S., Medved, A., & Mierin, L. (2018). Economic news and household decisions. *Contemporary Economics, 12*(1), 57–68.

Zhang, Z., Keasey, K., Lambrinoudakis, C., & Mascia, D. (2024). Consumer sentiment: The influence of social media. *Economics Letters,* 237.

Exploring the Domains of Consumer Perceived Ethicality in Developing Countries. Comparative Qualitative Research on Romania and China

Laszlo Alexandru Gyori, Flavia Dana Oltean, and Manuela Rozalia Gabor

Abstract This paper explores the domains that contribute to Consumer Perceived Ethicality (CPE) in developing countries. Brunk (AMA Winter Educators' Conference Proceedings. Brussels: American Marketing Association, pp. 32–33 (2010a)) original framework of six CPE domains (consumers, employees, environment, overseas community, local community and economy, and business community) is used as a foundation. In-depth interviews with consumers from Romania and China (developing countries) were conducted to understand their perspectives on ethical business practices. The findings suggest that all six of Brunk's domains are relevant to CPE in developing countries, but there may be nuances in how these domains are perceived compared to developed countries. For example, consumers in developing countries may be more concerned about the impact of business activities on local economies and the exploitation of labor overseas. This research contributes to the understanding of CPE by focusing on a previously understudied population (consumers in developing countries). The findings can be used by businesses to develop more ethical practices that resonate with consumers in these markets.

Keywords Consumer perceived ethicality · Ethics · Developing countries · Qualitative analysis

JEL Classification M14 · M31 · D12

L. A. Gyori
Department of Translation Services, Amazon Spain Services, Madrid, Spain

F. D. Oltean · M. R. Gabor (✉)
Department of Economic Science, Faculty of Economics and Law, "G. E. Palade" University of Medicine, Pharmacy, Science and Technology of Târgu Mureş, Gh Marinescu 38, 540142 Târgu Mureş, Mureş County, Romania
e-mail: manuela.gabor@umfst.ro

Department of Economic Research, Centre for Law, Economics and Business Studies, "G. E. Palade" University of Medicine, Pharmacy, Science and Technology of Târgu Mureş, Gh Marinescu 38, 540142 Târgu Mureş, Mureş County, Romania

IOSUD-Doctoral School, "G. E. Palade" University of Medicine, Pharmacy, Science and Technology of Târgu Mureş, Gh Marinescu 38, 540142 Târgu Mureş, Mureş County, Romania

A. L. Negruşa et al. (eds.), *Sustainable and Resilient Businesses in the Global Economy*, Springer Proceedings in Business and Economics,
https://doi.org/10.1007/978-3-031-75883-6_13

Introduction

Ethics refers to a set of moral norms, principles or values that guide people's behavior (Sherwin, 1983). The terms unethical or ethical describe an individual's subjective moral judgment of right/wrong or good/bad (Brunk, 2010b). Thus, starting from this general definition, "business ethics" is the extension of the ethics construct which can be defined as the critical, structured examination of how people and institutions should behave in the world of trade in following the laws, rules, moral. The concept of business ethics is not a distinct and separated element from ethics in general, but it takes part in and is a sub-element of it (Kumar et al., 2023). As today there are several sets of issues and expectations in doing business activities, which differ widely from one culture to another and from one country to another, business ethics represents the evaluation and application of moral standards in the business area (production, sales, communication, marketing, financial, accounting, etc.) according to its own right or wrong behavior perception). (Shah et al., 2020; Utkutug, 2022).

Consumer perceived ethicality (CPE) refers to the consumer's perspective of corporate social responsibility (Brunk, 2010b). Majority of existing literature relating to corporate social responsibility (CSR) and business ethics focuses on decision-making process in companies (Freeman & Hasnaoui, 2011) while there are a few particular research which is investigating the ethical judgments of marketing professional (Eryandra et al., 2018; Nill & Schibrowsky, 2007). When confronting with ethical dilemmas, it is important to understand how professional react, but an investigation of how corporates decisions are received by consumers should be equally interesting. There is a lack of understanding of consumers' ethical beliefs. The consumers' perspective of what constitutes un/ethical may not be in line with a company's definition (Kumar et al., 2023; Tsalikis et al., 2019).

Singh et al. (2012) affirms that corporates should use the ethical dimension as a strategy, not as a factor that may influence consumers' behavior. Brunk's (2010b) findings come in addition to Singh's affirmation and the results show that once a negative CPE is founded, it sounds to be rather static and resistant to change over time. Understanding this is important for companies because a misunderstanding can led to loss in profits or market shares. A relevant example is Nestlé in 1977 which introduced a new baby formula in Africa which was supposed to be mixed with water. The company did not consider that the water was polluted in most African countries, and it resulted in a boycott. Consumers had formed a negative behavior regarding Nestlé and according to Muller (2013) the company is still suffering from the past boycott. Perception of the consumers is based on their morals and judgments, which may be different depending on the economic status of the country. (Tsalikis et al., 2019; Utkutug, 2022).

The main studies we can rely on are those from Brunk (2010b) who studied domains of consumer perceived ethicality (CPE) and identified a wide range of business and social activities that may evoke un/ethical company/brand perceptions among consumers: consumers, employees, environment, overseas community, local economy and community, and business community. The study proposed 6 areas of

CPE to be analyzed as they were presented in the previous statement. Thus, this qualitative study analyzed data from consumers from UK and Germany (developed countries), so considered the international aspect.

To get a better understanding of the potential domains that forms the CPE, Awasthi and Singhal (2014) had continued Brunk's (2010b) study and explored whether a set of domains of consumers' ethical beliefs is exhaustive, and if different domains affect consumers perception and relationship with the company differently. The further research suggests some additions to the CPE framework, (Awasthi & Singhal, 2014). Another similar research was performed by Kumar et al., 2023 who attempted to analyze the brand ethicality by in relation to brand marketing communication. The studies regarding the formation of CPE and the potential sources target the developed countries. Other researchers like Brunk (2010b), Awasthi and Singhal (2014), Eryandra et al. (2018), Manansala et al. (2024) suggest further research from the "consumers side" and emphasize on developing countries. How consumers perceive an activity as being un/ethic is a matter of judgment, therefore a matter of subjectivism (Zeithaml, 1988). There are considerable differences of judgments between ethnicity which can affect a consumers' behavior differently (Shah et al., 2020).

There is a lack of understanding of consumers' ethical beliefs. The consumers' perspective of what constitutes un/ethical may not be in line with a company's definition (Kumar et al., 2023; Manansala et al., 2024; Utkutug, 2022). In developing countries, for instance, the imports are usually more expensive and scarcer than local products. Also, consumers in developing countries are relatively less affluent than those in developed countries and this can create a sense of insecurity among them. Thus, how consumers perceive a corporate activity as being ethical or not based on their judgments may differ depending on the country's economic status. Even if past studies analyzed the ethical dimension, some in developing countries (Hyder et al., 2007), none of the existing literature paid attention to the sources that forms the CPE in developing countries. Therefore, an empirical analysis is required.

The purpose of this paper is to explore the domains of CPE in developing countries. The framework is based on the 6 domains of origin analyzed by Brunk (2010b). To have a better understanding and focus on this topic, a research question has been formulated: What are the domains which contribute to the formation of CPE in developing countries?

The conceptual framework explains the key concepts used by the authors in this study. The 6 domains of origin are explained in this part. The methodology part explains how the authors of this study conducted the research and why a qualitative approach was taken and considered as the most suitable. Analysis and results will be based on in-depth interviews with consumers from Romania and China. In the conclusion part, the research question will be answered. How other organizations can use this research and what further research is suggested will be shown on limitations and managerial implication part.

Theoretical Framework

Based on Brunk (2010a, b) studies on qualitative method with an average of 20 interviews of general consumers in different developed countries. Consequently, this method implies naturally an international and cultural aspect of CPE behavior where the objective was to analyze 6 separated areas and 36 sub-domains of CPE origin, linked to the effects company behavior could have. In these domains, the way of how a company/brand handles with consumers, employees, environment, overseas community, local community and economy and business community are represented.

Consumers

The company's management practices affect widely and directly end-consumers as the target of the company's products and services, especially in the case of ethical activities. There are a lot of examples of negative CPE. The first typical unethical marketing activity is immoral advertising practices: companies which take advantage of socially disadvantaged (i.e., children, the elderly, disabled people) for profit (Brunk & Bluemelhuber, 2010; Utkutug, 2022). They can engage negative actions to the unethical company, independently of the product quality (Tsalikis et al., 2019). The second one is the sale of bad quality products and poor customer services. In particular, to hide the details of products such as technical weaknesses, side effects, and ingredients through labeling are perceived as unethical (Eryandra et al., 2018). Additionally, consumers want to have the right of freedom of choices and rational consumption. But when companies do unethical action, such as price fixing and sale strategies of consumer lock-in, they can cause negative CPE (Tsalikis et al., 2019).

Employees

The way in which employees are treated awakens strong feelings among consumers. In the Cowe and Williams's study (2000) and the 2001 IPSOS–Novethic poll (Galavielle, 2004), welfare for employees is ranked top of the list of consumers' ethical choice. Improper compensation, drive for cheap labor, unacceptable working conditions, such as unsatisfactory safety procedures and unusually long working hours can negatively affect CPE (Shah et al., 2020). Following several high-profile scandals (i.e., Nike, GAP, Shell and Nestlé), where subsequent consumer boycotts not only caused a decrease in sales revenues but also damaged brand image and reputation (De Pelsmacker et al., 2005). On the other hand, paying above average compensation and offering additional social benefits, such as pension insurance or providing childcare through company owned kindergartens, can positively affect CPE (Brunk, 2010c; Werner, 2009). Also, employee motivation can have a significant impact on the working condition. For example, respecting workers' rights and encouraging communication with management can be a factor of motivation for employees (Kumar et al., 2023). According to different countries and companies' culture, there are also various types of discrimination for workers within the company (age, race, religion, gender). Creating a work environment of the company does not discriminate against these workers raises ethical awareness (Brunk, 2010b).

Environment

Universal agreements predominate on environment issues. Consumers are looking for a company which does not cause any unethical environmental actions, like pollution and damages. When consumers discover an unethical company, they show their negative feeling to this company, such as boycott (Rana et al., 2020). Also, consumers expect that company will prevent environmental pollution. Therefore, under such circumstances, failing to conform to consumer expectations may also result in a negative CPE. Thus, companies must assume complete financial responsibility for environmental pollution caused by business activities for their image and reputation (Fransson & Garling, 1999; Cordano et al., 2010). The study is also concentrated on preventing damages of environment caused by a company's operation and production process. Consumers admit equally that needlessly often airline travel can cause damages of environment and disagree with the travel policies of many companies, calling for increased use of advanced telecommunication technology (e.g., video conferencing) (Brunk, 2010c). So over time, environment issues are more and more considered as an important area in business activities. So, this is the reason why consumers take into account more and more about this topic. Moreover, ethical consumption has a great impact on people and the environment (e.g., some types of sustainable and green product or organic food) (Rana et al., 2020).

Indeed, the animal protection issue is consistently polarizing. Most people allowed the necessity of animal testing for safety, but others opposed animal testing. They claim that use of animal for testing has to end (Rowan, 2015) and many cosmetics companies stop using animals for tests, such as Johnson & Johnson, L'Oreal (Garlow, 2014).

Overseas community

The consumers' perception of how a company acts on its foreign business can bring out strong unethical connotations (Brunk, 2010d; Kumar et al., 2023). According to them, companies must be aware about anyone of their stakeholders, so as far away they are. Therefore, the exploitation of labor, such as child labor, and bad working conditions, such as sweat-shop-like conditions, dangerous work environment are the highest criteria of unethical business actions (De Pelsmacker et al., 2005). In a second hand, customers rank the exploitation of natural resources abroad, and especially companies which profit of underdeveloped countries without appropriate financial compensation (Shah et al., 2020). Another area perceived as unacceptable business practices is dealing with countries under questionable political regimes, dictatorships or countries known for their violation of human rights. Also, human trafficking and illegal trade, such as arms dealing, should not be allowed in business custom (Brunk, 2010c). This is why it is rather important and useful to get involved with/in a high integrity in external business dealings which go hand in hand with creating greater transparency and increased integrity in internal relationships.

Local community and economy

Another business behavior considered as socially responsible and ethical by consumers is the community involvement by companies (Freeman and Hasnaoui (2011). Companies' behavior received attention from local communities. And this topic is more and more regarded as a core of many CSR programs and can be observed by creating foundations, contributing to charity organizations, and supporting local institutions such as libraries or schools (Utkutug, 2022). Also, investments for local community development are verified and charity activities help company in local community to get merit such as tax savings and qualitative improvement of local work (Park et al., 2014). However, indiscreet in moves of the company and outsourcing for low wage can be irresponsible for the local community and economy (Shah et al., 2020). This involvement goes beyond the strict local community and also touches the link between the company's home country and the local economy.

In addition, if one company has a relationship with organizations considered as dubious or harmful to the community, this company cannot avoid blame by consumers. Consumers do not accept those kinds of relationships.

Business community

Recently, consumer's interest in ethical and socially conscious product is increasing. However, when the product is unethical, such as goods made with sweatshop labor and unfair-trade products, it causes negative impact (Kumar et al., 2023). Also, corruption causes social and economy problems around the world and has negative impact (e.g., a drop in brand image) on trust. Business manager tried to check the material assets (e.g., property, equipment, patent…) as well as relational assets (e.g., corporate reputation, consumers' goodwill, and brands) before M&A—(Merger and Acquisitions) deal. When the company focused on the material assets, it can fail due to not consider the consumers (Papavasileiou, 2009).

Likewise, Brunk (2010b) has also been proven that consumers convict hostile corporate behavior, like corporations which use their purchasing power and financial strength to their advantage. It involves unethical connotations because deliberately putting smaller businesses and competitors out of business. This negative perception also works for the poor treatment of suppliers, corruptions, non-respect of the state and/or the country laws, etc.

Consumers think incorrect accounting practices as corruption. To be considered ethical by consumers, which is one of the important factors to survive in the market, fair accounting practices are essential for a company. For example, Enron went into bankruptcy after being disclosing their incorrect accounting practices (Brunk, 2010c) (Table 1).

Table 1 Consumer perceived ethicality (CPE)—domains of origin

Domains of CPE origin	Consumers	Pricing	Labelling	Advertising	Targeting	Product quality and customer services	Freedom of choices lock-in
	Employees	Health safety production	Working hours	Benefits and compensation	Self-righteous management behavior	Discrimination and integration	Care respect motivation
	Environment	Pollution	Corporate travel policies	Sustainable farming i.e., organic	Animal protection	Recycling	Pro-active search for alternatives
	Overseas community	Exploitation of labor, i.e., sweatshops, child-labor	Exploitation of natural resources	Violations of basic human rights	Human trafficking	Supporting questionable political regimes	Illegal trade, i.e., arms dealing
	Local community and economy	Location of head-office or place of production		Community support	Involvement with dubious organization	Charities foundations	Importing, i.e., sourcing from abroad
		Employment	Corporates taxes				
	Business community	Fair trade and sourcing	Pushing competitors out of business	Mergers and acquisitions	Intellectual rights	Accounting and reporting practices	Corruption, i.e., bribery

Source Authors' own processing based on Brunk (2010a, b, c, d, e)

Research Methodology

Qualitative method of research was a natural choice considering that the aim of this paper is to analyze the thoughts of consumers. The purpose of gathering data in qualitative research is to provide evidence for the experience it is investigating (Polkinghorne, 2005). It is not the evidence itself which marks in the paper, but the meanings represented in these interviews. The ideas and thoughts expressed by participants need to be analyzed, not the words or counting how many times a particular word appears in the text (Pokinghorne, 2005). Moreover, a qualitative approach is believed to provide a deeper understanding of the phenomena than would be obtained from purely quantitative methods, such as questionnaires (Gill et al., 2008).

Sample

The concern when choosing the sample in qualitative data is not how much data are gathered or from how many sources, it is weather the collected data are rich enough and refining enough to understand and experience (Polkinghorne, 2005). Hence, Gill et al. (2008) consider that the goal of sampling in qualitative study cases have to contain reach information for the purpose of the study. For many qualitative research projects, a sample of eight respondents is perfectly sufficient (Creswell, 2012). To enhance the validity of the interviews and to facilitate close association with the respondents, a small sample (less than 20) is recommended (Crouch, 2006). Keeping this guideline in mind, a sample of 14consumers was found to be sufficient for this study (8 respondents from Romania and 6from China). The sample consists of 50% males and 50% females with the age ranging from 21 to 50 years old. The sample consisted of students, academicians, and higher-level managers. Chinese respondents were selected using snowball method. Firstly, a Chinese student was asked to be part of the interview and then the person recommended us other persons to be part in the research.

Data collection

The consideration of consumers' evaluation of ethics or morality is considered a personal, subjective, and sensitive topic (Crane, 1999). Therefore, in depth interviews were considered the best approach. Interviews with the Romanian respondents were conducted through Skype and the interviews with the Chinese respondents were conducted at the Linnaeus University's library, face-to-face. A semi-structured interview format was provided to encourage discussions and open dialogue. A pre-test was sent to 2 respondents, 1 Chinese and 1 Romanian to ensure that the questions are understandable and that there are no misunderstandings regarding the topic. The format of the questions was open, and all participants covered the same topics, starting with a few general questions about the ethical concept and its importance, and then becoming increasingly specific. The data collection phase included 14 interviews, 8 Skype interviews and 6 face-to-face interviews.

Data Analysis

The purpose of this study intends to identify and conceptualize the domains of CPE origin from developing countries customers concerning companies' activities which give a favorable or unfavorable ethical perceptions. The analysis was conducted with a sample of 14 respondents from developing countries, using a semi-structured interview format. The main goal is to observe if there are differences with results from developed countries, which found the basis of all studies led about the CPE origin and perception.

In this study, 7 Romanian (4 men and 3 women) and 7 Chinese (3 men and 4 women) people were interviewed, following an opening questionnaire to let them tell more and deeply their opinion about all questions they were asked. The authors have analyzed each interview and identified keywords which allowed to separate interviewers' answers on themes/domains. This analysis is organized in explaining domain by domain what interviewed customers think about the CPE and corporate activities related to ethics. To illustrate the different thoughts and opinions expressed, several quotes were used throughout this analysis.

Concept of ethics

Analyzing the interviews, it appeared that all respondents have already heard about the ethical concept, as Chinese people as Romanian people, and as men as women. This observation reveals that this construct is now worldwide and touches developing countries in addition to developed countries.

When people were asked to explain what ethic is for them, the explanation which came up the most often was: a right behavior, a serious thinking about taking good decisions and a willing to act right. Andreea's opinion, a Romanian woman, is a relevant example of this thought: "*In my opinion, ethics is a set of principles according to which a human group manages its behavior, so that they can define what is good and what is wrong, or better, what is moral and what is immoral.*" Moreover, acting in ethical way can also mean respecting regulations at the risk of becoming illegal, as it can be seen by Jiang's words (Chinese man), "*In my opinion ethics is a good/fair behavior, regarding to respect the rules/laws.*"

For many of them, answering in a consumer status, companies have to behave according to ethical ways: they have to respect all shareholders because their activities imply impacts on them, "*it is the way of doing things and how it is affecting the society, the environment and people in a large way of saying.*" (Diana, Romanian woman). Furthermore, doing ethical business activities offers companies a good perception, feeling, judgment and thus, willing to buy their products from consumers. This is what Hsiung (Chinese man), says through this quote: "*I think a company must be ethical to their clients and employees, because that makes a good publicity for the company and attracts more clients.*"

Moreover, for some respondents, the ethicality in a company is an influent factor on the sustainability of the company. Indeed, if a company wants to be prosperous, it has to be ethical. This was mentioned by Choy, (Chinese woman), "*Yes, because a sustainable company is an ethical company.*" Thus, for all interviewees of this study, ethical domain is a known subject and even if they do not have a well-defined explanation of it, they all confess this is an important part in business activities.

Domain 1: Consumers

In the interviews, advertising, labeling, and pricing areas where the most focused concepts in respondents' answers.

Concerning advertising aspect, interviewees think that when companies promote their products, some of them consider the ethical side, like putting some stamps (organic food, ecological on clothes, local production...). But even if brands act in this way, it does not seem to be an important criterion for them but more for consumers involved in ethical way-of-life, "*may be a good aspect for a type of consumers that care about these things.*" (Rozsa, a Romanian man). And this is typically the case of Pang, Chinese woman interviewed, "*I buy in generally just products which I know that their producers respect the laws and the business rules.*"

However, concerning labeling, some of the respondents believe that some companies obscure or hide information to appear better, appealing, or right on the customer's point of view. "*The company can lie about the ingredients just because the product is more attractive that way.*" is a relevant example from a Romanian woman (Andreea) explaining the immoral and unethical act of some brands.

Moreover, talking about pricing area, there appeared that overpricing products are perceived as immoral for customers as for competitors, especially if companies practicing this activity do dumping price as well, "*I consider that overpricing products are immoral for consumers [...] companies which have dumping price and in these scenario, they can be immoral regarding other competitors.*" (Feng, Chinese man)

Domain 2: Employees

When the employees' domain was addressed, respondents argue that employees' safety is the most important part of this area, "*I think, in the first place, for a company, must be the employees.*" (Saba, Romanian man). More than important, ethicality regarding employees is a subject which must follow rules and regulations ordered by State, "*Employees safety is a subject of law more than ethics.*" (Jiang, Chinese man)

Besides, a kind of virtuous circle between employees' consideration, productivity and relationship regarding to customers and thus, the benefits on the turnover and profits has been observed. A Romanian woman, Moni, clearly explains this concept: "*If the company takes care of employees, the employees take care of clients. And the clients mean profit. In conclusion, I think safety employees are safety company.*"

Domain 3: Environment

Nowadays, environment aspect is becoming more and more important, protected, and promoted. Thus, it can explain the reason why respondents of this study feel affected when a company does not respect environment. Indeed, taking care of this domain is revealed rather important for them, "*It is crucial for a company to take care of the environment.*" (Diana, Romanian woman). This is why, adapting its consumer behaviors can be perceived seen by some respondents, as Wu (Chinese woman), like trying to make sure that purchases follow and respect ethic on environmental products, "*I try to buy products form companies that seems to care about protecting the environment.*" This example goes beyond the simple product purchase but reaches the companies choice and selection according to their involvement in environmental ethics. So, this commitment can be considered as a competitive advantage compared to other companies which don't take care or failed in the environment respect. And it can be seen in the Raul's behavior (Romanian man), as he takes part almost only in companies respecting environment, "*I would reject the services from a company which doesn't respect the environment.*"

Domain 4: Overseas Community

Doing international business is seen as a good way to make business. But it must respect the local market, economies and people (workers, inhabitants, salesmen…). This is the main ideas emerging from the respondents of this study: "*In my opinion it will be nice to try to encourage the local economies and try to not cross over the local market sellers.*" (Moni, Romanian woman). Moreover, being ethical and respecting the entry market is advised for a company because this gets new market shares and to be recognized in the whole world, as explain by Hsiung (Chinese man), "*Yes, because they have the opportunity to grow and become a worldwide recognized company*". Indeed, going abroad is one of the best external strategies to spread a company's power, reputation, attractiveness, and thus chances to get more customers and market shares.

In analyzing deeper, the importance of respecting the community, some actions appeared as completely immoral, detrimental, and unethical. Respondents gave relevant examples of business behaviors they see as wrong. Starting with exploiting labor, it is one of the most mentioned and most disapproved, "*First of all, trying to exploit the force of labor just to grow your volume of production is a wrong way to do things. I am thinking about Chinese market—where working overtime is not a matter of choice.*" (Adrian, Romanian man). Similarly, the environmental exploitation and destruction was also addressed and strengthens the importance given to this domain, "*Another example that comes to my mind is the exploitation of natural resources in Romania. There are big companies that are extracting oil from Romania exporting it to other developed European countries and then they sell it back to Romania more expensive that it should be. This is a good example of immorality when it comes to international business.*" (Andreea, Romanian woman)

However, although people feel affected and involved in ethical overseas business, they admit that it is not always possible and easy to buy products/services from ethical companies and avoiding the other ones. Indeed, there are some situations where the need is here and cannot be replaced by another one. Thus, even if the product/service which meets the need is not perceived as ethical, people will still buy it, "*Yes, I will feel affected. But I am not sure that I will not buy their product. Sometimes it is hard to avoid that specially, when it comes to buy oil—it is necessary, and you will buy it if you do not have another choice.*" (Xiao, Chinese woman)

Domain 5: Local Community and Economy

Be involved in local community and economy, like charity and donations, is seen as an ethical business action. But some of respondents seem to be disenchanted about this kind of behavior in the sense they see it as a strategy to appear ethical and not really engaged in the business ethics process: "*It is an imagine strategy for sure.*" (Diana, Romanian woman); "*This makes part of a marketing policy.*" (Feng, Chinese man). Also, as mentioned by Choy, (Chinese woman) consumers should be aware of the real meaning of the charity action to be sensitive or not, "*I think it's ethical for a company to use charity and donations, but I wouldn't especially be sensitive if the area doesn't interest me.*" This last example pinpoints the interest research and the selection of fields' intervention in which companies should act in order to draw its customers' attention and thus, convey a positive image and assessment from them. These examples show the real transparency, legitimacy, and trustworthiness that companies should behave with to convey confidence towards their customers, and more broadly towards all stakeholders. But, even if it seen as strategic, it stays well perceived and promoted, as argued by Jiang' words, "*I believe this is the best marketing action. If a company helps a local community, local people certainly will love the company and the company's name.*" (Chinese man)

Another sub-domain of local community and economy mention was the importation. When respondents were asked what the appropriateness of this practice could be, answers were rather balanced. This practice is perceived ethical in one side, and unethical in another side, according to a country situation and the type of product/service imported: "*It depends if this product exists or not in the local community. If we do not have that product in our country, it is logically to import it but if you can find plenty of it could be inappropriate.*" (Andreea, Romanian woman). For other respondents, the importation is a good opportunity for consumers from a country to know new products coming from abroad, "*They help the community to make contact with other products from abroad*". (Hsiung, Chinese man)

Domain 6: Business Community

In this part, mergers and acquisitions were the most focused areas addressed regarding the remaining time of the interviews. Taking over other companies is an extern growing business activity that companies are doing more and more. Even if this strategy is profitable for the biggest one, it has to be done in respecting those which are taken over. This is in this way that customers will judge this activity ethical or not, "*Mergers and acquisitions are both a way of doing business in my opinion if this was an act of consent between parts it is totally fine. But if we are talking about unmoral practices, like blackmail, it is not right.*" (Moni, Romanian woman)

Discussions and Conclusions

Since the last decade, and even more nowadays, the increase of ethical issues in the media over the world has resulted in a greater interest and consideration from the most of stakeholders about this concept. Starting from this observation, several researchers have led different studies about CPE (Consumer Perceived Ethicality), and more precisely about the sources of CPE. As this construct is now more analyzed, known and understood, the new framework of CPE origin brings out the fact that the perception of a brand's ethics from customers are more diverse and varied than previously, and not as easy to set up by companies. Moreover, the CPE differs from a country to another one, and deeper, change concerning the level of the country. But no previous research has been done about developing countries. Thus, the purpose of this research was to explore the importance of CPE in those countries which are concerned at several levels.

Through the results given by this research, it can be affirmed they are in line with those analyzed in past studies led on developed countries. Indeed, the totality of research about the sources of CPE ever done had based their interviews and thus, their findings, on developed countries (Brunk & Bluemelhuber, 2009, 2011; Brunk, 2010b, 2012, 2013). The resulting six domains of CPE origin found in these previous studies were also found and analyzed in this research: consumer, employees, environment, local community and economy, business community, and overseas community. Therefore, the research questions "What are the domains which contribute to the formation of CPE in developing countries?" is answered and addressed. These different areas constitute the main sources which can have an influence on the ethical perceptions of a company or brand from customers' point of view.

Unlike 36 sub-categories in the 6 sources explored by Brunk in her research, this study has analyzed the main sub-domains (11) which appeared the most occurred by the authors, regarding the relevance from all of them.

By analyzing the origins of un/favorable CPE from developing countries, China and Romania, this study gives useful and relevant information about the worldwide customers perception about un/ethical companies activities into the current literature. Indeed, like studies led on developed countries, this research exposes the same results. It demonstrates several perception and differences between the consumer and corporate perspective about ethics. Moreover, it enhances the fact that sources of CPE are the same from a country to another one, even if they are not in the same development level. Thus, this study proves the importance to take into an account and to launch ethical strategies in all countries involved in company business activities.

By relying on Crane (2005), he explained that most of companies do not have an entire and well conceptualization and understanding of their consumers, and especially concerning the link with ethics. That's why, they feel and deal with some difficulties to have a clear conception of what can be and how managing the CPE in their policy. Besides from Creyer (1997) findings who demonstrated the importance of dimensions identification for which customers pay attention and assess positively or negatively company ethics, this study strengthens this observation by proving this identification is valid, not only for developed countries, but also for developing countries.

This research not only contributes new insights into CPE in developing countries but also provides practical implications for managers. It emphasizes the crucial role of understanding and addressing CPE for general managers, marketing and brand managers, and other professionals. Unethical perceptions, as revealed by interviewed consumers, can negatively impact a company's image, reputation, and consumer behaviour. The study highlights the necessity for companies to communicate and establish clear CPE policies, especially in developing markets like Romania and China. Ethicality is identified as a significant factor for customer satisfaction and overall business success, making ongoing research in this area highly recommended.

Limitation. Managerial Implication. Further Research

In addition to providing new and deeper data about CPE in developing countries in the international literature, this research affords much more relevant information for the real application. Thus, it gives a best understanding to general managers, marketing and brand managers, and obviously all professionals managers involved in CPE. These results can help them to analyze the exact activities domains that may create un/ethical perceptions and assessment of their company/brand from consumers, and more widely by stakeholders. Companies' managers should be aware of how consumers reflect, act and perceive what may or may not raise un/ethical behavior. So, CPE is a crucial construct they have to know and take into account for their customers' loyalty, attractiveness and prosperity. As it has been demonstrated in this study from interviewed consumers, unethical perceptions can lead to a negative company/brand image and reputation. It can go further by leading a potential detrimental impact on consumers' attitudes and purchase behavior, as it has been said by different respondents.

Thus, communicating about its CPE involvement is nowadays fundamental for all companies to stay on the market according to customers, and also to be aware of what competitors do to improve their own CPE actions and policy. Moreover, as developing countries represent a new market to conquer for companies, it is all the more important for companies to follow a relevant CPE policy and well established in order to give a good image of them in those new countries. Developing countries and CPE, both combined, depict an opportunity for the future of companies regarding to their worldwide expansion. Ethical consumption has become a real influent factor in business, and a factor which accounts a lot for customers, and by extension for stakeholders. As mentioned in the analysis, a healthy company comes from customers who feel good in their relationship with the brand. So it is obvious that ethicality is important, especially in those developing countries like Romania and China, which were analyzed in this research, where sometimes conditions of work (an important factor in CPE) are not good at all and do not follow ethical rules.

Further research on this topic is suggested. A qualitative study on a large sample of consumers will allow researchers to create a better understanding of the phenomena.

References

Awasthi, K., & Singhal, M. (2014). Exploring the domain of consumer perceived ethicality. *Journal of Management Research, 14*(2), 101–105.

Brunk, K. (2010a). Exploring origins of ethical company/brand perceptions. A consumer perspective of corporate ethics. *Journal of Business Research, 63*(3), 255–262.

Brunk, K. (2010b). Ethical reputation: Fact or fiction? A qualitative consumer exploration. In *AMA Winter Educators' Conference Proceedings* (pp. 32–33). Brussels: American Marketing Association.

Brunk, K. (2010c). Exploring origins of ethical company/brand perceptions. A consumer perspective of corporate ethics. *Journal of Business Research, 63*(3), 255–262.

Brunk, K. (2010d). Exploring origins of ethical company/brand perceptions: Reply to Shea and Cohn's commentaries. *Journal of Business Research, 63*(12), 1364–1367.

Brunk, K. (2010e). Reputation building: Beyond our control? Inferences in consumers' ethical perception formation. *Journal of Consumer Behaviour, 9*(4), 275–292.

Brunk, K. (2012). Un/ethical company and brand perceptions: Conceptualising and operationalising consumer meanings. *Journal of Business Ethics, 111*(4), 551–565.

Brunk, K. H. (2013). *Consumer perceived ethicality (CPE) scale.* American Psychological Association.

Brunk, K., & Bluemelhuber, C. (2010). The impact of un/ethical corporate conduct on consumers ethical perceptions-a multidimensional framework. *Advances in Consumer Research, 37*, 368–373.

Brunk, K., & Bluemelhuber, C. (2011). One strike and you're out: Qualitative insights into the formation of consumers' ethical company or brand perceptions. *Journal of Business Research, 64*(2), 134–141.

Brunk, K. H., & Bluemelhube, C. (2009). The impact of un/ethical corporate conduct on consumers'ethical perceptions–a multidimensional framework. *Advances in Consumer Research, 37*(Special Issue), 368–373.

Cordano, M., Welcomer, S., Scherer, R., Pradenas, L., & Parada, V. (2010). A cross-cultural assessment of three theories of pro-environmental behavior: A comparison between business students of Chile and the United States. *Environment and Behavior, 43*(5), 634–657.

Cowe, R., & Williams, S. (2000). *Who are the ethical consumers*? (Vol. 2, pp. 1–44) UK: The Cooperative Bank.

Crane, A. (1999). Are you ethical? Please tick yes or no on researching ethics in business organizations. *Journal of Business Ethics, 20*(3), 237–248.

Crane, A. (2005). Meeting the ethical gaze: Challenges for orienting to the ethical market. In R. Harrison, T. Newholm, & D. Shaw (Eds.), *The ethical consumer* (pp. 219–232). Sage.

Creswell W. J. (2012). *Qualitative inquiry and research design: Choosing among five approaches* (3rd ed.). Sage Publication.

Creyer, E. (1997). The influence of firm behavior on purchase intention: Do consumers really care about business ethics? *Journal of Consumer Marketing, 14*(6), 421–432.

Crouch, M. (2006). The logic of small samples in interview-based qualitative research. *Social Science Information, 45*, 483–499.

De Pelsmacker, P., Driesen, L., & Rayp, G. (2005). Do consumers care about ethics? Willingness to pay for fair-trade coffee. *Journal of Consumer Affairs, 39*(2), 363–385.

Eryandra, A., Sjabadhyni, B., & Mustika, M. D. (2018). How older consumers' perceived ethicality influences brand loyalty. *SAGE Open, 8*(2), 1–5.

Fransson, N., & Grling, T. (1999). Environmental concern: Conceptual definitions, measurement methods and research findings. *Journal of Environmental Psychology, 19*(4), 369–382.

Freeman, I., & Hasnaoui, A. (2011). The meaning of corporate social responsibility: The vision of four nations. *Journal of Business Ethics, 100*, 419–443.

Galavielle, J. (2004). Business ethics is a matter of good conduct and of good conscience? *Journal of Business Ethics, 53*(1/2), 9–16.

Garlow A. (2014). Companies that should stop testing on animals (and what to do in the meantime). *One green planet*. Retrieved December 10, 2023, from http://www.onegreenplanet.org/animalsandnature/companies-that-should-stop-testing-on-animals-and-what-to-do-in-the-meantime/. (Online)

Gill, P., Stewart, K., Treasure, E., & Chadwick, B. (2008). Methods of data collection in qualitative research: Interviews and focus groups. *BDJ, 204*(6), 291–295.

Hyder, A., Harrison, R., Kass, N., & Maman, S. (2007). A case study of research ethics capacity development in Africa. *Academic Medicine, 82*(7), 675–683.

Kumar, V., Kaushal, & Shashi, V. (2023). Role of customer perceived brand ethicality in inducing engagement in online brand communities. *Journal of Retailing and Consumer Services, 71*(C).

Manansala, G., Arasanmi, C. N., & Ojo, A. O. (2024). Does customer's perception of ethicality influence brand affect, image and equity in the banking sector. *International Journal of Ethics and Systems, 40*(1), 68–84.

Muller, M. (2013). *Nestlebaby milk scandal has grown up but not gone away*. The Guardian. Retrieved December 11, 2023, from https://www.theguardian.com/sustainable-business/nestle-baby-milk-scandal-food-industry-standards. (Online)

Nill, A., & Schibrowsky, J. (2007). Research on marketing ethics: A systematic review of the literature. *Journal of Macromarketing, 27*(3), 256–273.

Papavasileiou, E. (2009). Corporate syntheses: Consumers' role in mergers and acquisitions. *Advances in Consumer Research, 36*, 1054–1056.

Park, B., Chidlow, A., & Choi, J. (2014). Corporate social responsibility: Stakeholders influence on MNEs' activities. *International Business Review, 23*(5), 966–980.

Polkinghorne, D. (2005). Language and meaning: Data collection in qualitative research. *Journal of Counseling Psychology, 52*(2), 137–145.

Rana, W., Zia, S. R., & Khan, M. I. (2020). The impact of green marketing strategies on consumer-based brand equity in the clothing retail industry of Pakistan. *Journal of Fashion Marketing and Management, 24*(2), 189–204.

Rowan, A. (2015). Ending the use of animals in toxicity testing and risk evaluation. *Cambridge Quarterly of Healthcare Ethics, 24*(04), 448–458.

Shah, S. H. H., Lei, S., Hussain, S. T., & Mariam, S. (2020). How consumer perceived ethicality influences repurchase intentions and word-of-mouth? A mediated moderation model. *Asian Journal of Business Ethics, 9*, 1–21.

Sherwin, D. (1983). The ethical roots of the business system. *Harvard Business Review, 61*(6), 183–192.

Singh, J., Iglesias, O., & Batista-Foguet, J. (2012). Does having an ethical brand matter? The influence of consumer perceived ethicality on trust, affect and loyalty. *Journal of Business Ethics, 111*(4), 541–549.

Tsalikis, J., Van Solt, M., & Seaton, B. (2019). Measuring consumers' perceptions of business ethicality on price, product, and service domains. *Cross Cultural and Strategic Management, 26*(4), 451–466.

Utkutug, C. P. (2022). The mediating effect of consumers' perceived ethicality in the relationship between consumer cynicism/material values and affective commitment. *Journal of Marketing Analysis*.

Werner, W. (2009). Corporate social responsibility initiatives addressing social exclusion in Bangladesh. *Journal of Health Population and Nutrition, 27*(4), 545–562.

Zeithaml, V. (1988). Consumer perceptions of price, quality, and value: A means-end model and synthesis of evidence. *Journal of Marketing, 52*(3), 2–22.

Stress and External Influence: A Comparison Between Working from Home and Working from the Office in the Context of the Digital Economy

Iulia Dăuş Ogoreanu, Alina-Maria Vieriu, Andreea Fortuna Şchiopu, and Ana Mihaela Pădurean

Abstract Does the expansion of the digital economy generate more remote working? Some think that is the case since working from home reduces carbon emission and energy consumption. People are now confronted with a connected environment through digital technologies that change how they live, behave, and work. In this context, the purpose of this paper is to establish the importance of authenticity in relation to two scenarios, respectively working from home or in the office. Authenticity is a tripartite concept defined by (1) authentic living, (2) self-alienation, and (3) accepting external influence. The focus of the study is on the third dimension of authenticity—accepting external influence. Additionally, stress in the workplace is also studied, as it plays an important part for authenticity. Gender is also considered when analyzing external influence and stress in the workplace, as women and men may perceive external influence and stress differently. A total of 258, among which 68% were women and 51% worked in the private sector, responded to a self-reported questionnaire. A paired t-test showed that participants had a moderately low level of accepting external influence and work-stress when working from-home, compared to the office. Furthermore, for women, there was a significant difference in accepting external influence when working from the office than when working from home. However, the place of work did not affect the perceived stress for this group. In the case of men, there was no significant difference between work environments in regard to neither accepting external influence nor stress. A Pearson's correlation test suggested a moderately positive correlation between work-stress and accepting external influence, in both scenarios (from home and from the office).

Keywords Authenticity · External-influence · Stress · Working environments in the digital economy · Work-from-home

I. D. Ogoreanu (✉) · A.-M. Vieriu · A. F. Şchiopu · A. M. Pădurean
The Bucharest University of Economic Studies, Bucharest, Romania
e-mail: dausiulia17@stud.ase.ro

A. L. Negruşa et al. (eds.), *Sustainable and Resilient Businesses in the Global Economy*, Springer Proceedings in Business and Economics,
https://doi.org/10.1007/978-3-031-75883-6_14

Introduction

Over the past two decades, researchers have emphasised the significance of authenticity in relation to well-being, particularly within the context of the workplace (van den Bosch & Taris, 2014a). Acting and expressing according to personal values and beliefs, is called authenticity. But how often are people authentic? van den Bosch and Taris (2014b) argued that authenticity is a state, as opposed to a trait (Wood et al., 2008) and it changes according to the environment. Peer pressure and social contexts vastly influence someone's behaviour or thoughts, as certain conducts are frowned upon, or not in line with the collective norms of a particular group. Although people want to be themselves, sometimes the environment may not allow it, and if they go against the wave, they must be ready to bear the consequences—rejection. Authenticity is comprised of authentic living, self-alienation and accepting external influence (Wood et al., 2008). The latter is the degree to which a person behaves according to the expectations of others. The more someone caters to external expectations, the less authentic they are. Therefore, considering that authenticity is conditional upon its environment, it can be inferred that the degree to which one accepts external influence also depends on the environment.

In the context of the digital economy, the working environment is transformed (Shkarlet et al., 2020) in the sense that there is a trend toward the digital office instead of the physical one. Using the hub management model, companies can use remotely the work of their employees from diverse locations, sometimes from different countries (Shkarlet et al., 2020). Mukherjee and Narang (2023) comment on a nascent 'remote work culture' in the current digitalised world where space is becoming disturbed and digitised. In this new setting, it is important to understand how people relate to the working environment, how it influences their behaviour and how they act and express themselves (i.e., how they experience authenticity) within the remote or physical arrangement.

In addition, stress is often a by-product of work. Demanding managers, long hours and tight deadlines, are factors contributing to stress at work. In the current setting, work stress can be seen from other angles: the stress from remote work and the stress at the office. It is not yet clear which of the two is greater. When working from home, people experience high levels of stress due to a perceived lack of guidance (Irawanto et al., 2021; Ortiz-Gómez et al., 2020), while going to the office means adhering to a less flexible programme, which also translates to increased stress. Within the realm of work-related authenticity, the work environment impacts both stress and the level of accepting external influence. Unsurprisingly, stress can also emerge as a consequence of inauthenticity (van den Bosch & Taris, 2014a). The workplace is an example of a social context where people spent significant time yet is the least social context in which people are authentic (Robinson et al., 2013).

To further establish the importance of authenticity in relation to the workplace, this paper widens its scope, by focusing on the third dimension of authenticity—accepting external influence. More than that, in light of the current working settings (home vs. office) and significance of stress, the first research question is: (1) *Does accepting external influence and stress differ when working remotely, compared to traditional office settings?* Secondly, gender is considered, thus the following question is: (2) *Are there differences in gender when it comes to accepting external influence and stress, in and between the work environments (home vs. office)*?

Literature Review

Work Environments in the Digital Economy

The pandemic has been a catalyst for working environments, and compared to 2015 (PayScale, 2024) remote work (often called teleworking or work from home) has increased worldwide. However, most companies still demand their employees to work from the office. This stems from a perceived negative impact on culture and collaboration (Gibson et al., 2023). Proximity bias (Gibson et al., 2023) could reinforce the idea that if employees are all together in the same location (i.e., the office), this will result in increased collaboration. In-office working is not a panacea for neither collaboration, nor productivity. The same applies to remote working. There are advantages and disadvantages either way, and remote working may not be suitable for everyone (Ferrara et al., 2022)—as each individual has their own dispositions and stress sensitivity. That is why, remote or in-office options should be flexible (Shirmohammadi et al., 2022), offered in accordance with the nature of the job and an individual' resources, in a way that is a win–win situation for the employer and the employee.

The surge of remote working also propelled research to explore the implications of it. Decreased commuting, flexible time and adoption of new technologies (Kłopotek, 2017; Saura et al., 2022; Shirmohammadi et al., 2022) are positive aspects of remote working. Employees report improved work-family balance (Charalampous et al., 2022; Ferrara et al., 2022; Shirmohammadi et al., 2022), as they spend more time at home, and an elevated sense of connection with their close peers (Srivastava et al., 2024). Moreover, from a mental health perspective, less commuting reduces the negative emotions associated with it (Charalampous et al., 2022; Shirmohammadi et al., 2022), therefore employees are experiencing a decrease in stress. Moreover, as the saying goes there is no place like home; Charalampous et al. (2022) show that employees feel more comfortable in their personal space. This should not come as a surprise; Srivastava et al. (2024) suggest that environmental stimuli contribute to having a comfortable working space; for example, temperature can affect levels of focus. When working from home, employees are able to control their environment and they have access to a healthier diet (Charalampous et al., 2022), as in-office employees are limited by the food served at the nearby cafeteria, unless they pack their lunch.

Working from home also has drawbacks. Although, remote work offers family-friendly arrangements, it may be more difficult to deal with household duties and children (Shirmohammadi et al., 2022), in addition to the tasks from work. Moreover, remote work has been associated with intense, long working hours (Shirmohammadi et al., 2022; Smirnykh, 2023), as delimiting the job and personal life becomes challenging. Employees also experience feelings of isolation and frustration (Charalampous et al., 2022) and increased emotional exhaustion due to the lack of peer support (Charalampous et al., 2022). Another issue is the sedentary lifestyle tied to remote working. Going to the office also means walking more (Charalampous et al., 2022; Srivastava et al., 2024); whether is a result of commuting, moving around the office or going to the cafeteria.

Accepting External Influence. Authenticity at Workplace

Being authentic can bring positive changes on wellbeing, job satisfaction (Ostermeier et al., 2022) and outlook on life. The workplace, a substantial part of our social environment, impacts the perceived authenticity of employees. Individuals whose actions are consistent with their beliefs are considered authentic (Wood et al., 2008; van den Bosch & Taris, 2014b) and functioning in an environment that appeals to one's true self increases the feeling of authenticity (van den Bosch & Taris, 2014b). More so, the fact that employees are looking for a workplace that encourages individuals' true self (Reis et al., 2017) furthers the importance of authenticity at work. As literature suggests (Wood et al., 2008), authenticity is a tripartite concept, which includes (1) authentic living, (2) self-alienation, and (3) accepting external influence. In their studies, van den Bosch and Taris (2014a, b) found that employees are more engaged and positive about their personal work achievements when they work in harmony with their true self. These findings were supported by others studies (Chinelato et al., 2015; Ostermeier et al., 2022; Rathi & Lee, 2021; Sutton, 2020) where life satisfaction, flourishing and emotional social support at work were seen positively impacted by authenticity. Individuals are more likely to express their emotions if they feel more authentic (Rathi & Lee, 2021), leading to better communication inside the team. As Sutton (2020) concluded from their meta-analysis research, if businesses want to improve job outcomes, efforts should be concentrated to foster healthier work settings where employees can be themselves. Although authenticity is important for both men and women (Sutton, 2020), employers that provide an authentic workplace are essential for women (Reis et al., 2017).

Oc et al. (2020) points that social context is a determinant of authenticity, especially in the workplace. One's authenticity depends largely on accepting external influence (AEI), which is the degree to which one conforms to the expectations of others. Employees with higher degrees of AEI experience increased emotional exhaustion (van den Bosch & Taris, 2014b; Chinelato et al., 2015). A high AEI may result in a surge in the: workload, responsibilities and tasks. An interesting finding, however, shows that employees accepting greater external influence also experience greater levels of satisfaction within their job (van den Bosch & Taris, 2014b). Contradictory, Lutz et al. (2023) found that satisfaction directly affects the acceptance of external influence, the higher the satisfaction the lower the acceptance of external influence. Catering to the interests of others instead of self (Hewlin et al., 2020) is a sign of inauthenticity, and as consequence, people feel more stress and neurotic (van den Bosch & Taris, 2014a; Rathi & Lee, 2021). In the workplace, feelings of nonauthenticity can occur when employees are suppressing their personal values because are not aligned with those of the company (Chou et al., 2019). When workers experience workplace ostracism, they tend to create a facade of conformity (Jiang & Zhang, 2021), that is stronger when there is a high level of collectivism value. In other words, employees that see their work and home persona more similarly are more authentic (Ebrahimi et al., 2020). Once someone annuls its own values on the outside, giving false impression of conformity, it leads to dissonance (van den Bosch & Taris, 2014a). The strong impact of social influence is seen in the number of breaks taken at work. Oliver et al. (2021) found that the number of breaks may be influenced by the employee's relations with their leader and colleagues; if the team does not engage in this behaviour, an individual may feel uncomfortable taking breaks. Therefore, these actions contribute to the formation of cultural norms for the team. One does not want leave earlier while others are working until late, regardless if their tasks are finished for the day.

Stress

Stress, a negative emotion that can take many forms, and has an adverse effect on people's wellbeing can impair job performance, personal satisfaction and family-friends' relations. A large source of stress arises from the workplace; whether working from home is more stressful than in-office, is not clear. Some research (Irawanto et al., 2021) suggests it may, as employees experience higher levels of stress at home, due to a perceived lack of guidance. For office workers, frequent commuting (Shirmohammadi et al., 2022), lack of enough designated space for lunch (Oliver et al., 2021) and unsuitable temperature and noise (Srivastava et al., 2024) in the office are sources of stress. While for remote workers, the isolation and long working hours (Shirmohammadi et al., 2022), (Charalampous et al., 2022) can increase their negative emotions.

Although stress is a common denominator in the workplace, its effects can diverge along gender lines. Factors such as societal expectations, organisational culture, and personal experiences contribute to stress differently between genders. In-office women have less autonomy in determining when to pause and are more prone to interruptions with unexpected tasks (Kouki, 2023). On the other hand, when working from home (even though they enjoy more freedom and independence) women face pressure to take care of the household and family (Kłopotek, 2017). These expectations, in addition to the tasks from the job can cause a lot of stress, and ultimately lead to severe problems, such as anxiety and depression. More so, women may experience higher levels of emotional exhaustion and burnout due to the cumulative impact of intersecting stressors (Mensah, 2021; Rivera-Torres et al., 2013). For men, the pressure to show traditional masculinity, acting with stoicism, competitiveness and dominance can lead to a stressful environment: constraining emotional expression and fostering a culture of silent suffering, leaving little room for vulnerability or self-care (Gunkel et al., 2007). Men tend to be more affected by heart disease, due to internalised stress (Rivera-Torres et al., 2013). This pressure puts the same weight on men, no matter the place they work from, home or the office. In all cases, stress impacts work outcomes such as productivity (Hoendervanger et al., 2019), turnover intentions, job satisfaction (Thorsteinsson et al., 2014) and wellbeing. Working in the wrong environment can exacerbate feelings of inauthenticity, when an employee conforms to the expectations of others and suppresses its own values (van den Bosch & Taris, 2014a; Rathi & Lee, 2021), ultimately resulting in greater stress.

So far, this paper highlighted the significance of authenticity at work and its outcomes—by accepting external influence a person's authenticity is negatively affected. It further argued that work environments are also social contexts: places where people are most likely to be influenced by their peers, thus less authentic. In addition, the work landscape has been considered (working from home and in-office) alongside stress. Taken together, the following hypotheses are put forward:

- H_1. The level of accepting external influences is different depending on the work environment.
- H_2. The perceived work stress is different depending on the work environment.
- H_3. There is a correlation between the level of accepting external influence and level of stress in the workplace.

As a secondary aim, given that gender may play an important role, the following hypotheses are proposed in addition.

- H_4a. The perceived work stress when working from the office/working from home is different depending on gender.
    - *H_4a_1. The perceived work stress for women is different depending on the work environment (office vs. from home).*
    - *H_4a_2. The perceived work stress for men is different depending on the work environment (office vs. from home).*
- H_4b. The level of accepting external influences when working from the office/working from home is different depending on gender.
    - *H_4b_1. The level of accepting external influences of women is different depending on the work environment (office vs. from home).*
    - *H_4b_2. The level of accepting external influences of men is different depending on the work environment (office vs. from home).*

Research and Methodology

A questionnaire was used to evaluate the perceived levels of (1) accepting external influence (AEI) and (2) work-stress (WS) depending on the work environment, work-from-home (H) or in-office (O) scenarios.

The construct of Accepting External Influence (AEI-H, AEI-O), sub dimension of authenticity, was measured for both scenarios, working from home and working in the office. This construct included four items such as "I am strongly influenced by what others think (when I work from home/office)" and was adapted after van den Bosch and Taris (2014a). The scale of Work-Stress (WS-H, WS-O) included six items such as "I feel like many things are out of my control…" contextualized after (Irawanto et al., 2021). The variables for stress had Cronbach alpha coefficients of 0.92 and 0.89, both positive and greater than 0.7 (Bujang et al., 2018), showing internal consistency and reliability (WS-H = 0.92, WS-O = 0.89). In addition, the variables for accepting external influence had Cronbach alpha coefficients positive and greater than 0.7 (AEI-H = 0.84 and AEI-O = 0.83) (Bujang et al., 2018).

All items were rated on a 7 points Likert-type scale, where 1 = highly disagree and 7 = highly agree. Because the first language of participants is Romanian, the questionnaire was translated to Romanian. The translation was approved by two experts. In addition, as many researchers such as Stockemer (2018) suggest, a pre-test was also done. 11 people completed the questionnaire during the pre-test, offering important insights regarding not only the translation, but also how easy to understand the questions were.

The questionnaire, created in Google Forms, was spred online, from 20th of January until second of February. The respondents were encouraged to send the questionnaire to other people, creating a snowball effect. The participants were filtered, so only those that travelled in the last two years and had work experience in both environments (from the office and from home) could answer the questions. In this manner, 258 people, ages between 18 and 68 responded to the questionnaire. Participation was voluntary and entirely confidential. The 258 respondents had the freedom to complete the questionnaires at their convenience, without any time pressure and could withdraw anytime. The random sample (Stockemer, 2018), obtained through a snowball effect is considered a medium sample and appropriate for conducting statistical analysis ensuring more accuracy in the results. As Wilson Vanvoorhis and Morgan (2007) suggest, a sample size of 200–300 participants it is perceived as a fair sample for vigorous results.

The collected data (N = 258) was analysed in SPSS. The number of respondents remained the same, with no outliers being removed. A paired t-test was applied to determine whether there is a significant difference between work environments. Testing the normality, Kolmogorov–Smirnov showed p-value < 0.05. Although the data follows a non-normal distribution, the parametric tests (Pearson's correlation and t-test) were chosen thanks to their robustness (Norman, 2010).

Results and Discussion

The data collected shown several noteworthy trends among respondents, such as the majority falling within the age group of 18–24 (34,1%). It is important to note that the gender distribution indicated a significant majority of females, accounting for 68% of the respondents. Notably, the majority of the respondents (38%) reported a wage between 2000 and 4000 RON/month.

External Influence and Stress While Working from Home

Analysis of work-from-home scenarios indicated moderate levels of acceptance of external influence and work stress among participants. Mean indices for AEI (3.03 $\pm$ 1.5) and WS (2.77 $\pm$ 1.51) suggested a moderately low level of both factors when working from home. Employees usually perceive working remotely as guaranteeing them more freedom, independence and flexibility, leading to a higher satisfaction with their jobs (Kłopotek, 2017), and lower levels of stress. A more relaxed environment that resonates with one's authentic self also enhances the feeling of authenticity (van den Bosch & Taris, 2014a, b). The absence of colleagues and supervisors (Charalampous et al., 2022) also translates into lower levels of accepting external influence. Regarding stress, the main stress factors for individuals working in the office are regular commuting (Shirmohammadi et al., 2022), insufficient allocated lunch spaces (Oliver et al., 2021), and inappropriate temperature and noise levels (Srivastava et al., 2024). None of these factors is present when working from home, which leads to lower stress levels (Table 1).

In Table 2 the general sentiment observed in employees working from home is that they are less likely to accept external influence and to give up their job. Items 2 and 3 show that 34%, respectively 36% of the respondents, strongly disagree with the statements that they can be influenced by their colleagues. Regarding item no 8, almost half of the respondents, 47% are firmly convinced they do not feel like quitting their job when working from home. This result is in accordance with what the participants responded at item 10 where only 16% felt frustrated with their work. Work frustration and overload are generally main reasons for which people leave their job. From Table 2, it can be concluded that employees tend not to feel a great pressure in accepting external influence when they are working from home, nor feel like quitting their job.

Table 1 Demographics

		Number	Percent (%)
Total no. of respondents		258	100
Gender	Female	175	67,8
	Male	81	31,4
	Other	2	0,8
Age	18–24	88	34,1
	25–34	42	16,3
	35–44	45	17,4
	45–54	61	23,6
	Over 54	22	8,5
Education level	School	4	1,6
	High school	46	17,8
	Post high school courses	26	10,1
	Bachelor's degree	127	49,2
	Mastre's degree	53	20,5
	Ph.D.	2	0,8
Income RON/month	Less than 2000	29	11,2
	2001–4000	88	34,1
	4001–5000	53	20,5
	5001–6000	29	11,2
	Over 6000	32	12,4
	Not saying	27	10,5
Sector	Public	118	45,7
	Private	131	50,8
	Both	9	3,5

Source Authors' own work

External Influence and Stress While Working from the Office

There was a slight increase in the mean for AEI (3.53 ± 1.57), sense people are more prone to accept external influence when working from the office. Work environments mainly influence people's behaviour and as Oc et al. (2020) noted, authenticity is significantly shaped in the workplace, particularly in terms of individuals' AEI. Moreover, working in an inappropriate environment can heighten feelings of inauthenticity, as employees may feel compelled to conform to others' expectations and suppress their own values (van den Bosch & Taris, 2014a; Rathi & Lee, 2021) in order to fit in, resulting in increased stress levels. However, WS-O (2.95 ± 1.54) showed minimal variation, compared to working from home. Stress profoundly impacts various work-related outcomes such as productivity (Hoendervanger et al.,

Table 2 Results' frequency per item for working from home

Item	Response						
When working from home	1 (%)	2 (%)	3 (%)	4 (%)	5 (%)	6 (%)	7 (%)
1. I feel the need to do what others expect me to do	25	17	13	14	16	9	6
2. I am strongly influenced by the opinions of others	34	26	14	12	8	4	2
3. Other people influence me greatly	36	22	16	10	9	6	1
4. I behave in a manner that people expect me to behave	24	11	10	16	14	14	11
5. I am discouraged about my work	33	23	16	12	8	3	5
6. I feel many things are beyond my control	28	21	14	15	8	9	5
7. I feel overwhelmed by completing work	34	19	20	11	6	7	3
8. I feel like giving up on my job	47	16	13	7	10	5	2
9. I feel unable to get out from my work	27	19	14	12	12	9	7
10. I feel frustrated with my work	35	22	15	11	8	6	3

Note—1 = strongly disagree and 7 = strongly agree
Source Authors' own work

2019), intentions to leave a job, job satisfaction (Thorsteinsson et al., 2014), and overall well-being.

When looking at Table 3, it can be observed that there is no general consensus about the perceived acceptance of external influence, when working from the office. For item 11, regarding the need to do what others are expecting, the respondents are split almost equally, with 41% of them leaning towards agreeing with this affirmation, while 42% disagree and 17% being neutral. Item 12 shows that 63% of the respondents are inclined to disagree with the stance that they are influenced by others' opinions. However, almost half of the respondents, 47%, agree with the affirmation presented at item 14 and recon they behave in a manner that other people expect them to behave. For item 18 (I feel like giving up my job), only 5% strongly agree with this affirmation, while 43% strongly disagree. What is more, only 5% of the people feel like their work is beyond their control (item 16). Even though the general consensus regarding the work stress is that respondents do not feel significantly more stressed while working from the office than from home, an item is in disparity. Item 19, regarding the ability to get out from work, shows 38% of the respondents agree with this affirmation, while less than half (49%) disagree. By contrast, for the same item, but in the context of working from home, 28% agree that they cannot get out from work.

From Tables 2 and 3, it can be concluded that people feel the same amount of stress when working, no matter the place they are doing their job from, home or the office. In the case of accepting external influence it is clear that there is a slight

Table 3 Results' frequency per item for working from the office

Item	Response						
When working from the office	1 (%)	2 (%)	3 (%)	4 (%)	5 (%)	6 (%)	7 (%)
1. I feel the need to do what others expect me to do	15	12	15	17	14	13	14
2. I am strongly influenced by the opinions of others	26	24	13	13	12	7	5
3. Other people influence me greatly	27	20	16	12	12	8	5
4. I behave in a manner that people expect me to behave	17	11	12	13	17	18	12
5. I am discouraged about my work	34	23	13	11	7	7	5
6. I feel many things are beyond my control	30	21	14	11	9	10	5
7. I feel overwhelmed by completing work	25	20	19	14	8	8	6
8. I feel like giving up on my job	43	20	10	10	7	5	5
9. I feel unable to get out from my work	22	13	14	13	9	15	14
10. I feel frustrated with my work	39	23	9	11	8	6	4

Note—1 = strongly disagree and 7 = strongly agree
Source Authors' own work

difference between people's perceived acceptance of external influence. In the case of working from home, AEI seems to be lower.

Differences Between Work Environments Regarding Influence and Stress

The paired t-test (Table 4) showed a statistically significant difference between respondents' *AEI-O* and *AEI-H* [$t = 5.05$, $p < 0.05$]. This result proves that the first hypothesis *H1. The level of accepting external influences is different depending on the work environment*, is valid. This means people felt the perceived acceptance of external influence greater while working from the office, than from home. These findings align with previous research (Chou et al., 2019) that states employees suppress their personal values due to differences with the company's ideology. When individuals compromise their values externally it generates an internal conflict (van den Bosch & Taris, 2014b). Employees who perceive similarities between their work and home personas experience higher levels of feelings of authenticity (Ebrahimi et al., 2020). This means that people working from home feel they are more authentic to themselves, which leads to a lower level of susceptibility to external influence compared to the scenario when they are working from the office.

Contrary to expectations, as previous studies (Irawanto et al., 2021; Ortiz-Gómez et al., 2020) suggested working from home may be even more stressful for employees,

Table 4 Paired samples t-test for accepting external influence and stress

		Mean	Std. Deviation	t	Sig. (2-tailed)
Pair 1	AEI-H—AEI-O	0.497093	1.57952	5.055	0.000*
Pair 2	WS-O—WS-H	0.17765	1.95894	1.457	0.146

* $p < 0.05$
Source Authors' own work

our research shows no significant mean difference between *WS-O* and *WS-H* [t = 1.45, p > 0.05]. This means stress is perceived similarly in both scenarios and the second hypothesis *H2. The perceived work stress is different depending on the work environment* is rejected. Stress can impair job performance, personal satisfaction and family-friends relations. When working from the office, frequent commuting (Shirmohammadi et al., 2022), not enough designated space for lunch (Oliver et al., 2021) and unsuitable temperature and noise (Srivastava et al., 2024) are among the main stressors. However, people working from home are also facing different factors that can contribute to high levels of stress, such as isolation and long working hours (Shirmohammadi et al., 2022), (Charalampous et al., 2022) that can increase their negative emotions.

Further analysis stratified by gender, revealed significant differences in acceptance of external influence between office and home environments for women, but not for men. In Table 5, for *AEI-H*, the gender mean difference is statistically significant [t = −2.937, p < 0.05]. This means that women and men perceive the acceptance of external influence differently, based on the work place. The result supports the hypothesis *H4b. The level of accepting external influences when working from home is different depending on gender*. This finding can be linked to the fact that women feel safer and more autonomous when working from home (Fagan & Burchell, 2002). Srivastava et al. (2024) shows that environmental stimuli are important factors for a comfortable working space, which ultimately leads to increased feelings of authenticity, thus, lower levels of accepting external influence. In Table 5, in the case of *AEI-O*, the mean difference across genders is not statistically significant [t = −1.17, p > 0.05]. This shows that men and women have similar levels of accepting external influence when working from the office. The results align with previous studies (Charalampous et al., 2022) that show employee are more prone to accepting external influence when working from the office due to the permanent presence of colleagues and supervisors (van den Bosch & Taris, 2014b).

When considering *WS-O*, there is a significant difference between men and women [t = −1.916, p < 0.10], but only at 10% level. Even though there is a slight difference between genders, men and women still perceive the work stress at the same intensity when working from the office. In the case of *WS-H*, there is also a significant difference between men and women [t = −1.881, p < 0.10], but only at 10% level, which means there is a narrow contrast between the two groups. When working from home, women may experience heightened levels of emotional exhaustion and burnout due to long working hours and managing the household at the same time

Table 5 Independent samples test women versus men

	Mean (Std. Dev)		t	Sig. (2-tailed)
	F [0]	M [1]		
AEI—O	3.469 (1.581)	3.713 (1.495)	−1.170	0.243
AEI—H	2.861 (1.491)	3.441(1.422)	−2.937	0.004*
WS—O	2.833 (1.542)	3.228 (1.519)	−1.916	0.057**
WS—H	2.647 (1.502)	3.021 (1.439)	−1.881	0.061**

AEI = Accepting External Influence, WS = Work Stress, O = Office, H = Work from home
$^{*}p < 0.05$, $^{**}p < 0.10$
Source Authors' own work

(Kouki, 2023), while men may feel an even greater pressure to conform to the traditional male model (Gunkel et al., 2007). These results show that hypothesis *H4a. The perceived work stress when working from the office/from home is different depending on gender* should be rejected. Previous research also states that both men and women feel similar stressors in the office, such as regular commuting (Shirmohammadi et al., 2022) and inappropriate temperature and noise levels (Srivastava et al., 2024).

The results in Table 6, highlight a significant disparity between the levels of *AEI-H* (M = 2,86; SD = 1,49) and *AEI-O* (M = 3,46; SD = 1,58) for women. This result shows that hypothesis *H4b1. The level of accepting external influences of women is different depending on the work environment (office vs home)*, should be accepted. The office environment fosters traditional hierarchical structures creating an uncomfortable environment for women, leading to their predisposition to accept external influence from superiors or colleagues (Fagan & Burchell, 2002) Moreover, women working from home often juggle multiple roles simultaneously, like household management, and professional responsibilities (Kouki, 2023). Consequently, when working from home, women may be more focused on managing their immediate surroundings and familial responsibilities, leading to a decreased receptiveness to external influence.

Women experience almost the same levels of stress in both cases *WS-H* (M = 2,64; SD = 1,50) and *WS-O* (M = 2,83; SD = 1,50). This means hypothesis *H4a1. The perceived work stress for women is different depending on the work environment (office vs. home)* should be rejected.

Stress and external influence experienced by men are often affected by various factors, however, the work environment does not seem to be one of them. From

Table 6 Women in the work place. T-test for accepting external influence and stress

		Mean	Std. Deviation	t	Sig. (2-tailed)
Pair 1	AEI-H—AEI-O	0.607	1.570	5.115	0.000*
Pair 2	WS-O—WS-H	0.187	2.029	1.218	0.225

$^{*}\ p < 0.05$
Source Authors' own work

Table 7 Men in the workplace. T-test for accepting external influence and stress

		Mean	Std. Deviation	t	Sig. (2-tailed)
Pair 1	AEI-H—AEI-O	0.272	1.602	1.526	0.131*
Pair 2	WS-O—WS-H	0.208	1.698	1.105	0.274

* $p < 0.05$
Source Authors' own work

Table 7 we can see that there is no significant difference in the case of external influence, *AEI-H* (M = 2,95; SD = 1,71) *AEI-O* (M = 3,19; SD = 1,77), nor in the case of stress, *WS-H* (M = 2,73; SD = 1,80) and *WS-O* (M = 2,78; SD = 1,88). This means that men experience the same levels of *WS* and *AEI* in both cases (working from home vs. the office), leading to the conclusion that both hypotheses *H4a2. The perceived work stress for men is different depending on the work environment (office vs. home)* and *H4b2. The level of accepting external influences of men is different depending on the work environment (office vs. home)* should be rejected. In the context of the workplace, men may face stressors related to showing traditional masculinity and acting with stoicism, in addition to competitiveness and deadlines, challenges that persists regardless of the work environment (Gunkel et al., 2007). When talking about external influence, most men, experience the same levels of accepting external influence even when working from home. Usually, men feel the same pressure to behave in a way the others expect them to behave, whether they work from home or from the office.

Pearson's correlation was computed between (AEI-H and WS-H, r = 0.31; AEI-O and WS-O, r = 0.34) and indicates a moderate positive correlation between work stress and the acceptance of external influence in both scenarios (home vs. office). The result means that hypothesis *H3. There is a correlation between the level of accepting external influence and level of stress in the workplace* can be accepted. The findings suggest that as stress at work increases, so does the propensity to accept external influence, regardless of the work environment.

In Table 8, it can be observed that only 4 out of 9 hypotheses were accepted, all being related to accepting external influence. Thus, it is shown that the level of accepting external influence is different for women, compared to men and it is affected by the work environment (from home or from the office).

Conclusions

As an asset in human resources, authenticity is important for the well-being of employees and it influences employees' relationships in the workplace. Additionally, being authentic is not only influenced by personal style, but also by external factors, which can be determined by including a person's actual place of work.

Table 8 Conclusions regarding the hypotheses

Hypothesis	Conclusion
H_1. The level of accepting external influences is different depending on the work environment	Accepted
H_2. The perceived work stress is different depending on the work environment	Rejected
H_3. There is a correlation between the level of accepting external influence and level of stress in the workplace	Accepted
H_4a. The perceived work stress when working from the office/working from home is different depending on gender	Rejected
H_4a_1. The perceived work stress for women is different depending on the work environment (office vs. from home)	Rejected
H_4a_2. The perceived work stress for men is different depending on the work environment (office vs. from home)	Rejected
H_4b. The level of accepting external influences when working from the office/ working from home is different depending on gender	Accepted
H_4b_1. The level of accepting external influences of women is different depending on the work environment (office vs. from home)	Accepted
H_4b_2. The level of accepting external influences of men is different depending on the work environment (office vs. from home)	Rejected

Source Authors' own work

Accepting external influence means for a person to behave according to the expectations of others. The more external influence accepted, the less authenticity. Considering that authenticity is also influenced by the environment, one can state that accepting external influence depends on the environment as well. Similarly, stress is also influenced by the working environment.

This paper is focused on one of the three dimensions of authenticity, which is accepting external influence. The authors show that perceived external influence is accepted to a greater extent by employees working in person compared to the ones working from home, and work-stress does not significantly differ depending on the working environment. At the same time, accepting external influence is significantly swayed by gender identity, since women's perception is different from that of men. However, the perception of stress is equally as intense when working in person, while, when working remotely, both accepting external influence, as well as work-stress, have registered a moderate low level. This fact is based on the freedom that employees perceive when working remotely. Furthermore, there is also a correlation between perceived external influence and work stress. Therefore, the study shows that employees who work from home are less susceptible to the influence of others. According to these results, the working environment influences an employee's authenticity.

Employee authenticity represents a key element in human resources, since it is important in obtaining performance for business entities, which justifies enhanced research efforts and preoccupations in the context of an ever-changing labour market.

Limitations and Future Directions

Although our paper has definite contributions to the literature on working environments, it has a few limitations. First, it is based only on quantitative data, recorded in one moment in time, in one country. Since we know that opinions change over time, this type of examination should continue in other periods to validate the results over time and in other geographical areas. Also, a qualitative approach might provide valuable details on employee internalising behaviours related to working in an office or at home. Another future direction for other studies might delve into the identification of the differences between other groups of employees. For example, an interesting path would be to examine how people react to dissimilar working environments (office vs. home) based on their personality characteristics (e.g., extroversion vs. introversion). Group comparisons are valuable from both a theoretical and a practical point of view (Voll & Pfnür, 2023). This would allow managers to adapt their decision-making to the particular characteristics of their subordinates, taking advantage of both types of environments and creating appropriate policies for remote work. Nevertheless, our paper offers new ideas about employees' compliment to the expectations of others and stress and about how gender affects these states. How males and females relate to external influence and stress has been sparingly discussed in the literature until now, becoming a strong point of our paper.

Acknowledgements This paper was co-financed by The Bucharest University of Economic Studies during the Ph.D. program.

References

Bujang, M. A., Omar, E. D., & Baharum, N. A. (2018). A review on sample size determination for Cronbach's alpha test: A simple guide for researchers. *Malaysian Journal of Medical Sciences, 25*(6), 85–99. https://doi.org/10.21315/mjms2018.25.6.9.

Charalampous, M., Grant, C. A., & Tramontano, C. (2022). "It needs to be the right blend": A qualitative exploration of remote e-workers' experience and well-being at work. *Employee Relations: The International Journal, 44*(2), 335–355. https://doi.org/10.1108/ER-02-2021-0058.

Chou, H.-H., Fang, S.-C., & Yeh, T.-K. (2019). The effects of facades of conformity on employee voice and job satisfaction. *Management Decision, 58*(3), 495–509. https://doi.org/10.1108/MD-04-2019-0492.

Chinelato de, R. S. C., Ferreira, M. C., Valentini, F., & Van den Bosch, R. (2015). Construct validity evidence for the individual authenticity measure at work in Brazilian samples. *Revista de Psicología del Trabajo y de las Organizaciones, 31*(2), 109–118. https://doi.org/10.1016/j.rpto.2015.03.005. (Online)

Ebrahimi, M., Kouchaki, M., & Patrick, V. M. (2020). Juggling work and home selves: Low identity integration feels less authentic and increases unethicality. *Organizational Behavior and Human Decision Processes, 158*, 101–111. https://doi.org/10.1016/j.obhdp.2019.02.005.

Fagan, C., & Burchell, B. (2002). *European foundation for the improvement of living and working conditions*. Dublin. www.eurofound.eu.int. (Online)

Ferrara, B., Pansini, M., De Vincenzi, C., Buonomo, I., & Benevene, P. (2022). Investigating the role of remote working on employees' performance and well-being: An evidence-based systematic review. *International Journal of Environmental Research and Public Health, 19*(19), 12373. https://doi.org/10.3390/ijerph191912373.

Gibson, C. B., Gilson, L. L., Griffith, T. L., & O'Neill, T. A. (2023). Should employees be required to return to the office? *Organizational Dynamics, 52*(2), 100981. https://doi.org/10.1016/J.ORGDYN.2023.100981.(Online).

Gunkel, M., Lusk, E. J., Wolff, B., & Li, F. (2007). Gender-specific effects at work: An empirical study of four countries. *Gender, Work & Organization, 14*(1), 56–79. https://doi.org/10.1111/j.1468-0432.2007.00332.x.

Hewlin, P. F., Karelaia, N., Kouchaki, M., & Sedikides, C. (2020). Authenticity at work: Its shapes, triggers, and consequences. *Organizational Behavior and Human Decision Processes, 158*, 80–82. https://doi.org/10.1016/j.obhdp.2020.01.010.

Hoendervanger, J. G., Van Yperen, N. W., Mobach, M. P., & Albers, C. J. (2019). Perceived fit in activity-based work environments and its impact on satisfaction and performance. *Journal of Environmental Psychology, 65*, 101339. https://doi.org/10.1016/j.jenvp.2019.101339.

Irawanto, D., Novianti, K., & Roz, K. (2021). Work from home: Measuring satisfaction between work-life balance and work stress during the COVID-19 pandemic in Indonesia. *Economies, 9*(3), 96. https://doi.org/10.3390/economies9030096.

Jiang, P., & Zhang, L. (2021). Does conformity lead to gains? The effect of workplace ostracism on performance evaluation from a self-presentational view. *Acta Psychologica Sinica, 53*(4), 400–412. https://doi.org/10.3724/SP.J.1041.2021.00400.

Kłopotek, M. (2017). *The advantages and disadvantages of remote working from the perspective of young employees*. www.oamquarterly.polsl.pl. (Online)

Kouki, A. (2023). Beyond the "Comforts" of work from home: Child health and the female wage penalty. *European Economic Review, 157*, 104527. https://doi.org/10.1016/j.euroecorev.2023.104527.

Lutz, P. K., Newman, D. B., Schlegel, R. J., & Wirtz, D. (2023). Authenticity, meaning in life, and life satisfaction: A multicomponent investigation of relationships at the trait and state levels. *Journal of Personality, 91*(3), 541–555. https://doi.org/10.1111/jopy.12753.

Mensah, A. (2021). Job stress and mental well-being among working men and women in Europe: The mediating role of social support. *International Journal of Environmental Research and Public Health, 18*(5), 2494. https://doi.org/10.3390/ijerph18052494.

Mukherjee, S., & Narang, D. (2023). Digital economy and work-from-home: The rise of home offices amidst the COVID-19 outbreak in India. *Journal of the Knowledge Economy, 14*(2), 924–945. https://doi.org/10.1007/s13132-022-00896-0.

Norman, G. (2010). Likert scales, levels of measurement and the "laws" of statistics. *Advances in Health Sciences Education, 15*(5), 625–632. https://doi.org/10.1007/s10459-010-9222-y.(Online).

Oc, B., Daniels, M. A., Diefendorff, J. M., Bashshur, M. R., & Greguras, G. J. (2020). Humility breeds authenticity: How authentic leader humility shapes follower vulnerability and felt authenticity. *Organizational Behavior and Human Decision Processes, 158*, 112–125. https://doi.org/10.1016/j.obhdp.2019.04.008.

Oliver, M., Rodham, K., Taylor, J., & McIver, C. (2021). Understanding the psychological and social influences on office workers taking breaks; a thematic analysis. *Psychology & Health, 36*(3), 351–366. https://doi.org/10.1080/08870446.2020.1764954.

Ortiz-Gómez, M., Ariza-Montes, A., & Molina-Sánchez, H. (2020). Human values and work engagement: The mediating role of authenticity among workers in a Spanish religious organization. *Frontiers in Psychology*, 11. https://doi.org/10.3389/fpsyg.2020.00076.

Ostermeier, K., Cooper, D., & Caldas, M. (2022). Can I Be Who I Am? Psychological authenticity climate and employee outcomes. *Human Performance, 35*(1), 1–30. https://doi.org/10.1080/08959285.2021.1998060.

PayScale. (2024). Percentage of employees who work from home all or most of the time worldwide from 2015 to 2023. Statista. *Statista Inc*. Retrieved March 28, 2024, from https://www.statista.com/statistics/1450450/employees-remote-work-share/. (Online)

Rathi, N., & Lee, K. (2021). Does It Pay to Be Authentic? Implications of authenticity for life satisfaction and psychological well-being in a collectivist culture. *Journal of Happiness Studies, 22*(1), 147–161. https://doi.org/10.1007/s10902-020-00223-x.

Reis, G. G., Braga, B. M., & Trullen, J. (2017). Workplace authenticity as an attribute of employer attractiveness. *Personnel Review, 46*(8), 1962–1976. https://doi.org/10.1108/PR-07-2016-0156.

Rivera-Torres, P., Araque-Padilla, R., & Montero-Simó, M. (2013). Job stress across gender: The importance of emotional and intellectual demands and social support in women. *International Journal of Environmental Research and Public Health, 10*(1), 375–389. https://doi.org/10.3390/ijerph10010375.

Robinson, O. C., Lopez, F. G., Ramos, K., & Nartova-Bochaver, S. (2013). Authenticity, social context, and well-being in the United States, England, and Russia. *Journal of Cross-Cultural Psychology, 44*(5), 719–737. https://doi.org/10.1177/0022022112465672.(Online).

Saura, J. R., Ribeiro-Soriano, D., & Zegarra Saldaña, P. (2022). Exploring the challenges of remote work on Twitter users' sentiments: From digital technology development to a post-pandemic era. *Journal of Business Research, 142*, 242–254. https://doi.org/10.1016/J.JBUSRES.2021.12.052.

Shirmohammadi, M., Au, W. C., & Beigi, M. (2022). Remote work and work-life balance: Lessons learned from the Covid-19 pandemic and suggestions for HRD practitioners. *Human Resource Development International, 25*(2), 163–181. https://doi.org/10.1080/13678868.2022.2047380.

Shkarlet, S., Dubyna, M., Shtyrkhun, K., & Verbivska, L. (2020). Transformation of the paradigm of the economic entities development in digital economy. *WSEAS Transactions on Environment and Development, 16*, 413–422. https://doi.org/10.37394/232015.2020.16.41.

Smirnykh, L. (2023). Working from home and job satisfaction: Evidence from Russia. *International Journal of Manpower*. https://doi.org/10.1108/IJM-02-2023-0089.

Srivastava, C., Murnane, E. L., Billington, S. L., & Samuelson, H. W. (2024). Impact of workplace design on perceived work performance and well-being: Home versus office. *Journal of Environmental Psychology, 95*, 102274. https://doi.org/10.1016/j.jenvp.2024.102274.

Stockemer, D. (2018). Quantitative methods for the social sciences: A practical introduction with examples in SPSS and Stata. *Quantitative Methods for the Social Sciences: A Practical Introduction with Examples in SPSS and Stata*. https://doi.org/10.1007/978-3-319-99118-4.

Sutton, A. (2020). Living the good life: A meta-analysis of authenticity, well-being and engagement. *Personality and Individual Differences, 153*, 109645. https://doi.org/10.1016/j.paid.2019.109645.

Thorsteinsson, E. B., Brown, R. F., & Richards, C. (2014). The relationship between work-stress, psychological stress and staff health and work outcomes in office workers. *Psychology, 05*(10), 1301–1311. https://doi.org/10.4236/psych.2014.510141.

Van den Bosch, R., & Taris, T. W. (2014a). Authenticity at work: Development and validation of an individual authenticity measure at work. *Journal of Happiness Studies, 15*(1), 1–18. https://doi.org/10.1007/s10902-013-9413-3.

Van den Bosch, R., & Taris, T. W. (2014b). The authentic worker's well-being and performance: The relationship between authenticity at work, well-being, and work outcomes. *The Journal of Psychology, 148*(6), 659–681. https://doi.org/10.1080/00223980.2013.820684.

Voll, K., & Pfnür, A. (2023). Is the success of working from home a matter of configuration?–A comparison between the USA and Germany using PLS-SEM. *Journal of Corporate Real Estate*. https://doi.org/10.1108/JCRE-03-2023-0010.

Wilson Vanvoorhis, C. R., & Morgan, B. L. (2007). *Understanding power and rules of thumb for determining sample sizes*. Tutorials in Quantitative Methods for Psychology.

Wood, A. M., Linley, P. A., Maltby, J., Baliousis, M., & Joseph, S. (2008). The authentic personality: A theoretical and empirical conceptualization and the development of the authenticity scale. *Journal of Counseling Psychology, 55*(3), 385–399. https://doi.org/10.1037/0022-0167.55.3.385.

The Role of Non-monetary Benefits in Motivating and Improving Employee Performance

Anastasia-Cristiana Dumitru and Andrei Popescu

Abstract The research investigates the relationship between fringe benefits, employee satisfaction, motivation, and their impact on performance in the workplace. By statistically analysing survey data, this study aims to explore how providing fringe benefits influences the levels of satisfaction and motivation among employees in Romania. The research employs quantitative methodologies for data collection and analysis, using IBM SPSS software to conduct correlation analyses, multiple regression tests, chi-square tests of independence, and t-tests for hypothesis testing. Based on the existing literature, four general hypotheses were formulated, demonstrating that the number of fringe benefits is linked to employee satisfaction and motivation, that the utilization of benefits leads to performance enhancement, and that there are age-related differences in the use of specific benefits. In light of the results obtained in this research, a deeper understanding emerges of how the preferences and needs of different age groups of employees influence the use of fringe benefits. These differences underscore the need for organizations to adopt a personalized approach regarding the structure of fringe benefits to address the diversity of their employees' needs and preferences.

Keywords Non-monetary benefits · Employee motivation · Job performance · Job satisfaction

Introduction

The benefits for employees are additional rewards and compensations outside of salary (Siyal & Garavan, 2020). These can take various forms depending on the size of the organization or budget. According to (Anon, 2023), the most common types of benefits are flexible work programs, wellness programs, educational programs, life and health insurance, and retirement plans. Employee benefits can be provided based

A.-C. Dumitru (✉) · A. Popescu
The Bucharest University of Economic Studies, Bucharest, Romania
e-mail: dumitruanastasia16@stud.ase.ro

A. L. Negrușa et al. (eds.), *Sustainable and Resilient Businesses in the Global Economy*, Springer Proceedings in Business and Economics,
https://doi.org/10.1007/978-3-031-75883-6_15

on performance, efficiency, individual requirements and needs, or the organization's economic growth, profitability, or revenue (Siyal & Garavan, 2020).

Non-wage benefits, also known as non-monetary advantages, are an essential component of the compensation package offered by organizations to their employees. These benefits include a diverse range of facilities and services that do not directly involve cash payments but significantly contribute to employee satisfaction and engagement (Sittenthaler & Alwine, 2020).

Companies that implement programs for the benefit of employees enhance productivity, reduce staff turnover rates, and have positive effects on employees by providing a balance between their personal and professional lives (Kalev & Dibbin, 2022).

According to (Li, 2023) employees are influenced by how they perceive the importance of benefits by internal factors that motivate them, which can be explained by Maslow's hierarchy of needs, as well as external factors, such as changes in the social environment.

As concluded by (Hardiyansyah et al., 2023) performance represents the results of employees' work based on their level of experience, skills, and the time dedicated to fulfilling specific tasks. Workplace performance reflects the extent to which employees fulfil their responsibilities and contribute to achieving organizational objectives.

The purpose of this research is to analyse and highlight the importance of non-wage benefits in terms of motivating and improving employee performance across a variety of organizational contexts. This study aims to examine the impact that non-wage benefits have on employee morale and engagement, with the goal of evaluating how these benefits can contribute to enhancing their satisfaction and productivity within organizations. The ultimate aim of this article is to provide organizations with useful information and practical recommendations for improving compensation and benefits strategies to increase employee motivation and performance.

The research objectives are as follows: to review the literature to understand the role and importance of non-wage benefits, to investigate employee preferences and expectations regarding these benefits and assess how they influence job satisfaction, motivation, and engagement, to analyse data collected through a questionnaire and provide practical recommendations for organizations regarding the development and implementation of effective compensation and benefits strategies that support employee motivation and performance.

The structure of the article consists of five distinct sections, each playing a specific role in the development and presentation of our research. The introduction section provides the general context of the study, the importance of non-wage benefits in motivating and improving employee performance, as well as our research objectives. The literature review explores previous research and relevant theories in the field of non-wage benefits and their impact on employees and organizations. The methodology details the research process, including the study design, data collection tools, and study population. The results and discussions section will present and interpret the results obtained from the data analysis and highlight their implications in the context of research hypotheses and literature. Finally, the conclusions synthesize the main findings of the study, discuss their implications, and suggest directions for future research in the field of non-wage benefits and employee performance. Through this structure, we aim to provide a coherent and comprehensive presentation of our research, facilitating understanding and appreciation of the results and their contribution to the field of human resources and organizational management.

Our research makes significant contributions to the field of human resources and organizational behavior. Firstly, the study provides updated empirical data on employees' preferences for different types of fringe benefits, based on diverse samples from the public and private sectors. Secondly, the research examines in detail how these benefits influence employee satisfaction, commitment, and performance, thereby contributing to a deeper understanding of the mechanisms through which fringe benefits can enhance organizational efficiency.

The originality of our study lies in its differentiated approach to the impact of fringe benefits on employees from various age groups. Most previous studies have focused on general benefits without considering how employees' preferences and needs may vary by age. Additionally, our study employs a rigorous methodology, including advanced statistical analyses such as MANOVA and t-tests, to ensure the validity and reliability of the results. This allows for a more nuanced evaluation of the relationships between variables and provides a more detailed perspective on the effects of fringe benefits.

Literature Review

To begin with, it is opportune to clarify the terms used in the presented research. According to (Chauhan, 2015), non-wage benefits refer to tangible and intangible means of motivating and rewarding employees. These may include, among others, flexible working hours, salary contributions or bonuses, training, savings accounts, private health insurance, or even longer paid leave.

Employee motivation and increasing workplace performance are fundamental aspects of human resource management and organizational performance improvement. Motivation is defined as an internal force that stimulates and controls human behaviour, according to (Robbins & Judge, 2017). Employee motivation and their commitment to the company are essential for achieving organizational goals and increasing workplace performance. By understanding these concepts and applying appropriate human resource management strategies, organizations can create a work environment in which employees are motivated, engaged, and productive, contributing to their long-term success and sustainability (Chabé-Ferret et al., 2023).

Work performance, according to (Burhanudin et al., 2023), is the result of activities carried out by employees or teams according to their duties and responsibilities, measured by the quantity and quality of work, timeliness, attendance, efficiency, and effectiveness in completing tasks, compared to pre-established standards and periodically communicated to improve overall employee performance.

Non-wage benefits represent a vital aspect of modern organizations' compensation and motivation strategies, playing a significant role in attracting, retaining, and satisfying employees (Abdulkhaleq, 2021). These benefits can take various forms, including health benefits, paid time off, professional development opportunities, vacation vouchers, meal vouchers, reimbursement of certain personal expenses, etc. In light of these aspects, exploring relevant theories becomes essential to understand how non-wage benefits can be efficiently used to motivate employees and improve workplace performance:

- **Maslow's Hierarchy of Needs theory**: People have different levels of needs, which must be satisfied in a certain order, starting from basic needs and moving up to higher ones (Mendez et al., 2023). Regarding non-wage benefits, organizations can use this theory to develop programs that meet employees' diverse needs. For example, offering benefits such as private health insurance, life insurance, or access to wellness programs can satisfy employees' physiological and safety needs, while personal development programs or advancement opportunities can contribute to fulfilling their esteem and belongingness needs within the organization.
- **Adams' Equity Theory**: Employees compare the rewards they receive from the company with the effort they put in and the results they achieve compared to other employees (Tseng & Kuo, 2014). Applied to our research topic, it is crucial for organizations to allocate all these benefits in a fair and transparent manner, without differentiation, so that employees perceive they are treated equally and to avoid feelings of unfairness and dissatisfaction. For example, in a health benefits program, employees should receive the same options and benefits, regardless of their position in the organization or performance level.
- **Vroom's Expectancy Theory**: his theory states that employees are motivated by their expectations regarding the relationship between the effort they put into the organization (expectancy), the performance achieved (instrumentality), and the reward received (valence) from the employer (Lee, 2007). In other words, they are motivated to improve their performance at work when they believe their efforts will lead to certain outcomes, and that these outcomes are valued and will be rewarded. Thus, organizations should identify and offer non-wage benefits that meet these criteria. For example, a performance-based bonus program can be implemented in a company, where employees are clearly informed about performance criteria and how they can achieve their goals to obtain these bonuses. By implementing such a reward system, the organization can stimulate employees' motivation and performance, providing them with a concrete and attractive incentive to achieve their individual and organizational objectives.
- **Herzberg's Two-Factor Theory of Motivation**: This theory proposes that there are two distinct sets of factors that influence motivation and job satisfaction: hygiene factors, which are represented by working conditions, salary, relationships with colleagues, and motivational factors, which are related to the nature of the work itself, such as recognition, responsibility, advancement opportunities, and professional growth (Ozsoy, 2019). To explain the implication of this theory in this research topic, let's consider two companies, Company A and Company B, which have the same basic benefits: private health insurance

and meal vouchers. Company A chooses to invest in new non-wage benefits such as gym memberships, team-building sessions, recreation facilities, while Company B ignores these benefits and focuses strictly on salary and wage benefits. According to Herzberg's theory, non-wage benefits represent motivational factors that contribute to employee satisfaction and performance improvement by creating a pleasant and stimulating work environment. Thus, employees who have access to such facilities will be more motivated to fulfill their duties and improve their professional performance. On the other hand, the absence of non-wage benefits in Company B could lead to feelings of dissatisfaction and lack of motivation among employees. Even if the salary is similar to that offered by Company A, the lack of other benefits that stimulate employee satisfaction and engagement could affect their performance within the organization. Additionally, another motivating factor that companies can exploit is offering advancement or professional development opportunities.

- **Organizational Commitment Theory**: According to this theory, organizational commitment refers to the degree to which employees identify with the goals and values of the organization and are dedicated to contributing to its development and success (Ghosh & Swamy, 2014). Regarding non-wage benefits, companies should use these rewards to strengthen employees' sense of belonging to the organization and to encourage their loyalty and involvement. An example of such a benefit is the opportunity to participate in organizational decision-making.

In addition to these theories mentioned above, we can also analyse the difference in how non-wage benefits impact employee motivation compared to monetary ones. Thus, in the research conducted by (Dewhurst et al., 2009), there are strong arguments supporting the effectiveness of non-monetary incentives compared to monetary ones in terms of employee motivation. Their study, based on a global McKinsey survey, found that three non-monetary incentives—praise and recognition, attention from managers, and opportunities to take on new tasks—are even more effective in motivating employees than evaluated salary incentives.

Furthermore, to verify the objectives proposed at the beginning of the research, namely investigating employees' preferences and expectations regarding these benefits and assessing how they influence job satisfaction, motivation, and engagement, we will rely on the specialized literature. Additionally, based on the specialized literature, we aim to formulate general hypotheses that will later be confirmed or disproven.

Fringe benefits, such as wellness programs, flexible scheduling, or other perks, are considered essential for increasing employee engagement and performance (Sittenthaler & Alwine, 2020). Studies indicate that employees who feel supported by their organization through these benefits tend to be more engaged and motivated in performing their daily tasks. For example, previous research has shown that companies offering appropriate fringe benefits can experience a decrease in turnover rates and an increase in job satisfaction, highlighting the importance of tailoring these benefits to the individual needs of employees to enhance engagement and performance (Sittenthaler & Alwine, 2020).

Based on these studies, we can formulate the first hypothesis of our research: *There is a significant and positive correlation between offering fringe benefits within organizations and the level of employee satisfaction and motivation* (Sittenthaler & Alwine, 2020).

Recent studies, such as those presented by (Srna & Dinc, 2017), highlight that fringe benefits can be a crucial factor in increasing employee loyalty and productivity. Specifically, these benefits can contribute to improving employees' health and overall well-being, thus positively impacting their workplace performance. Previous research suggests that employees who receive such benefits are less likely to leave the organization and tend to have a more active presence within teams, underscoring the strategic importance of managing and providing these benefits effectively (Srna & Dinc, 2017).

Based on these studies, the second hypothesis is formulated: *The utilization of fringe benefits leads to an increase in employee engagement and performance within the organization* (Srna & Dinc, 2017).

In the specialized literature, (Wardiansyah et al., 2024) have highlighted significant variations in employee preferences based on organizational context and individual characteristics. This diversity in perceptions and preferences among employees can influence how fringe benefits are perceived and utilized within organizations. For example, employees with different needs and distinct expectations regarding the benefits offered may have varying perspectives on the relevance of these benefits to their well-being and job satisfaction, thus underscoring the complexity and necessity of adapting compensation and benefits strategies to meet the diversity of employees (Wardiansyah et al., 2024).

Thus, we formulate the general hypothesis number three, as follows: *There is no significant association between employees' job satisfaction levels and their preferences for fringe benefits* (Wardiansyah et al., 2024).

Relevant studies suggest that employees' preferences and needs can vary significantly based on age, influencing how benefits such as health insurance versus lifestyle and fitness-related perks are perceived and utilized. For instance, younger employees may be more receptive to lifestyle and wellness benefits, whereas older employees might prioritize benefits that support their overall health and financial well-being (Sittenthaler & Alwine, 2020; Srna & Dinc, 2017). This hypothesis underscores the need to tailor compensation and benefits strategies according to employees' demographic characteristics to enhance their effectiveness and relevance within organizations.

These insights form the basis of general hypothesis number four: Employees aged between 25 and 35 will show a higher utilization of lifestyle and fitness-related fringe benefits, whereas employees aged over 45 will demonstrate a higher utilization of health-related fringe benefits (health insurance).

Research Methodology

Research methodology represents the set of methods, techniques, and tools used to investigate and address the research questions formulated within a study. It includes details regarding the research plan, data collection and analysis methods, as well as procedures for validating and interpreting the results (Swarooprani, 2022).

Therefore, in this chapter of the research, we will describe the methodology used to achieve the outlined objectives. To evaluate the impact of non-wage benefits on employee motivation and performance, we opted for a quantitative research design, using a structured questionnaire as the data collection instrument. The choice of this approach was justified by our research objectives, which required quantitative measurements to identify relationships and trends in an objective and repli cable manner. A representative sample of employees from various fields of activity, different ages, and with varying levels of education was selected to ensure generalizability of responses. The method chosen for sample selection was random, thus ensuring a fair representation of the various categories of employees.

To collect data, the online survey was distributed to a sample of employees from various organizations. The questionnaire was distributed through online survey platforms and social networks, ensuring that participants came from a diverse range of sectors. This method allowed us to reach a large number of respondents in a relatively short time, thereby maximizing the diversity and representativeness of our sample.

The questionnaire was designed to measure various aspects related to fringe benefits and their impact on employee satisfaction and performance. Before distributing the questionnaire, it underwent a pilot test with a small group of employees to ensure clarity and relevance of the questions. The feedback received was used to adjust and improve the formulation of questions, aiming to accurately reflect the research objectives.

To ensure the anonymity and confidentiality of respondents, the questionnaire did not request personally identifiable information. Participation in the survey was completely voluntary, and respondents were informed about the purpose of the research and how their data would be used. Clear instructions were provided regarding how to complete the questionnaire, ensuring that participants understood the questions and responded honestly and accurately.

The collected data were stored and analyzed using appropriate statistical methods to test the research hypotheses. We utilized statistical analysis software to compute frequencies, means, and other descriptive statistics. Additionally, we conducted significance tests such as t-tests and MANOVA to evaluate relationships between the variables of interest. This approach allowed us to rigorously analyze the data and draw meaningful conclusions based on statistical evidence.

The questionnaire developed for this research consists of questions aimed at creating a demographic profile of the respondents and structured questions aimed at analyzing employees' perceptions and experiences regarding the non-wage benefits offered by the organizations they belong to. Additionally, within the questionnaire, questions were included regarding the level of job satisfaction and the factors influencing it, as well as questions related to the factors influencing employee motivation and how these factors can impact workplace performance. Thus, to describe in more detail the measures and questions used in the questionnaire, it is important to present the rationale behind each question and how they are meant to measure the study's variables of interest.

In the demographic section, we will begin with essential demographic questions to understand the basic characteristics of the respondents. The gender question helps identify the gender distribution within the study population, which can influence perceptions and preferences regarding fringe benefits. Categorizing by age allows analysis of differences in the utilization and appreciation of fringe benefits based on life stage. Determining the industry sector (public or private) is crucial for analyzing structural and cultural differences that may influence the availability and utilization of fringe benefits.

In the section concerning the use and perception of fringe benefits, we focus on identifying the types of fringe benefits employees currently enjoy and assessing their perceptions regarding the impact of these benefits. The question about current benefits provides an overview of the available benefit packages, while the question about the influence of benefits on engagement measures employees' perception of how these benefits affect their level of engagement within the organization. Evaluating the value and frequency of benefit utilization allows quantifying the level of appreciation and identifying potential improvements needed in benefit packages.

The section on job satisfaction and performance includes questions designed to measure employees' overall satisfaction levels and their perception of their performance. Questions about overall satisfaction and self-assessment of performance provide essential data to investigate the correlation between benefit utilization and job satisfaction or performance. Identifying preferences for fringe benefits helps tailor and personalize the benefit packages offered based on employees' needs and desires.

Regarding the specific usage of different fringe benefits, this section details the frequency of use across various categories of fringe benefits, providing a more precise analysis of the relevance of each type of benefit. Questions about frequency of use and preferences for specific benefits are intended to identify the most popular and frequently used benefits, indicating the actual preferences of employees. Detailing the frequency of use for specific categories such as healthcare, lifestyle and fitness, education and professional development, recreation and entertainment, and work-life balance services provides granularity in the collected data.

In the section on importance and satisfaction with work-life balance, we explore the perceived importance of flexible work schedules and additional days off, which are essential elements for achieving a balance between personal and professional life. Assessing satisfaction with this balance is crucial for understanding how it influences employees' well-being and motivation in the workplace. Questions about the impact of work-life balance on motivation and performance investigate the relationship between satisfaction with work-life balance and key aspects of job performance.

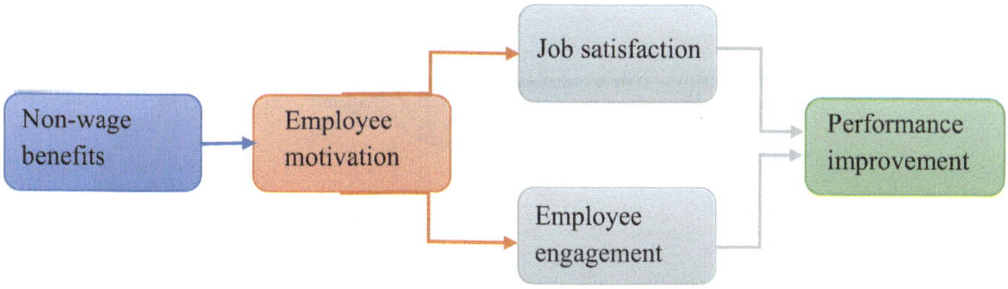

Fig. 1 Conceptual model of the study. *Source* Created by the author

In the last section, training and professional development, the questions focus on employees' participation in training and professional development programs and the impact of these programs on their skills and self-confidence. The frequency of participation in training programs provides information about employees' access to and involvement in professional development, while evaluating the impact of training helps determine the effectiveness of the training programs offered. The question about the link between skill development and performance explores employees' perceptions of how their skill development contributes to improving job performance (Fig. 1).

For a better explanation of the concept to be used, we also chose to create a conceptual model.

After obtaining the data resulting from the questionnaire, we will create a demographic profile of the respondents based on age, gender, and field of activity. Then, to confirm or refute the hypotheses stated above, with the help of specialized software such as SPSS, we will conduct statistical tests: such as correlation analyses, multiple regression tests, chi-square tests of independence, and t-tests.

Results and Discussions

To initiate discussions based on the research, it is deemed appropriate to present the administered questionnaire. It consisted of 20 questions, of which three aimed to create a profile of the respondents in terms of age, gender, and sector of activity. One multiple-choice question was intended to determine the non-wage benefits that respondents currently enjoy, while the remaining questions aimed to analyse employees' perceptions regarding non-wage benefits, how they motivate them, and improve their professional performance, and to demonstrate the hypotheses outlined in the previous chapter.

Thus, following the distribution of the questionnaire, a total of 165 unique responses were obtained, out of which 76 were from male respondents, representing 46.1%, and 89 were from female respondents. Among all respondents, 21.8% are employed in the public sector, while the rest, totalling 129, are active in the private sector (Fig. 2).

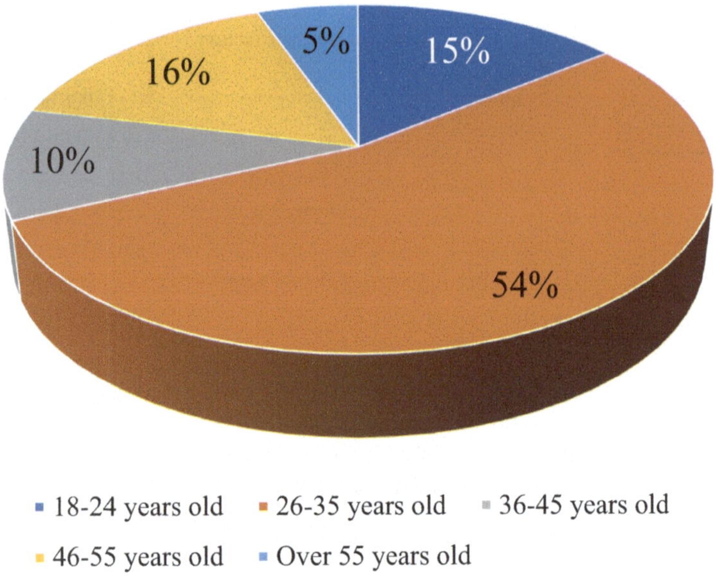

Fig. 2 Representation of respondents by age categories. *Source* Conducted by the author, based on the results obtained from research using SPSS

The total of 165 respondents were distributed into 5 age categories as can be observed in the previous figure. The majority, 54%, belong to the age category of 26–35 years old, followed by a balance between the categories of 18–24 years old and 46–55 years old with 24 and 26 respondents, respectively, followed by the categories of 36–45 years old, 10%, and over 55 years old with 5% (Fig. 3).

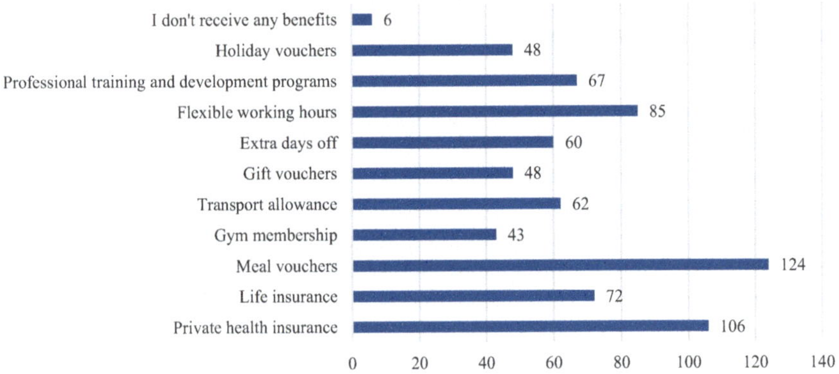

Fig. 3 Non-wage benefits enjoyed by respondents. *Source* Conducted by the author, based on the results obtained from research using SPSS

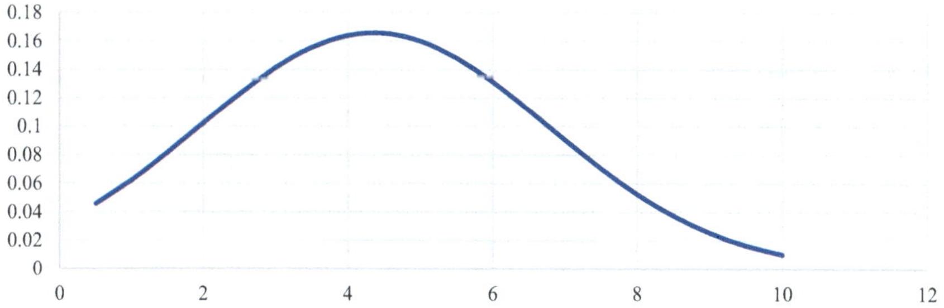

Fig. 4 Graphic representation of Std. Deviation regarding the number of non-wage benefits. *Source* Conducted by the author, based on the results obtained from research using SPSS

Analysing the existing benefits in the Romanian labour market, the ten benefits mentioned in the previous chart are the most common. Within the questionnaire, out of the total of 165 respondents, it can be observed that 124 representing 75.15% of the total respondents benefit from meal vouchers, 106 of them from private health insurance, and 85 from flexible working hours. Additionally, 6 of them do not benefit from any non-wage benefits.

Running a frequency test using SPSS on the number of benefits each respondent has access to, from no benefits to ten benefits, it can be observed that 17% have access to two benefits, 16.4% to 4 benefits, 15.2% to 3 benefits, and a percentage of 4.2% have access to all ten identified benefits. Also, based on this test, the results indicate an average number of non-wage benefits of 4.36, with a standard deviation of 2.404. This suggests that, on average, respondents benefit from approximately 4–5 non-wage benefits, with considerable dispersion around this mean. A higher standard deviation indicates a greater variation among participants regarding the number of non-wages benefits they receive (Fig. 4).

Another question in the questionnaire aimed to assess the respondents' desire to have access to certain non-wage benefits available on the labour market in Romania. The results indicate that most respondents wish to have access to private medical insurance, with 115 expressing this preference. Additionally, 105 respondents expressed a desire for meal vouchers, while 103 expressed a desire for additional days off or flexible working hours (Fig. 5).

In our research, alongside questions aimed at profiling respondents and identifying the fringe benefits employees receive or desire, we utilized a series of structured questions to assess employees' perceptions and behaviors regarding fringe benefits and job satisfaction. In this context, it is crucial to gain a clear and detailed understanding of the distribution of responses and the level of variability associated with them. To achieve this, we calculated means and standard deviations for key variables, providing a comprehensive picture of how employees perceive and utilize fringe benefits, as well as their overall job satisfaction level (Table 1).

To assess the reliability and consistency of the scale used in the questionnaire, Cronbach's alpha analysis was applied to the variables of interest. Table 2 presents the results of these analyses, including the specific values of Cronbach's alpha for

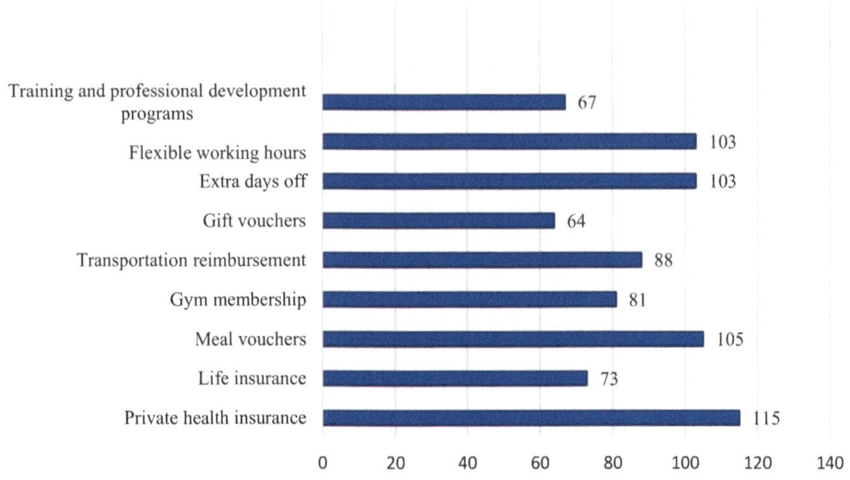

Fig. 5 The non-wage benefits desired by respondents. *Source* Conducted by the author, based on the results obtained from research using SPSS

Table 1 Descriptive statistics of key variables

Question	Mean	Std. Dev
To what extent do you appreciate the current range of fringe benefits offered by your organization?	3.710	0.859
How often do you use the fringe benefits offered by your organization?	3.975	0.954
Please rate your overall job satisfaction level	3.736	0.852
How do you rate your job performance recently?	4.270	0.690
On a scale of 1–5, how important is it for you to have a flexible work schedule?	4.515	0.786
On a scale of 1–5, how important is it for you to have additional days off outside regular vacation time?	4.478	0.794
On a scale of 1–5, how satisfied are you with the balance between your personal and professional life?	3.987	0.849

Source Conducted by the author, based on the results obtained from research using SPSS

each variable assessed in the questionnaire. These results are crucial to ensure that the questions used are reliable and align with the research objectives regarding measuring employee satisfaction, use of fringe benefits, their preferences, and other relevant aspects for the conducted study.

To analyse the first general hypothesis, according to which: There is a positive and significant correlation between the provision of non-wage benefits within organizations and the level of employee satisfaction and motivation (Sittenthaler & Alwine, 2020), we will conduct a correlation analysis that will provide us with a Pearson

Table 2 Cronbach's alpha coefficients

Construct	Alfa Cronbach
General job satisfaction	0.73
Use of fringe benefits	0.81
Importance of flexible work schedule and additional days off	0.78
Perception of the influence of work-life balance	0.80

Source Conducted by the author, based on the results obtained from research using SPSS

coefficient and associated p-values. Thus, the independent variable is represented by the total number of non-wage benefits received by the respondent, while the dependent variable is represented by the level of employee satisfaction at the workplace, on a scale from 1 to 5, where 1 means very dissatisfied and 5 means very satisfied (Table 3).

From the previous table, a p-value associated with the correlation coefficient of 0.003 is obtained, which is smaller than the significance level of 0.01 (2-tailed), indicating a significant correlation. Since the Pearson coefficient has a value of 0.230, being positive, it means that this correlation is positive, meaning that as the number of non-wage benefits offered to employees increases, their overall job satisfaction tends to increase as well. Thus, the results of the correlation analysis indicate a significant positive correlation between the number of non-wage benefits employees receive and their overall job satisfaction. It is important to emphasize that this correlation does not necessarily imply a direct causal relationship between the number of non-wage benefits and employee satisfaction, but only a statistical association, as there may be other factors, such as organizational culture, work environment, and other factors, that can influence employee satisfaction.

To demonstrate the second hypothesis of the research: *The use of non-wage benefits leads to an increase in employee commitment and performance within the organization* (Srna & Dinc, 2017) we will remove from the sample those individuals who

Table 3 Results of the correlation test for general hypothesis 1

		How many fringe benefits do you currently receive?	Evaluate your overall job satisfaction
How many fringe benefits do you currently receive?	Pearson correlation	1	0,230**
	Sig. (2-tailed)		0,003
	N	165	165
Rate your overall job satisfaction	Pearson correlation	0,230**	1
	Sig. (2-tailed)	0,003	
	N	165	165

** Correlation is significant at the 0.01 level (2-tailed)
Source conducted by the author, based on the results obtained from research using SPSS

do not have access to such benefits, leaving us with a total of 158 responses to be subjected to a multiple regression analysis. To verify this hypothesis, we will define the number of non-wage benefits employees use as the independent variable, and the evaluation of job performance of the respondents on a scale from 1 to 5 as the dependent variable (Table 4).

In the model summary of the multiple regression test, we can observe a value of the multiple correlation coefficient, R, with a value of 0.222, indicating a relatively weak positive correlation between the use of fringe benefits and the dependent variable. The coefficient of determination, R Square, is 0.049, indicating that approximately 4.9% of the variation in employee performance can be explained using fringe benefits (Table 5).

The ANOVA analysis indicates that there is a significant relationship between fringe benefits and job performance. The values in the ANOVA table, Sum of Squares and Mean Square, provide information about the total variation in performance that can be attributed to fringe benefits. The F and Sig. values indicate whether the observed difference between the groups of fringe benefits is statistically significant. In this case, F is significant, 8.151, suggesting that the difference between the groups of fringe benefits is greater than expected from random fluctuations alone. The significant p-value of 0.005 indicates that this difference is attributed to the effects of fringe benefits. This suggests that fringe benefits explain a significant proportion of the variation in job performance.

Next, to test the third general hypothesis according to which: *There is no significant association between employees' job satisfaction and preferences for fringe benefits* (Wardiansyah et al., 2024), we will rely on 2 questions included in the questionnaire, one that assesses the preferred fringe benefit and the other one that assesses employees' job satisfaction. Thus, the independent variable will be represented by the preference for a specific benefit, while the dependent variable will be employees' job satisfaction. To analyse the hypothesis, we will use the chi-square test of independence, which will help us determine if there is a significant association between the two variables or if they are independent (Table 6).

Table 4 Model summary of the multiple regression test for hypothesis 2

R	R Square	Adjusted R Square	Std. Error of the estimate
0,222	0,049	0,043	0,676

Source Conducted by the author, based on the results obtained from research using SPSS

Table 5 ANOVA of the multiple regression test for hypothesis 2

	Sum of squares	df	Mean square	F	Sig.
Regression	3,720	2	3,720	8,151	0,005
Residual	71,651	157	0,456		
Total	75,371	158			

Source Conducted by the author, based on the results obtained from research using SPSS

Table 6 The chi-square test for hypothesis 3

	Value	df	Asymptotic significance (2-sided)
Pearson chi-square	60,073	24	<0,001
Likelihood ratio	54,630	24	<0,001
Linear-by-linear association	0,001	1	0,981
N of valid cases	159		

Source Conducted by the author, based on the results obtained from research using SPSS

The results of the Chi-Square test indicate a significant association between preferences for extra-salary benefits and the level of employee satisfaction, as the Pearson Chi-Square test statistic is 60.073, with 24° of freedom and a significance level (p-value) less than 0.001, indicating a significant association between the analysed variables. Additionally, the Likelihood Ratio test provided a test statistic of 54.630 with 24° of freedom and a significance level less than 0.001, confirming the significant association between variables. It is important to examine and consider the actual strength of the association, not just the significance level, thus we will also analyse the Cramer's V coefficient (Table 7).

Analysing Cramer's V coefficient, which has a value of 0.307, and the associated significance value less than 0.001, indicates a moderate association between the established variables. In conclusion, employees' preferences for extra-salary benefits are related to their level of job satisfaction.

To verify the last hypothesis formulated, that Employees aged between 25 and 35 will have a higher usage of lifestyle and fitness-related extra-salary benefits, while employees over 45 will have a higher usage of health-related extra-salary benefits (health insurance), we will rely on 3 questions from the questionnaire: one that divides respondents into age categories, the second that establishes the frequency of using health-related benefits, and the last one that checks the frequency of using lifestyle and fitness-related benefits. Thus, we will perform a multivariate MANOVA statistical test, in which the independent variable is represented by the age category, and the dependent variables are represented by the frequency of usage for health-related benefits and lifestyle and fitness-related benefits (Table 8).

Following the test execution, we can observe that the values of the effect measures (Pillai's Trace, Wilks' Lambda, Hotelling's Trace, Roy's Largest Root) are relatively high, indicating that age category has a significant effect on the use of benefits.

Table 7 Symmetric measures for the chi-square test for hypothesis 3

		Value	Approximate significance
Nominal by nominal	Phi	0.615	<0.001
	Cramer's V	0,307	<0,001
N of valid cases		159	

Source Conducted by the author, based on the results obtained from research using SPSS

Table 8 Multivariate test for hypothesis 4

Effect		Value	F	Hypothesis df	Error df	Sig.
Intercept	Pillai's trace	0.892	632.013	2.000	153.000	„<0.001
	Wilks' lambda	0.108	632.013	2.000	153.000	<0.001
	Hotelling's trace	8.262	632.013	2.000	153.000	<0.001
	Roy's largest root	8.262	632.013	2.000	153.000	<0.001
@ Age	Pillai's trace	0,328	7,546	8,000	308,000	<0.001
	Wilks' lambda	0,695	7,630	8,000	306,000	<0.001
	Hotelling's trace	0,406	7,713	8,000	304,000	<0.001
	Roy's largest root	0,295	11,345	4,000	154,000	<0.001

Source Conducted by the author, based on the results obtained from research using SPSS

The low p-values, less than 0.001, show that this difference is significant, meaning that there are significant differences in the use of fringe benefits among different age groups. To verify the research hypothesis and to see if the use of these benefits differs significantly by age, we will rely on a T-test for comparison between two groups, 25–35 years old and over 45 years old (Table 9).

Running the T-test and focusing only on the two age categories, we can observe that for the dependent variable "Private Health and Life Insurance", Lovene's test for

Table 9 Independent sample test for hypothesis 4

		Lovene's test for equality of variances		t-test for Equality of Means			
						Significance	
		F	Sig.	t	df	One-sided p	Two-sided p
Private health and life insurance	Equal variances assumed	25.162	<0.001	−5.178	116	<0.001	<0.001
	Equal variances not assumed			−4.218	41.675	<0.001	<0.001
Lifestyle and Fitness	Equal variances assumed	0.515	0.475	−0.223	116	0.412	0.824
	Equal variances not assumed			−0.224	59.13	0.412	0.824

Source Conducted by the author, based on the results obtained from research using SPSS

Equality of Variances indicates that the variances in the two groups are significantly different, with an F value of 25.162 and a very low significance level, less than 0.001. This represents significant differences between the variances of the two age groups regarding the use of these benefits. For the t-test for Equality of Means, a t value of −5.178 is obtained with a number of 116° of freedom, df, and a significance level less than 0.001. Based on the statistical data, it can be concluded that employees aged over 45 use health-related benefits significantly more than those aged between 25 and 35.

Regarding the dependent variable "Lifestyle and Fitness," Lovene's test indicates that the variances in the two groups are not significantly different, with an F value of 0.515 and a significance level of 0.475, indicating that the variances are comparable between the two groups. For the t-test for Equality of Means, a t value of −0.223 is obtained, with a number of degrees of freedom, df, of 116 and a significance of 0.412, suggesting that there is no significant difference between employees aged between 25 and 35 and those aged over 45 regarding the use of lifestyle and fitness benefits.

In conclusion, the results suggest that there is a significant difference between age groups regarding the use of health-related benefits, but not regarding the use of lifestyle and fitness benefits. Thus, the fourth hypothesis is rejected (Table 10).

Based on the tests conducted earlier, the following conclusions can be drawn.

The questionnaire also included two questions analysing the importance of flexible working hours and the possibility of having days off outside of annual leave, using a scale where 1 represented "very unimportant" and 5 represented "very important". As can be seen, for the respondents, these are two extremely important benefits, with an average response of 4.52 for flexible working hours, with 64.8% of respondents choosing the "very important" option, and an average of 4.48 for having days off, with 64.6% of respondents choosing option 5 (Fig. 6).

Conclusions

This research investigated the impact of fringe benefits on employees, starting with profiling the respondents based on several factors, continuing with analysing their preferences regarding benefits, their level of job satisfaction, and motivation. The results were obtained through administering a questionnaire and analysing it using a comprehensive set of statistical methods, interpreting them in the context of the relevant literature.

Thus, a significant positive correlation was found between the number of employee benefits and their level of job satisfaction, the utilization of benefits leading to increased employee performance, employees' preferences for fringe benefits being linked to their job satisfaction, and a significant difference existing between age groups regarding the utilization of health-related benefits, but not for benefits related to lifestyle and fitness.

The findings of this research make a significant contribution to the literature on human resource management and organizational psychology. Firstly, this study

Table 10 Testing hypotheses results

Hypothesis	Results	Interpretation
Hypothesis 1: There is a significant positive correlation between offering fringe benefits within organizations and the level of employee satisfaction and motivation (Sittenthaler & Alwine, 2020)	Significant positive correlation (p-value associated with correlation coefficient of 0.003 is less than the significance level of 0.01 (2-tailed). Pearson coefficient = 0.230)	This hypothesis was confirmed, indicating that employees who receive more fringe benefits tend to have higher levels of satisfaction and motivation at work
Hypothesis 2: The use of fringe benefits leads to increased employee commitment and performance within the organization (Srna & Dinc, 2017)	Positive correlation (R, with a value of 0.222, R Square is 0.049) and significant relationship (F is significant, 8.151)	This hypothesis was confirmed, suggesting that the use of fringe benefits is associated with higher employee commitment and performance
Hypothesis 3: There is no significant association between employee job satisfaction and preferences for fringe benefits (Wardiansyah et al., 2024)	Significant association (Pearson Chi-Square test statistic is 60.073 and Cramer's V coefficient is 0.307, with significance value less than 0.001)	This hypothesis was rejected, indicating that employees' preferences for fringe benefits are linked to their job satisfaction
Hypothesis 4: Employees aged 25–35 will show higher utilization of lifestyle and fitness-related fringe benefits, while employees over 45 will show higher utilization of health-related fringe benefits (health insurance)	Significant differences ($p < 0.001$)	This hypothesis was rejected, indicating that preferences for fringe benefits vary significantly depending on age group, with significant differences observed in health-related benefits but not in lifestyle and fitness-related benefits between the two age groups

Source Conducted by the author, based on the results obtained from research using SPSS

validates and extends previous findings regarding the importance of fringe benefits in enhancing employee satisfaction and performance. The study confirms the general hypothesis that offering fringe benefits is positively correlated with employee satisfaction and motivation (Sittenthaler & Alwine, 2020).

Secondly, the research provides a new perspective on how age influences preferences for different types of fringe benefits. These findings contribute to a deeper understanding of how organizations can customize benefit packages more effectively to meet the needs of diverse employee segments.

Moreover, the results refute the hypothesis that there is no significant association between employees' job satisfaction levels and preferences for fringe benefits (Wardiansyah et al., 2024). This suggests that benefit preferences are an important indicator of job satisfaction, adding a new dimension to the literature on employee motivation.

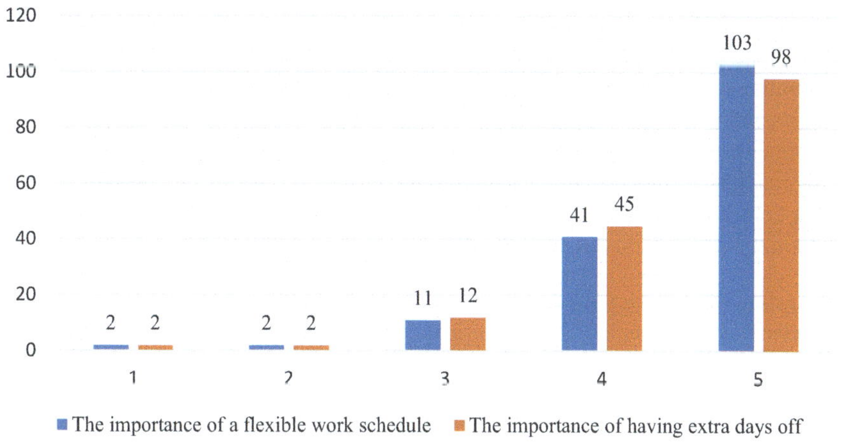

Fig. 6 The importance of flexible working hours and additional days off. *Source* Conducted by the author, based on the results obtained from research using SPSS

Our study has multiple practical implications for human resource management. Firstly, customizing fringe benefits based on employees' individual preferences and needs can significantly enhance their satisfaction and commitment. Human resource managers should consider implementing flexible benefit platforms where employees can choose from a diverse range of options, including health insurance, meal vouchers, or gym memberships.

Secondly, improving work-life balance through benefits such as flexible work schedules or additional days off can positively influence employee satisfaction. Organizations could implement policies that provide more flexibility in working hours and more options for leave to support overall employee well-being.

Also, promoting health and wellness through wellness and fitness programs can support the health and well-being of employees of all ages. Companies can develop such programs to meet the needs of different age groups, thereby contributing to better employee retention.

Although this study provides valuable insights, there are several limitations that need to be acknowledged. First, the sample of respondents may not be representative of all industries and economic sectors in Romania, which could limit the generalizability of the findings. Second, using a self-administered questionnaire can introduce response errors due to subjective perceptions of the respondents. Additionally, the study relies on cross-sectional data, which means causality between the studied variables cannot be determined. Third, despite employing rigorous statistical methods for data analysis, there may be other uncontrolled factors that could have influenced the results. For example, the specific economic and cultural context of Romania could impact employees' perceptions of extra-salary benefits.

Future research should consider using larger and more diverse samples that include employees from various industries and economic sectors. Longitudinal studies would be beneficial to examine how perceptions and utilization of extra-salary benefits change over time. Another research direction could be a detailed exploration of contextual factors that influence the relationship between extra-salary benefits and employee satisfaction. This could involve specific economic, cultural, and organizational factors unique to each country or region.

In conclusion, this research has highlighted the importance of extra-salary benefits in enhancing employee satisfaction and performance, emphasizing the need to customize these benefits to meet the diverse needs of employees.

Acknowledgements This paper was co-financed by The Bucharest University of Economic Studies during the Ph.D. program.

References

Abdulkhaleq, N. Q. (2021). The effect of non-monetary incentives & work environment. *Studies of Applied Economics, 39*(7).

Anon. (2023). 25 Types of employee benefits to look for in a new job, (Interactiv). Retrieved January 10, 2024, from https://www.indeed.com/career-advice/career-development/types-of-employee-benefits.

Burhanudin, B., Deliah, F. P., & Yayuk, A. (2023). The effect of work motivation and work discipline on employee performance. *International Journal of Humanities Education and Social Sciences (IJHESS), 2*(5), 1453–1467.

Chabé-Ferret, S., Le Coent, P., David-Legleye, V., & Delannoy, V. (2023). Non-monetary incentives to increase enrollment in payments for environmental services. *European Review of Agricultural Economics, 50*(4), 1401–1427.

Chauhan, V. (2015). Relevance of non-monetary incentives in an organization. *SSRN Electronic Journal, 28*, 10.

Dewhurst, M., Guthridge, M., & Mohr, E. (2009). Motivating people: Getting beyond money, (Interactiv). Retrieved March 10, 2024, from https://www.mckinsey.com/capabilities/people-and-organizational-performance/our-insights/motivating-people-getting-beyond-money.

Ghosh, S., & Swamy, D. R. (2014). A literature review on organizational commitment–a comprehensive summary. *International Journal of Engineering Research and Applications, 4*(12), 04–14.

Hardiyansyah, I., Razak, I., & Silitonga, E. S. (2023). The influence of motivation and compensation on employee performance at Pt Unisadhuguna with job satisfaction mediation. *Journal of Economics, Finance and Management Studies, Issue, 6*, 2.

Kalev, A., & Dibbin, F. (2022). The surprising benefits of work/life support, (Interactiv). Retrieved January 02, 2024, from https://hbr.org/2022/09/the-surprising-benefits-of-work-life-support.

Lee, S. (2007). Vroom's expectancy theory and the public library customer motivation model. *Library Review, 56*(9), 788–796.

Li, X. (2023). Employee benefits and its impacts on business performance-a systematic review. *SHS Web of Conferences, 170*(03021), 3.

Mendez, A., Rojas, M., & Watkins-Fassler, K. (2023). *The hierarchy of needs empirical examination of Maslow's theory and lessons for development* (p. 165). Word Development.

Ozsoy, E. (2019). An empirical test of Herzberg's two-factor motivation theory. *Marketing and Management of Innovations, 1*, 11–20.

Robbins, S. P., & Judge, T. A. (2017). *Organizational behavior* (17th ed.). s.l.: Pearson Education Limited.

Sittenthaler, H. M., & Alwine, M. (2020). Cash, non-cash, or mix? Gender matters! The impact of monetary, non-monetary, and mixed incentives on performance. *Journal of Business Economics, 90*, 1253–1284.

Siyal, S., & Garavan, T. (2020). *Employee benefits* (pp. 1–2).

Srna, E., & Dinc, M. S. (2017). The significance of non-monetary incentives and its relationship with employee motivation: A case of civil service employees in Bosnia and Herzegovina. s.n.: Sarajevo

Swarooprani, K. (2022). An study of research methodology. *International Journal of Scientific Research in Science, Engineering and Technology, 9*(3), 537–543.

Tseng, L.-M., & Kuo, C.-L. (2014). Customers' attitudes toward insurance frauds: An application of Adams' equity theory. *International Journal of Social Economics, 41*(11), 1038–1054.

Wardiansyah, D. R., Indrawati, N. K., & Kurniawati, D. T. (2024). The effect of employee motivation and employee engagement on job performance mediated by job satisfaction. *International Journal of Research in Business and Social Science, 13*(1), 220–231.

Preliminary Investigation of Entrepreneurial Traits of Business Students

Elisabeta Butoi and Paul-Sorin Lazăr

Abstract Given its critical role as a cornerstone of a developing and transforming economy, entrepreneurship continues to capture the attention of scholars. In this context, analyzing the entrepreneurial traits of students becomes a valuable endeavor to assess the potential trajectory of future entrepreneurs and the business environment. As a fundamental pillar of the developing and transforming economy, entrepreneurship still captivates scholars. Analyzing the entrepreneurial inclination of students deserves the examination for the estimation of further development of the business environment. The primary objective of this research was to examine the entrepreneurial character traits among students in the Faculty of Business and to reveal the relationship between these entrepreneurial psychological attributes. By identifying the degree of their inclinations toward an entrepreneurial attitude, the study seeks to add valuable insights into the education of future entrepreneurs and the potential impact on economic development. The research is based on primary quantitative data collected using a questionnaire and analyzed through descriptive and inferential statistics. The research examines 15 entrepreneurial character strengths of a sample of 350 business students. The results showed that the entrepreneurial character traits of students differ in gender and studying program. The results indicate significant gender differences in open-mindedness, love of learning, wisdom, perseverance, social intelligence, spirituality, equity, and courage. Other important differences were identified based on the study program. The results showed significant differences between undergraduate and master's students. The statistical calculations reveal positive inter-correlations between all 15 entrepreneurial traits. Some interesting scores of the measured variables are very relevant for the education and business representatives.

Keywords Entrepreneurial traits · Skill · Character strength · Business students · Entrepreneurial education

---

E. Butoi (✉) · P.-S. Lazăr
Babeş-Bolyai University, Cluj-Napoca, Romania
e-mail: elisabeta.butoi@ubbcluj.ro

A. L. Negruşa et al. (eds.), *Sustainable and Resilient Businesses in the Global Economy*, Springer Proceedings in Business and Economics,
https://doi.org/10.1007/978-3-031-75883-6_16

Introduction

Entrepreneurship permeates every corner of the world. It manifests itself in the modest family-run shops of developing nations and multinational corporations around the globe. The foundation of the economic system lies in the spirit of entrepreneurship. This field has captivated academics for decades, counting vast literature and research studies. However, the interest and attention directed toward entrepreneurship is still increasing in the transformational and technologically adapting society.

The traditional view of entrepreneurial success often correlates with the success of the venture itself, typically measured in financial terms. However, researchers have expanded this narrow scope to include both monetary and non-monetary factors as sources of entrepreneurial success. Today, entrepreneurs having a good image in society (Md Yasin et al., 2011), are a key driver of economic growth, innovation, and social change. Their entrepreneurial activity promotes job creation, competitiveness, and technological advancement. Given its importance, researchers have studied various factors influencing entrepreneurial activity, including cultural, social, economic, political, technological, legal, and psychological aspects. Entrepreneurial skills are a combination of abilities and knowledge that allow individuals to identify and evaluate business opportunities, develop and manage ventures, and navigate the challenges of the marketplace. Entrepreneurs require robust practical skills and knowledge to effectively produce goods or services and run a company successfully. Their ability to set goals and be committed to accomplishing them, the capacity to decide, plan, organize, and coordinate people to work efficiently and effectively are the rational competencies of an entrepreneur. Aspiring entrepreneurs must possess a combination of hard and soft skills, ranging from technical knowledge to interpersonal abilities (Leon, 2017).

The psychological dimensions of entrepreneurship play a critical role in entrepreneurial success. The potential of education to cultivate and stimulate entrepreneurship is gaining increased recognition. Furthermore, fostering an entrepreneurial mindset equips students to navigate the ever-changing business environment (Sun, 2023). Researchers suggest that an entrepreneurial mindset can be nurtured through formal education (Bayar et al., 2022; Bejinaru, 2018; Bilén et al., 2005). Further emphasize the potential for acquiring the necessary personality traits, competencies, and skills to become an entrepreneur through learning. These studies advance that aspiring entrepreneurs can develop their knowledge through both "learning by doing" and formal entrepreneurship education (Chang & Rieple, 2013; Lyons et al., 2019a; Oosterbeek et al., 2010).

This research seeks to elucidate the entrepreneurial traits of university students who intend to become entrepreneurs. The impetus for the present study emerges from a critical analysis of existing literature that underscores a predominant focus on the acquired skills of students, as opposed to an in-depth exploration of their character traits, both of which are pivotal in shaping the future of students on their path to entrepreneurship. The study is preliminary and representative at the faculty level. The analysis focuses on the identification of character traits that reveal the entrepreneurial inclinations of students. The structure of this study is organized as follows: the literature review provides a comprehensive analysis of existing research examining entrepreneurial abilities, followed by the presentation of methodology, and then the results and discussion of the findings of this research are presented through descriptive and inferential statistics. Finally, the conclusions summarize the contributions of this study.

Literature Review

"Given the proper motivation, training, and support, anyone with reasonable mental capacity has the potential to become a successful entrepreneur" (Lyons et al., 2019a, 2019b, p. 117).

In society, it is considered that entrepreneurs have the role of driving economic progress and development. In psychology, entrepreneurs possess distinct personality traits contributing to their success. Socio-behavioral theories, on the other hand, emphasize how an entrepreneur's experiences, knowledge, environment, and capacity for learning shape their success, in addition to personality factors (Pokrajčić, 2004). Successful entrepreneurship is the result of applied entrepreneurial capabilities.

Entrepreneurial skills refer to the abilities and competencies required for individuals to recognize and pursue opportunities to create and grow a successful business. Entrepreneurial skills are task and people-oriented (Chell, 2013), technical skills (Salun et al., 2021), managerial skills (like management, finance, legal, marketing), entrepreneurial skills (business concept, environmental scanning, and networking), and personal skills (like creativity, accountability, emotional coping with problems, self-awareness,) (Lyons et al., 2019a, 2019b). Later approach considered skills from the perspective of transformational management skills including problem-solving (Raine & Pandya, 2019), persistence, passion, adaptability, knowledge, and creativity (Bejinaru, 2018; Gatner et al., 2004; Koe et al., 2018; Oosterbeek et al., 2010; Štrukelj et al., 2019), innovation, leadership, resilience (Gatner et al., 2004; Lee, 2023), and resourcefulness. Furthermore, considering the importance of relationships, teaming (Salun et al., 2021; Štrukelj et al., 2019), community involvement, building reputation, and networking are essential. Otherwise, decision-making, communication (Štrukelj et al., 2019), process design, conflict management, and performance are necessary to manage organizational processes (Lyons et al., 2019a). When explaining the cognitive characteristics of the entrepreneur, Gatner and et al., (2004) added, opportunity recognition (Koe et al., 2018), planning, commitment, and intentions. Another study with a wider focus examined the following: entrepreneurial focus, perceived barriers, self-control, pro-entrepreneurial attitude, self-efficacy (Lee & Koh, 2023; Oosterbeek et al., 2010), and perceived support factors (Md Yasin et al., 2011). Others consider important these skills: creative thinking, leadership, and time management (Salun et al., 2021). The following was presented as the most important entrepreneurial skills: need of power, achievement, autonomy, social orientation, risk-taking, persistence, market awareness, and flexibility (Oosterbeek et al., 2010). Purposeful work, achievement desire, spirituality, and resilience are the driving factors of women entrepreneurs' perseverance (Lee, 2023). Another study considered entrepreneurial skills: critical thinking, decision-making, active learning, and judgment (Bejinaru, 2018). Successful entrepreneurship is contingent upon three critical factors: curiosity, creativity, and commitment to one's venture (Raine & Pandya, 2019). One approach posits that entrepreneurial skills are multi-dimensional, encompassing behavior, emotions, and know-how (Chell, 2013).

As some of the entrepreneurial skills presented in these cited resources are in fact competencies in the curricula of the faculty of the studied sample of this research, those competencies are not chosen as part of this study. Furthermore, in the literature were identified mixed ways of expressing entrepreneurial skills, being aptitudes, abilities, behaviors, capabilities, competencies, or characteristics, also some critics concerning the proper word (Leon, 2017). Thus, the choice for this research was to consider studying these entrepreneurial skills that were also identified in human character strengths defined by the specialist in human personalities and values (Bluemke et al., 2021; Park et al., 2017; Peterson & Seligman, 2004). Based on rational analyses, there were selected 15 entrepreneurial character traits, as follows.

Curiosity, defined as a strong desire to explore, to know or to learn something, plays a crucial role in various aspects of life, including education. It is a fundamental characteristic that drives individuals to seek knowledge, ask questions (Raine & Pandya, 2019), and explore new ideas or experiences (Peterson & Seligman, 2004). In the context of education, curiosity is essential for student engagement, learning enhancement, and problem-solving approach (Bluemke et al., 2021; Raine & Pandya, 2019).

Open-mindedness is the readiness and capacity to consider opposing beliefs and perceptions impartially, setting aside one's own commitments (van Brussel et al., 2023). It acts as a corrective virtue against confirmation bias, aiding in reasoning and evidence evaluation (Alsharif & Symons, 2020). Open-mindedness manifests through interpersonal openness, promoting tolerance and productive interactions, but also encouraging the acceptance of new ideas, diverse opinions, tolerance, and respect for different cultural values (van der Westhuizen, 2023).

Love of learning is deriving enjoyment from deepening understanding and acquiring new abilities (Bluemke et al., 2021). It is the longing to know more, reading, and understanding (Zábó et al., 2023). The higher education institutions have the possibility and necessity to cultivate appropriate competencies in students during the learning process, aligning with practical requirements (Štrukelj et al., 2019).

Creativity, characterized by divergent thinking, imagination, and cognitive flexibility, is a vital catalyst for economic growth and a cornerstone of education. Its applications transcend the realms of art and music, proving highly valuable to innovative organizations. As innovation is pivotal to success, and human capital is recognized as the most precious resource for driving company performance, the generated ideas by human minds contribute to cultivating an innovative culture, enabling the organization to thrive (Butoi, 2014). The capacity to generate multiple ideas to identify the best solution for a problem or create something different encompasses the creative personal strength (Lyons et al., 2019a, 2019b).

Perspective is the capacity to see things from different viewpoints or contexts. It is the wisdom to understand complex issues, recognize patterns, and apply knowledge across diverse situations (Peterson & Seligman, 2004).

Perseverance, persistence, or commitment comprehends the continuous and persistent nature of effort (Gatner et al., 2004; Peterson & Seligman, 2004). It is a psychological concept that describes an individual's strong desire for significant accomplishment, mastery of skills, and personal fulfillment through setting and meeting high standards. Persistence reflects an internal drive for action and a preference for achieving success through one's efforts, being strongly correlated with the need for achievement. Perseverance is considered an important factor of entrepreneurial resilience for success (Raine & Pandya, 2019).

Courage is generally understood as the ability to face fear and act despite it. Courage is linked to positive outcomes such as goal achievement, well-being, and satisfaction with life (Santisi et al., 2000). Entrepreneurial courage involves taking calculated risks (Leon, 2017) (Leon, 2017). Entrepreneurs don't blindly jump into ventures. They assess risks, plan, and act strategically while acknowledging the possibility of failure (Bockorny & Youssef-Morgan, 2019). Positive psychology researchers Peterson and Seligman, known for their Values in Action character strengths framework, identified courage as a component of psychological capital, which together with optimism and hope, are strengths that can contribute to greater well-being (Peterson & Seligman, 2004), a measured quality of life indicator in the sustainable business environment (Eurostat, 2020).

Enthusiasm is the ability to engage with zest, vibrant energy flowing from mind and body, synergizing the physical and mental well-being (Bluemke et al., 2021). The desire for success has the driving force, the inexhaustible vitality, an insatiable desire that can only continue. The vitality of a person is the capacity to go outwards and express in deeds and achievements (Dorson, 2023).

Emotional intelligence, referred also as emotional coping, implies the ability to cope with problems (Lyons et al., 2019a), to understand the self and others' feelings and be able to adapt socially to each situation (Bluemke et al., 2021).

Optimism or hope is the positive belief that someone will succeed in the future. Entrepreneurs who are optimistic trust that they can reach their goals (Bockorny & Youssef-Morgan, 2019).

The research generally emphasizes the transcendent aspects of *spirituality* and the practical implications it may have on business practices. One important element of the spiritual mind is the imagination, the pivotal force of creativity and further innovation (Chiles et al., 2021). It's considered that entrepreneurs' religious beliefs influence their persistence in entrepreneurial careers, with altruism serving as a mediating psychological mechanism, seeing that the moral values rooted in entrepreneurs' beliefs infuse into their entrepreneurial mindset (Chen et al., 2023).

Teamwork involves collaboration among team members to achieve common goals. It also implies loyalty to others (Zábó et al., 2023). In today's interconnected world, teamwork is a vital skill for students. By working together, they not only contribute their knowledge but likewise gain valuable insights from their peers, preparing them for success in the classroom and later in the collaborative environment of modern workplaces (Lee & Koh, 2023).

Equity is the act of treating everyone the same, it derives from the principles of ethics, specifically fairness. Unethical actions in business damage the trust of partners and investors. Maintaining trust is useful especially for private firms to attract investors. A study found that the perception of ethical activities at the country level positively influenced the performance and outcomes of companies (Knill, 2012). The fairness attitude of entrepreneurs could be affected by cultural, legal, and socio-demographic factors (Crane, 2009).

Leadership is the capacity to guide, motivate, and influence others (Leon, 2017), also to organize the "collective success of a group and fostering good working relationships among members" (Bluemke et al., 2021, p. 7). Self-leadership and spirituality are considered resources for entrepreneurs to navigate the demands and stressors inherent in entrepreneurship. Spirituality and self-leadership exert a reciprocal influence on each other, impacting the thought patterns and self-efficacy of the entrepreneur (Godwin et al., 2016).

Self-regulation or self-control is the ability to take a thoughtful approach, considering the options before making a decision (Peterson & Seligman, 2004). There is also the locus control term, referring to the capacity to work hard and make the plans work without creating problems (Gatner et al., 2004).

All these entrepreneurial character strengths are not just determinants of entrepreneurs' success but also have a robust relation with life satisfaction (Park et al., 2017) and well-being (Bluemke et al., 2021). These entrepreneurial traits were investigated among business students most likely to become tomorrow's entrepreneurs.

Research Methodology

The Faculty of Business is entrusted with a pivotal role in shaping the future of the economy by nurturing the most proficient and talented human capital. This responsibility does not only encompass the preparation of individuals for existing industries but also extends to the cultivation of innovative and entrepreneurial attitudes capable of generating new and robust ventures. Considering this, the present research endeavors to examine the entrepreneurial traits of students. This analysis aims to ascertain the extent to which students are predisposed towards an entrepreneurial mindset.

For the 2023–2024 academic year, the faculty's total enrollment comprised 1,671 students. Utilizing a 5% margin of error and a 95% confidence level, the calculated required sample size was determined to be 313 participants. However, this study successfully engaged 350 respondents, thereby exceeding the minimum threshold for statistical representativeness. This consistency in sample representation underscores the reliability of the research outcomes.

Students' entrepreneurial traits were assessed using quantitative data obtained through a questionnaire in Romanian and English language distributed online to undergraduate and master's students. The instrument comprised demographic variables and 15 entrepreneurial personality strengths variables, measured using a 5-point Likert scale. The entrepreneurial traits were characterized based on the study of Bluemke and et al., (2021) defining human character strength scales. They developed a short version of the International Personality Item Pool—Values in Action (IPIP-VIA) having just 4 items for each scale, instead of 10 items per scale, as the VIAs were initially defined by Peterson and Seligman, (2004). The results were examined using the Statistical Package for the Social Sciences program (SPSS) version 28 for Windows.

Results and Discussions

The faculty offers studying programs in Romanian and in English language, thus having foreign students. The respondents are represented by 323 Romanian students and 27 other nationalities. In terms of gender, the sample was divided into 60.9% females, 38.3 males, and 0.9% of non-disclosed gender. 94.3% of respondents were between 18 and 25 years of age. The sample is composed of 331 undergraduate students and 19 master's students. Additionally, 68% of them do not work, 6% are freelancers, and the rest have full time (10%), part-time (12.6%), or hourly pay (3.4%) jobs (Table 1).

At the beginning of the results analyses, the descriptive statistics were examined for each item of the entrepreneurial traits variables. The Skewness of the items didn't exceed the absolute value of +2, also the Kurtosis values of the items didn't exceed the absolute value of +2 with one item exception, (having the value of 2.11) (Table 1). Furthermore, the internal consistency of the research variables, composed of 4 items each, was measured using Cronbach's Alpha. The alpha values lapse between 0.5 and 0.8, with one exception below 0.5, is indicating acceptable internal reliability of the variables (Table 2).

The mean values of the entrepreneurial character strengths variables had the lowest rate for optimism (3.26) and self-control (3.27), while the highest rate was for spirituality (3.94) and love of learning (3.91), followed by perseverance (3.89) (Table 2). The relatively low mean values for optimism and self-control warrant a critical examination. As the respondents are children of persons who lived their childhood under communism, probably some inheritance from them (Sălcudean, 2015) may have been a contributing factor to their optimism. However, this needs further research, as other environmental factors, like economic, politic, or cultural could be considered. Optimism is often highlighted as a key trait for entrepreneurs (Sun, 2023), as it drives them to pursue their vision despite potential obstacles (Bockorny &

Table 1 Demographic data

Variable	Items	Frequency	Percent
Nationality	Romanian	323	92.3
	Others	27	7.7
Gender	Feminine	213	60.9
	Masculine	134	38.3
	Prefer not to say	3	0.9
Age	18–25 years	330	94.3
	26–35 years	15	4.3
	36–45 years	4	1.1
	46–55 years	1	0.3
Studying program	Bachelor's	331	94.6
	Master's	19	5.4
Studying year	Bachelor's first year	169	48.3
	Bachelor's second year	107	30.6
	Bachelor's third year	55	15.7
	Master's first year	10	2.9
	Master's second year	9	2.6
Employment	I don't work	238	68.0
	I am a full-time employee	35	10.0
	I am a part-time employee	44	12.6
	I am employed on hourly pay	12	3.4
	I am self-employed (freelancer)	21	6.0

Source Authors' calculations

Youssef-Morgan, 2019). Self-control, similarly, is essential for maintaining focus and discipline (Chell, 2013; Lyons et al., 2019a, 2019b), yet it appears to be less dominant. This could indicate that entrepreneurs prioritize flexible and dynamic approaches over rigid self-restraint (Dorson, 2023), which may be necessary in the ever-changing business environment (Salun et al., 2021).

Spirituality may contribute to an entrepreneur's sense of purpose (Lee & Koh, 2023) and resilience, providing a broader context for their enterprises beyond mere financial success (Godwin et al., 2016). Love of learning, on the other hand, reflects the entrepreneurial drive to continuously acquire knowledge, innovate, and adapt (Lyons et al., 2019a; Raine & Pandya, 2019; van der Westhuizen, 2023). The respondents' level of love of learning is additionally justified by their choice of student status. Perseverance, following closely behind spirituality and love of learning, is a critical strength for entrepreneurs (Chell, 2013). It encapsulates the determination and boldness required to endure the trials of entrepreneurship (Salun et al., 2021). A study of entrepreneurs in Australia identified that the entrepreneurs with higher

Table 2 Descriptive statistics

Variable	Item	N	Mean	Std. Deviation	Skewness	Kurtosis	Variable's mean	Cronbach's Alpha
Curiosity	CUR1	350	4.04	0.97	−0.77	−0.01	3.83	0.5
	CUR2	350	3.81	1.01	−0.55	−0.30		
	CUR3R	350	3.73	1.31	−0.65	−0.77		
	CUR4R	350	3.75	1.25	−0.72	−0.55		
Judgment/ open−mindedness	JUD1	350	4.10	0.96	−0.96	0.50	3.78	0.51
	JUD2	350	3.94	0.97	−0.69	0.03		
	JUD3R	350	3.40	1.20	−0.28	−0.81		
	JUD4R	350	3.66	1.27	−0.61	−0.70		
Love of learning	LER1	350	4.17	1.01	−1.07	0.38	3.91	0.52
	LER2	350	4.23	0.97	−1.23	0.99		
	LER3R	350	4.32	1.04	−1.64	2.11		
	LER4R	350	2.92	1.50	0.00	−1.40		
Originality/ creativity	ORI1	350	3.85	1.03	−0.62	−0.25	3.81	0.62
	ORI2	350	3.80	1.01	−0.44	−0.43		
	ORI3R	350	3.73	1.17	−0.66	−0.31		
	ORI4R	350	3.84	1.18	−0.69	−0.51		
Perspective/ wisdom	PER1	350	4.20	0.98	−1.20	1.00	3.84	0.53
	PER2	350	3.85	1.03	−0.75	0.10		
	PER3R	350	3.75	1.17	−0.64	−0.45		
	PER4R	350	3.57	1.15	−0.47	−0.63		
Industry/ perseverance/ persistence	PEV1	350	3.76	1.14	−0.62	−0.46	3.89	0.72
	PEV2	350	4.02	1.00	−0.81	−0.03		
	PEV3R	350	3.88	1.19	−0.73	−0.50		
	PEV4R	350	3.91	1.16	−0.90	0.01		
Valor/ Bravery/ Courage	VAL1	350	3.55	1.17	−0.54	−0.44	3.54	0.52
	VAL2	350	3.75	1.08	−0.55	−0.44		
	VAL3R	350	3.83	1.17	−0.71	−0.40		
	VAL4R	350	3.03	1.23	0.05	−0.91		
Zest/enthusiasm/ vitality	ZES1	350	3.63	1.12	−0.48	−0.58	3.55	0.61
	ZES2	350	3.75	1.07	−0.54	−0.25		
	ZES3R	350	3.64	1.21	−0.62	−0.49		
	ZES4R	350	3.16	1.19	−0.04	−0.87		
Social/personal/ emotional intelligence	SIQ1	350	4.21	0.96	−1.23	1.18	3.80	0.5
	SIQ2	350	3.97	1.05	−0.82	−0.05		
	SIQ3R	350	3.46	1.26	−0.30	−1.01		
	SIQ4R	350	3.57	1.14	−0.36	−0.70		

(continued)

Table 2 (continued)

Variable	Item	N	Mean	Std. Deviation	Skewness	Kurtosis	Variable's mean	Cronbach's Alpha
Hope/optimism	HOP1	350	3.92	1.04	−0.81	0.05	3.26	0.51
	HOP2	350	3.90	1.05	−0.72	−0.21		
	HOP3R	350	2.88	1.23	0.09	−0.93		
	HOP4R	350	2.34	1.22	0.56	−0.69		
Spirituality/ religiousness	SPI1	350	3.34	1.23	−0.21	−0.93	3.94	0.5
	SPI2	350	4.21	1.08	−1.37	1.15		
	SPI3R	350	4.21	1.17	−1.42	0.99		
	SPI4R	350	4.01	1.26	−1.02	−0.09		
Teamwork/ citizenship	TEA1	350	4.45	0.80	−1.79	3.97	3.80	0.51
	TEA2	350	4.28	0.86	−1.08	0.69		
	TEA3R	350	3.80	1.22	−0.74	−0.43		
	TEA4R	350	2.67	1.19	0.28	−0.81		
Equity/fairness	EQU1	350	3.80	1.05	−0.52	−0.55	3.76	0.53
	EQU2	350	4.32	0.97	−1.44	1.52		
	EQU3R	350	3.96	1.19	−0.87	−0.31		
	EQU4R	350	2.97	1.23	0.07	−0.90		
Leadership	LEA1	350	3.82	1.04	−0.55	−0.48	3.62	0.64
	LEA2	350	3.40	1.20	−0.31	−0.68		
	LEA3R	350	3.67	1.08	−0.52	−0.29		
	LEA4R	350	3.59	1.29	−0.54	−0.79		
Self-regulation/ self-control	REG1	350	3.71	1.10	−0.61	−0.29	3.27	0.23
	REG2	350	3.29	1.17	−0.27	−0.63		
	REG3R	350	3.03	1.34	0.03	−1.17		
	REG4R	350	3.07	1.12	0.04	−0.63		

Source Authors' calculations

scores of perseverance and consistency had more successful businesses (Raine & Pandya, 2019).

To examine gender-based disparities among respondents, independent sample tests were conducted across the 15 entrepreneurial characteristics. The analysis revealed a predominant trend wherein female participants displayed higher mean scores than their male counterparts in 12 out of the 15 traits assessed. The exceptions to this pattern were observed in courage/valor, optimism, and self-control, where male respondents demonstrated higher scores. Female respondents showed higher mean scores in the following traits: curiosity, open-mindedness, love of learning, creativity, wisdom, perseverance, vitality, emotional intelligence, spirituality, teamwork, fairness, and leadership. However, not all these differences were statistically significant.

The independent sample tests (Levene's test and t-test) established statistically significant differences (Sig. (2-tailed) <0.05) in gender for these seven traits: open-mindedness, love of learning, wisdom, perseverance, social intelligence, spirituality, and equity. In each of these categories, female respondents consistently demonstrated significantly higher scores than males. Interestingly, while females scored lower in courage, optimism, and self-control, the difference in these traits reached statistical significance just for courage, the characteristic where males scored higher (Table 3).

Research has demonstrated that women face persistent discrimination and overt oppression across various domains. This discrimination extends beyond personal spheres into professional realms, compounding the challenges women entrepreneurs encounter (Lee, 2023). Even if, Romanian society still perceives the negative effects of communism and its legacy in the process of transformation (Sălcudean, 2015), the statistics for 2021 present almost equality in gender distribution of the population involved in business start-ups in Romania, with 9.8% males and 9.6% female (McEvoy, 2023). Furthermore, the results of this study identify a higher inclination towards entrepreneurial characteristics in females than males, for most entrepreneurial character strengths, as explained previously. Romania has emerged from its challenging post-communist transition experiencing renewed growth driven by foreign investment and supported by international institutions. Recognizing the importance of international relations, Romania aims to foster global political and economic ties being an attractive destination for international business expansion (Doaré & Coroş, 2017). Therefore, entrepreneurship-oriented education, which proved to have an impact on entrepreneurial economic activities, is fundamental. Developing countries should implement comprehensive entrepreneurship-focused education across all academic levels to stimulate the entrepreneurial tendencies of students (Bayar et al., 2022).

Also, there are statistically significant differences between the undergraduates and master students concerning open-mindedness, creativity, perspective, perseverance, courage, enthusiasm, and emotional intelligence (Table 3). In terms of age group, the one-way Anova test identified significant differences between students younger than 26 years and all the others for the variables: judgment, creativity, perspective, courage, enthusiasm, and teamwork, the oldest scored higher. Some of those latest differences overlap with the differences between undergraduates and master's students, also the age of master's students is naturally different than undergraduates.

Table 4 presents the interrelationships among the 15 entrepreneurial personality traits, by the correlation coefficient (r) values. The correlations are exclusively positive. The most significant associations were between open-mindedness and wisdom (0.567) followed very closely by the correlation of wisdom with social intelligence (0.565), and then the teamwork strongest correlation with leadership (0.558) and curiosity (0.551). On the other side of the spectrum, self-control demonstrated the lowest overlap with all the other scales. The low value of self-regulation mean value and weak correlations with the other variables were identified also in Bluemke and et al., (2021) study. However, self-regulation is most strongly associated with perseverance (0.374), spirituality (0.360), and persistence (0.350). Perceived behavioral control is a driver of entrepreneurship (Md Yasin et al., 2011).

Table 3 Independent samples tests (Gender, studying programs)

Variable	Gender	N	Mean	Levene's test	t-test	Studying program	N	Mean	Levene's test	t-test
				F	Sig. (2-tailed)				F	Sig. (2-tailed)
CUR	M	134	3.76	0.029	0.131	BA	331	3.8248	1.803	0.307
	F	213	3.88			MA	19	4.0000		
JUD	M	134	3.67	0.937	0.024	BA	331	3.7553	0.741	0.024
	F	213	3.84			MA	19	4.1316		
LER	M	134	3.77	0.069	0.005	BA	331	3.8927	2.928	0.080
	F	213	4.00			MA	19	4.1974		
ORI	M	134	3.71	0.334	0.056	BA	331	3.7863	1.961	0.043
	F	213	3.87			MA	19	4.1447		
PER	M	134	3.68	0.000	0.000	BA	331	3.8157	3.848	0.007
	F	213	3.95			MA	19	4.2632		
PEV	M	134	3.73	1.918	0.002	BA	331	3.8701	0.003	0.022
	F	213	4.00			MA	19	4.3158		
VAL	M	134	3.66	0.494	0.013	BA	331	3.5181	0.122	0.022
	F	213	3.46			MA	19	3.9211		
ZES	M	134	3.55	1.317	0.955	BA	331	3.5174	0.144	0.005
	F	213	3.56			MA	19	4.0263		
SIQ	M	134	3.70	0.116	0.026	BA	331	3.7810	1.671	0.022
	F	213	3.87			MA	19	4.1579		
HOP	M	134	3.30	1.470	0.462	BA	331	3.2470	0.330	0.185
	F	213	3.24			MA	19	3.4737		
SPI	M	134	3.84	0.509	0.030	BA	331	3.9350	0.108	0.375
	F	213	4.02			MA	19	4.0921		
TEA	M	134	3.76	0.039	0.446	BA	331	3.7840	1.176	0.101
	F	213	3.82			MA	19	4.0395		
EQU	M	134	3.52	1.217	0.000	BA	331	3.7508	2.751	0.141
	F	213	3.92			MA	19	4.0000		
LEA	M	134	3.52	0.021	0.058	BA	331	3.6042	0.102	0.123
	F	213	3.69			MA	19	3.8947		
REG	M	134	3.33	0.780	0.221	BA	331	3.2598	6.752	0.083
	F	213	3.24			MA	19	3.5263		

Note M = Male, F = Female, BA = Bachelor's study, MA = Master's study
Source Authors' calculations

Table 4 Variables' inter-correlations

	CUR	JUD	LER	ORI	PER	PEV	VAL	ZES	SIQ	HOP	SPI	TEA	EQU	LEA	REG
CUR	1	0.431**	0.515**	0.491**	0.443**	0.387**	0.342**	0.498**	0.418**	0.354**	0.381**	0.551**	0.284**	0.532**	0.143**
JUD	0.431**	1	0.339**	0.424**	0.567**	0.431**	0.340**	0.218**	0.362**	0.161**	0.310**	0.387**	0.257**	0.395**	0.179**
LER	0.515**	0.339**	1	0.511**	0.445**	0.364**	0.337**	0.363**	0.412**	0.210**	0.466**	0.337**	0.278**	0.295**	0.227**
ORI	0.491**	0.424**	0.511**	1	0.534**	0.447**	0.467**	0.367**	0.485**	0.288**	0.471**	0.354**	0.223**	0.458**	0.293**
PER	0.443**	0.567**	0.445**	0.534**	1	0.531**	0.435**	0.371**	0.565**	0.215**	0.459**	0.427**	0.306**	0.546**	0.350**
PEV	0.387**	0.431**	0.364**	0.447**	0.531**	1	0.355**	0.363**	0.452**	0.335**	0.412**	0.477**	0.356**	0.478**	0.374**
VAL	0.342**	0.340**	0.337**	0.467**	0.435**	0.355**	1	0.329**	0.505**	0.278**	0.294**	0.349**	0.135*	0.424**	0.303**
ZES	0.498**	0.218**	0.363**	0.367**	0.371**	0.363**	0.329**	1	0.454**	0.464**	0.428**	0.430**	0.272**	0.426**	0.308**
SIQ	0.418**	0.362**	0.412**	0.485**	0.565**	0.452**	0.505**	0.454**	1	0.291**	0.436**	0.422**	0.298**	0.513**	0.336**
HOP	0.354**	0.161**	0.210**	0.288**	0.215**	0.335**	0.278**	0.464**	0.291**	1	0.295**	0.382**	0.266**	0.345**	0.239**
SPI	0.381**	0.310**	0.466**	0.471**	0.459**	0.412**	0.294**	0.428**	0.436**	0.295**	1	0.351**	0.308**	0.360**	0.360**
TEA	0.551**	0.387**	0.337**	0.354**	0.427**	0.477**	0.349**	0.430**	0.422**	0.382**	0.351**	1	0.442**	0.558**	0.190**
EQU	0.284**	0.257**	0.278**	0.223**	0.306**	0.356**	0.135*	0.272**	0.298**	0.266**	0.308**	0.442**	1	0.263**	0.198**
LEA	0.532**	0.395**	0.295**	0.458**	0.546**	0.478**	0.424**	0.426**	0.513**	0.345**	0.360**	0.558**	0.263**	1	0.223**
REG	0.143**	0.179**	0.227**	0.293**	0.350**	0.374**	0.303**	0.308**	0.336**	0.239**	0.360**	0.190**	0.198**	0.223**	1

Note ** Correlation is significant at the 0.01 level (2-tailed). N = 350
Source Authors' calculations

Curiosity is considered the desire to learn and explore (Peterson & Seligman, 2004), thus the results reveal the link between curiosity and love of learning. Love of learning has the strongest correlation with curiosity (0.515) and creativity (0.511) (Table 4). Love of learning proved the most significant correlation with curiosity, like in a study conducted in Germany and the UK (Bluemke et al., 2021). The learning orientation is one of the drivers of entrepreneurial intention formation (De Clercq et al., 2012). A study on students in Iraq found that self-perceived creativity has a positive impact on e-entrepreneurship intention (Al Halbusi et al., 2022). This strong correlation between creativity and the desire to launch a new venture was identified by another study of Chinese students (Sun, 2023).

Courage was strongest correlated with social intelligence (0.505). Being a character strength, courage is also described as an element of entrepreneurial success. Courage helps entrepreneurs by giving them the confidence to take risks and try new things (Bockorny & Youssef-Morgan, 2019). Spirituality correlated strongest (0.471) with creativity (Butoi, 2014) enhances entrepreneurial altruism and afterward increases the probability of persisting in an entrepreneurial career (Chen et al., 2023). Open-mindedness proved the strongest correlation with wisdom in this study. The proper mindset for inter-disciplinary teams and cross-industry researchers requires a high level of open-mindedness approach and teamwork abilities for those attempting to develop new products and embrace innovation (Bilén et al., 2005).

Workplace occurrences can elicit either positive or negative emotional responses, which subsequently influence significant work-related attitudes and behaviors. Teamwork requires the engagement of people through curiosity, creativity, and social intelligence to predict positive interpersonal relationships (Gheorghe et al., 2022). Leadership and teamwork had the strongest correlation in-between (0.558) compared with their correlation with the other scales (Table 4). The correlation of teamwork with equity (0.442) is very important to develop trust among members of a team or organization (Knill, 2012). However, to create a learning organization, the role of leadership and teamwork is decisive in the changing business environment (Şahin & Bilir, 2024). Equity/fairness is most strongly correlated with teamwork. Fairness is the basic element of relating to others, and having a responsible attitude, in a collaborative teamwork environment (Chen et al., 2022). Social intelligence and leadership skills are needed by entrepreneurs and give them the wisdom to start a business (Md Yasin et al., 2011).

While an entrepreneur's personal characteristics provide an advantageous foundation, they alone do not guarantee success. These traits represent a predisposition towards entrepreneurship. It is the ability to learn that empowers entrepreneurs to capitalize on their inherent potential, talents, and predispositions, as well as to acquire and refine the essential business insight. Ultimately, an entrepreneur's success centers on the capacity for learning and growth (Pokrajčić, 2004). Therefore, educators in the field of entrepreneurship are tasked with equipping students with the essential skills required to thrive in a rapidly evolving environment. Fostering the development of entrepreneurial skills can be enhanced by creating a learning environment that facilitates students' interaction with real businesses through live projects (Salun et al., 2021).

Conclusions

Contrary to the belief that character strengths are a given, they are not static traits, they can be developed and nurtured over time (Lyons et al., 2019a, 2019b; Štrukelj et al., 2019). Everyone enters this world with an open eye for curiosity (Raine & Pandya, 2019), this being the basement for the love of learning considered crucial for entrepreneurs (Bayar et al., 2022; van der Westhuizen, 2023), as it underpins the ability to acquire and refine the necessary knowledge and competencies, enables continuous growth, and the ability to overcome challenges that arise while running a business. Objective evaluation of all pertinent evidence and being willing to change one's opinion accordingly defines open-mindedness (Santisi et al., 2000) essential in a transforming environment. Having original ideas (Dorson, 2023; Pokrajčić, 2004), being creative (Chell, 2013; Gatner et al., 2004; Leon, 2017) and being able to see the big picture, as a perspective, could ignite innovation. Courage, manifested as a willingness to confront risk and uncertainty, provides the incentive to overcome barriers (Bockorny & Youssef-Morgan, 2019; Santisi et al., 2000). Enthusiasm fuels the motivational drive necessary for the venture (Dorson, 2023; Gatner et al., 2004). Perseverance, defined as reliable persistence in a course of action despite obstacles (Salun et al., 2021; Santisi et al., 2000), ensures the resilience required to navigate the challenges of new business development. As the human capital is the highest value in every enterprise, being an optimistic (van der Westhuizen, 2023), self-controlled (Chell, 2013) and spiritual leader (Chen et al., 2023), approaching every human encountered with fairness (Bluemke et al., 2021) and proving social intelligence (Santisi et al., 2000), would develop successful teamwork, partnerships, and businesses.

As defined in this study, curiosity, open-mindedness (judgment), love of learning, creativity (originality), perspective (wisdom), perseverance (persistence), courage (valor, bravery), enthusiasm (zest, vitality), social (personal or emotional) intelligence, hope (optimism), spirituality (religiousness), equity (fairness), leadership, and self-control (self-regulation) encompass the 15 entrepreneurial traits examined. These traits were assessed within the representative sample of 350 business students and revealed interesting results. An analysis of the mean values across the entrepreneurial traits exposed the lowest values for optimism and self-control, while the highest values were recorded for spirituality, love of learning, and perseverance. A salient finding that warrants careful consideration by policymakers is the diminished prevalence of optimism, hope, and self-regulatory capacity among the students. This observation raises significant implications regarding the psychological resilience within the students, future entrepreneurs, partners in business or teamwork members. Thus, involvements addressed in these directions would be a proactive approach. Educational programs could prioritize interventions aimed at improving these specific traits to enhance students' psychological resilience.

Regarding the differences in terms of gender of the respondents, independent sample tests were performed. For 12 out of 15 entrepreneurial characteristics, females scored higher than males, except for courage/valor, optimism, and self-control. The statistically significant difference is just for courage. The females scored higher for curiosity, open-mindedness, love of learning, creativity, wisdom, perseverance, vitality, emotional intelligence, spirituality, teamwork, fairness, and leadership. Thus, findings surface statistically significant differences between females and males for open-mindedness, love of learning, wisdom, perseverance, social intelligence, spirituality, and equity. These results underscore the importance of considering gender as a potentially influential factor in the expression and development of entrepreneurial characteristics. Further research is warranted to explore the underlying causes of these gender-based disparities and their implications for entrepreneurial education and support programs. Given the significant differences found between males and females in certain entrepreneurial traits, educators might consider tailoring some aspects of their curriculum to address these disparities. For example, programs could focus on building courage, optimism, and self-control in female students, while emphasizing traits like open-mindedness, love of learning, and social intelligence for male students.

The results were also analyzed from the perspective of the studying program. The findings underline higher scores of mean values for master's students statistically different than undergraduate students for open-mindedness, creativity, perspective, perseverance, courage, enthusiasm, and emotional intelligence. These entrepreneurial strengths are better developed in master's students. This suggests that more advanced entrepreneurial education could be beneficial at the graduate level.

This research examines the interrelationships among 15 entrepreneurial personality traits, revealing only significant positive correlations between them. The most pronounced correlations were observed between open-mindedness and wisdom, followed by wisdom and social intelligence, and teamwork with leadership. Notably, curiosity demonstrated the highest associations with the love of learning, teamwork, and leadership. The investigation also highlighted the significance of creativity, which showed stronger correlations with the love of learning and spirituality. Courage displayed a stronger relationship with social intelligence, while emotional intelligence had the strongest correlations with wisdom, leadership and bravery. Spirituality relates strongest with creativity and love of learning. Optimism correlated higher with enthusiasm, while zest correlated stronger with curiosity. Equity as a prerequisite for teamwork had a higher correlation with it. Leadership proved the highest associations with teamwork, wisdom, curiosity, and social intelligence. Self-regulation had the highest associations with perseverance, spirituality, and perspective. All the interrelationships between the traits indicate their collective importance in fostering entrepreneurial aptitudes. These results are reinforcing entrepreneurial traits role as a critical element of entrepreneurial success. The research revealed positive correlations among all 15 entrepreneurial traits studied. This implies that entrepreneurship education should aim to develop these traits collectively rather than in isolation, as they seem to reinforce each other.

While recognizing the importance of developing entrepreneurial skills, educators face a significant challenge in delineating the right skills that should be cultivated. Entrepreneurial abilities encompass a range of aptitudes that are important for starting and growing a successful business. Entrepreneurs need to develop these skill sets also through experience (Leon, 2017). A study in which students worked on live projects with real entrepreneurs recommended a more structured and timely approach to integrating skills training into the curriculum. This could involve providing targeted interventions based on the specific needs students encounter throughout their projects (Chang & Rieple, 2013).

This study is the first analysis of the entrepreneurial personality characteristics of the students of the Faculty of Business. Furthermore, longitudinal studies can be particularly insightful, tracking how attitudes toward entrepreneurship evolve. This approach, along with research examining the link between student demographics and entrepreneurial inclinations, would provide more valuable data for policymakers, curriculum developers, and teachers. With such insights, they can refine current curriculums and delivery methods, fostering a more innovative and entrepreneurial culture. Ultimately, this will equip graduates with the skills and confidence to consider entrepreneurship a viable career pathway. More than that, to create a thriving entrepreneurial ecosystem, governments should actively promote policies that make it easier to start and run businesses (Koe et al., 2018). This implies that entrepreneurship education should be an ongoing process, with regular assessments and adjustments to the curriculum based on students' changing needs and attitudes.

For further statement confirmation of results of the business students, comparative analyses with students from other fields than business would be a confirmatory factor. One research identified a difference in the willingness to become an entrepreneur between the non-technical students and the technical ones. The non-technical students were open to business opportunities arising, while the technical students perceived the difficulty of starting a business and were not inclined to become entrepreneurs (Md Yasin et al., 2011). This suggests that entrepreneurship education could benefit from cross-disciplinary approaches.

In conclusion, the entrepreneurial character traits of students, including but not limited to those in the Faculty of Business, are of vital importance in fostering an innovative and entrepreneurial culture in a dynamic economy. Educational programs could focus on cultivating this culture within their institutions. The research provides a comprehensive understanding of entrepreneurial traits, thereby contributing to the effective preparation of future entrepreneurs. Nurturing entrepreneurial attitudes among students is not only beneficial for their future enterprise, but is also crucial for the labor market, economy, and for a sustainable business environment.

References

Al Halbusi, H., Soto-Acosta, P., & Popa, S. (2022). Entrepreneurial passion, role models and self-perceived creativity as antecedents of e-entrepreneurial intention in an emerging Asian economy: The moderating effect of social media. *Asia Pacific Journal of Management*.

Alsharif, H., & Symons, J. (2020). Open-mindedness as a Corrective Virtue. *Philosophy, 96*(1), 73–97.

Bayar, Y., Sart, G., Gavriletea, M. D., & Coroş, M. M. (2022). The impact of entrepreneurial education and educational attainment on entrepreneurial activity: Evidence from selected high-income economies. *Journal of Business Economics and Management, 23*(6), 1257–1279.

Bejinaru, R. (2018). Assessing students' entrepreneurial skills needed in the knowledge economy. *Management & Marketing, 13*(3), 1119–1132.

Bilén, S. G., Kisenwether, E. C., Rzasa, S. E., & Wise, J. C. (2005). Developing and assessing students' entrepreneurial skills and mind-set. *Journal of Engineering Education, 94*(2), 233–243.

Bluemke, M., Partsch, M., Saucier, G., & Lechner, C. (2021). *Human character in the IPIP: Towards shorter, more content-valid, and cross-culturally comparable IPIP-VIA character strength scales*. s.l.:s.n.

Bockorny, K., & Youssef-Morgan, C. M. (2019). Entrepreneurs' courage, psychological capital, and life satisfaction. *Frontiers in Psychology, 10*.

Butoi, E. (2014). Fostering creative thinking to spark innovativeness. In *Proceedings of Developing Entrepreneurship and Creativity in the Romanian Business Environment* (pp. 93–107). Oradea: Emanuel University Press.

Chang, J., & Rieple, A. (2013). Assessing students' entrepreneurial skills development in live projects. *Journal of Small Business and Enterprise Development, 20*(1), 225–241.

Chell, E. (2013). Review of skill and the entrepreneurial process. *International Journal of Entrepreneurial Behavior and Research, 19*(1), 6–31.

Chen, C.-Y., Hsu, P.-Y., & Vu, H.-N. (2022). Collaborative process tailoring in evolutionary software development: A teamwork-quality perspective. *Software Quality Journal, 31*(1), 89–119.

Chen, Y., Klyver, K., & Zhang, H. (2023). Altruistic entrepreneurs: How does religious belief affect entrepreneurial career persistence? *Journal of Small Business Management*, 1–32.

Chiles, T. H., Crawford, B., & Elias, S. R. S. T. A. (2021). *Mind, body, and soul: A spiritual perspective on the entrepreneurial imagination*. Organization Theory.

Crane, F. G. (2009). Ethics, entrepreneurs and corporate managers: A Canadian study. *Journal of Small Business & Entrepreneurship, 22*(3), 267–274.

De Clercq, D., Honig, B., & Martin, B. (2012). The roles of learning orientation and passion for work in the formation of entrepreneurial intention. *Researching Entrepreneurship, 31*(6), 652–676.

Doaré, V. Q., & Coroş, M. M. (2017). La Roumanie, peut-elle être une destination attractive pour les entrepreneurs français? *Studia Universitatis Babeş-Bolyai Negotia, 62*(1), 63–91.

Dorson, J. (2023). Caveman, genius, artist, entrepreneur: Success and self-realization from literary naturalism to advice literature. *Subjectivity, 30*(3), 273–296.

Eurostat. (2020). *Eurostat statistics-explained*. Retrieved June 2, 2024, from https://ec.europa.eu/eurostat/statistics-explained/index.php?title=Quality_of_life_indicators_-_overall_experience_of_life#Overall_life_satisfaction_in_the_context_of_quality_of_life. (Online)

Gatner, W. B., Shaver, K. G., Carter, N., & Reynolds, P. D. (2004). *Handbook of entrepreneurial dynamics: The process of business creation*. SAGE Publications.

Gheorghe, A., et al. (2022). The effect of humor and perceived social interdependence on teamwork engagement in student groups. *Current Psychology, 42*(27), 23830–23841.

Godwin, J. L., Neck, C. P., & D'Intino, R. S. (2016). Self-leadership, spirituality, and entrepreneur performance: A conceptual model. *Journal of Management, Spirituality & Religion, 13*(1), 64–78.

Knill, A. (2012). The value of country-level perceived ethics to entrepreneurs around the world. *The European Journal of Finance, 18*(3–4), 209–237.

Koe, W., Krishnan, R., & Utami, S. (2018). The influence of entrepreneurial skills on business start-up intention among bumiputra students. *Journal of Advanced Manufacturing Technology (JAMT), 53–64*, 2.

Lee, V. B. (2023). *The role of spirituality for successful established women entrepreneurs*. Georgia State University.

Lee, J., & Koh, E. (2023). *Teamwork dimensions classification using BERT* (pp. 254–259). s.l.: Springer.

Leon, R. -D. (2017). Developing entrepreneurial skills. An educational and intercultural perspective. *Journal of Entrepreneurship, Management and Innovation, 13*(2017), 97–121.

Lyons, T. S., Lyons, J. S., & Jolley, G. J. (2019b). Entrepreneurial skill-building in rural ecosystems. *Journal of Entrepreneurship and Public Policy, 9*(1), 112–136.

Lyons, T. S., Lyons, J. S., & Jolley, G. J. (2019a). The readiness inventory for successful entrepreneurship (RISE): A tool for university engagement in entrepreneurial learning. *The Journal of Economic Development in Higher Education, 2.*

McEvoy, O. (2023). *Gender distribution of early-stage entrepreneurs in Europe by country 2021.* Retrieved May 10, 2024, from https://www.statista.com/statistics/315773/gender-distribution-of-early-stage-entrepreneurs-in-europe/. (Online).

Md Yasin, A. Y., Nik Mahmood, N. A. A., & Nik Jassfar, N. A. (2011). Students' entrepreneurial inclination at a Malaysian polytechnic: A preliminary investigation. *International Education Studies, 4*(2).

Oosterbeek, H., VanPraag, M., & Ijsselstein, A. (2010). The impact of entrepreneurship education on entrepreneurship skills and motivation. *European Economic Review, 54*(3), 442–454.

Park, N., Barton, M., & Pillay, J. (2017). Strengths of character and virtues: What we know and what we still want to learn. In M. A. Warren, & S. I. Donaldson (Eds.), *Scientific advances in positive psychology* (pp. 73–101).

Peterson, C., & Seligman, M. E. P. (2004). *Character strengths and virtues: A handbook and classification.* Osford University Press.

Pokrajčić, D. M. (2004). The characteristics of successful entrepreneurs. *Economic Annals, 49*(162), 25–43.

Raine, A. L., & Pandya, M. (2019). Three keys to entrepreneurial success: Curiosity, creativity, and commitment. *Entrepreneurship Education, 2*(3–4), 189–198.

Şahin, N., & Bilir, F. P. (2024). The effect of transformational leadership and personal cultural values on creating a learning organization. *Humanities and Social Sciences Communications, 11*(1).

Sălcudean, N. I. (2015). Repere imagologice ale culturii româneşti. Aglutinări şi disocieri. In V. Dumitraşcu (Ed.), *Interculturalitate şi interdisciplinaritate în ştiinţele socio-umane* (pp. 110–128). Pro Universitaria.

Salun, M., Zaslavska, K., Vaníčková, R., & Šindelková, K. (2021). Formation of entrepreneurial skills in students in a changing world. *SHS Web of Conferences, 90*(02009).

Santisi, G., et al. (2000). A relationship between psychological capital and quality of life: The role of courage. *Sustainability, 12*(13), 5238.

Štrukelj, T., Zlatanović, D., Nikolić, J., & Zabukovšek, S. S. (2019). A cyber-systemic learning action approach towards selected students' competencies development. *Kybernetes, 48*(7), 1516–1533.

Sun, L. (2023). Do entrepreneurship education activities have an impact on entrepreneurial behavior? *An Application of Behavioral Entrepreneurial Intention., 30*(54), 115855–115869.

van Brussel, S., Timmermans, M., Verkoeijen, P., & Paas, F. (2023). Comparing instructional strategies to support student teachers' learning to prepare an open-minded citizenship education lesson. *Instructional Science, 51*(3), 451–473.

van der Westhuizen, T. (2023). Youth entrepreneur ecosystem. In *Youth entrepreneurship* (pp. 57–111) .s.l.:s.n.

Zábó, V., Oláh, A., Erát, D., & Vargha, A. (2023). Assessing your strengths-Hungarian validation of the 24-item values in action inventory of strengths on a large sample. *European Journal of Mental Health, 18*(e0012), 1–16.

Frontier Technologies and Economic Development in New EU Member Countries: An Application of the JKS Panel Causality Test

Oguzhan Yelkesen, Yilmaz Bayar, and Marius Dan Gavriletea

Abstract Economic development is a broad concept that incorporates social, political, and economic dimensions. Countries prioritize economic growth as a main tool that led to an increase in welfare to achieve economic development. From this perspective, the use of technology, which has been a common practice in growth models for many years, emerges as one of the most significant driving forces behind economic growth. However, it is not just the utilization of technology that matters; physical investments aimed at technology advancement, human capital that enables the use of technology, and national efforts towards the adaptation and integration of technology also serve as catalysts for economic growth. Consequently, a relationship between economic growth and 'frontiers technology,' which includes all these factors related to technology usage, becomes apparent. In this context, this study aims to analyze the relationship between frontier technology and economic development by deploying panel causality analysis for new EU member countries over the period 2008–2021. The empirical results reveal that frontier technologies granger cause economic development in new EU countries. That is to say, changes in frontier technologies explain changes in economic development in new EU countries. This result demonstrates that frontier technologies have significant impacts on economic development, making them crucial tools for new EU member countries in terms of their EU cohesion processes and economic developments. When designing policies, new EU members should consider the importance of these tools. This approach not only facilitates the integration and convergence processes within the EU but also supports these countries in leveraging the potential of frontier technologies to accelerate their economic development.

O. Yelkesen (✉) · Y. Bayar
Department of Public Finance, Bandirma Onyedi Eylul University, Bandirma, Turkey
e-mail: oyelkesen@bandirma.edu.tr

Y. Bayar
e-mail: ybayar@bandirma.edu.tr

M. D. Gavriletea (✉)
Faculty of Business, Babeş-Bolyai University, Cluj-Napoca, Romania
e-mail: marius.gavriletea@ubbcluj.ro

A. L. Negruşa et al. (eds.), *Sustainable and Resilient Businesses in the Global Economy*,
Springer Proceedings in Business and Economics,
https://doi.org/10.1007/978-3-031-75883-6_17

Keywords Frontier technologies · Economic development · New EU members · Panel causality test

Introduction

In traditional growth models, it was initially proposed that technology plays a critical role as an exogenous factor for economic activities (Solow, 1956). Subsequently, researchers have integrated technology as an endogenous factor within growth models by focusing on the interaction between technology, human capability, and productivity (Lucas, 1988). These discussions have explored the significant driving force of technology in the production process from various perspectives, suggesting that technology is not merely a factor that accelerates the production process, but it is a key to long-term economic development. Indeed, these debates have facilitated a shift in the focus of technology use from narrow-based economic growth perspective to broader economic development concept. As a result of economic development and progress, the understanding of technology has also evolved.

With economic development and growth, the need has arisen not only for information and communication technologies (ICT), which are the standard indicators of technology, but also for investment and financing for the existence of these tools, human capital that will make technology use advantageous, and integration with artificial intelligence (AI) to adapt technology to various fields. In this manner, AI-supported ICT proliferation increases productivity per employee (Papaioannou & Dimelis, 2007) as well as technological advancement and sustainable economic development (Li, 2021). In this context, a technological structure that evolves with new technological developments appears to be an important factor for development. Thus, not only the use of ICT tools but also their adaptation, integration with AI, support for R&D activities, and the development of capabilities to keep pace with this changing technology, collectively referred to as "frontier technologies," are emerging as key factors in economic development.

Frontier technologies, like AI, blockchain, Internet of Things (IoT), and big data seem to be key drivers of transformation in productivity and human capital (Fanoro et al., 2021; Lau et al., 2023). These technologies are expected to reach USD$ 3.2 trillion by 2025 (UN, 2021). These technologies also offer substantial opportunities for economic growth by promoting innovation, improving efficiency, and generating employment (Xu, 2022). Furthermore, they can improve countries' global competitiveness and attract foreign investments. Therefore, it seems essential for nations to implement supportive policies and foster an environment that would encourage innovation and technological adoption to fully benefit from these technologies.

In recent years, ICT are seen as one of the most crucial components of the information society, which signifies economic, social, and cultural transformation. From this perspective, the *eEurope* initiatives launched by European Union (EU) member countries in 1999 aim to be a part of this transformation (Alfaro Cortés & Alfaro

Navarro, 2011). EU member states expect candidate countries to have infrastructure so that they can achieve similar objectives and promote cohesion within the EU region. By doing so, they can also reduce economic and social disparities among member countries. Consequently, how new EU member countries' investments and integrations in new technologies could impact their economic development is a topic of interest in the literature.

From this perspective, this study aims at analyzing the association between frontier technologies and economic development in new EU member countries for the period 2008–2021 by using Juodis, Karavias and Sarafidis (JKS, 2021) causality test. The causality results indicate the presence of a unidirectional relationship from frontier technologies to economic development. Therefore, it can be stated that changes in economic development in new EU member countries can be explained by the variations in frontier technologies. In other words, frontier technologies are significant for the economic development of new EU member countries and play a crucial role in the cohesion process of these countries. In the literature, there are lots of studies investigating the relationship between ICT penetration and economic development. However, there is a relatively limited number of studies exploring the relationship between frontier technologies and economic development. In line with this, this study aims to contribute to the literature by examining the nexus between economic development and frontier technologies from the standpoint of new EU member countries, which are in the process of establishing their institutional infrastructures. This focus not only addresses a gap in the research but also underlines the potential impacts and importance of frontier technologies in shaping the economic development trajectories of these new EU economies.

The study is organized as follows. Section "Literature Review" reviews literature. Section "Data and Research Methodology" defines the data set and econometric methodology. Section "Results and Discussions" focuses on empirical analyses and results. Section "Conclusions" concludes.

Literature Review

Studies in the literature approach the relationship between technology and economic development from various angles. The studies generally imply that ICT penetration positively affects important elements of economic development. This outcome is not surprising as we consider the impact of technological advancements on production, trade, and capital, along with their role in facilitating the rapid dissemination of information, thereby increasing production and positively affecting economic development (Welfens, 2005). On the other hand, as Vu (2011) suggests, advancements in technology usage will enhance economic growth by fostering innovation, increasing the decision-making capacity of firms and institutions, and reducing production costs. Vu's (2011) perspectives are based on the foundational ideas previously expressed by Lucas (1988) and Romer (1990) and suggest that enhancements in human capital can inspire individuals to embrace creativity and leverage technology, thereby fostering growth. Moreover, ICT infrastructure enables countries to collect more resources by taxing servicing, income, and consumption, which, in turn, leads to higher GDP (Tcheng et al., 2007). Empirically testing these mechanisms of the ICT on economic growth and output, Nasab and Aghaei (2009) have found that an increase in technology investments in OPEC countries has improved individuals' welfare. However, it is not just the use of technology, but technological investment and its association with human capital, represented broadly by 'frontiers technology', appear significant for economic growth. This situation underlines the importance for countries aiming for economic growth as part of their development to consider and support these factors in their policy-making processes.

Empirical studies in the literature generally view the relationship between economic development and ICT from the perspective of economic growth. For instance, in their study examining the impact of ICT on economic growth and human development across 27 European Union (EU) member countries, Alfaro Cortés and Alfaro Navarro (2011) highlighted the significant effects of ICT on economic growth. In a follow-up study, Arendt (2015) emphasized the importance of the relationship between the digital economy and economic growth for 15 Central and Eastern European Countries (CEE). Similarly, Aghaei and Rezagholizadeh (2017) investigated the impact of ICT investments on economic growth between 1990 and 2014 in countries of the Organization of the Islamic Cooperation (OIC) and concluded that investments in ICT are crucial for economic growth. They also stressed the importance of creating an investment environment oriented towards technology in OIC countries.

Apart from these findings, there are also studies highlighting the negative effects of ICT penetration. For instance, Nimrod (2018) demonstrates the potential negative impacts of psychostress, a psychological side effect of technology, on people's lives. Morawczynski and Ngwenyama (2007) argue that ICT penetration alone may not be sufficient for economic development and that technological penetration needs to be supported by other complementary factors. These studies indicate that ICT can have both positive and negative effects on economic development, and these effects may vary depending on a country's level of development, its characteristic structures, and other factors. Additionally, these results emphasize the necessity of a certain institutional infrastructure to create the desired effects of technology. This suggests that while ICT can significantly contribute to economic growth and development, its full benefits can only be realized within an environment that would support its effective use that would, in turn, mitigate its potential adverse effects.

As previously stated, economic development is a broader concept that incorporates economic growth in addition to other factors. Therefore, in discussing the relationship between ICT penetration and development, it is essential not just to consider its interaction with economic growth but also to examine its connections with factors such as human capital and the environment. In this context, Bayar et al. (2023) conducted a study on 11 transition European countries for the years 1996–2021, using cointegration and causality analyses to explore the relationship between ICT dissemination and human development. They found that ICT has a positive effect on human capital. In more comprehensive analyses, Aksentijević et al. (2021) investigated the relationship between ICT and human development in 130 countries from

2007 to 2019 using the GMM estimator and suggested that this relationship is positive in low-income countries but not significantly meaningful in high-income countries. From an environmental perspective, Wang et al. (2023) examined the relationship between ICT dissemination and environmental factors in the BRICS countries for the period 1996–2020 and showed that ICT has positive environmental effects by reducing energy consumption. Similarly, Khan et al. (2020) analyzed the relationship between ICT and environmental sustainability in 91 countries from 1990 to 2017 using POLS and GMM methods. They found that ICT has positive effects in developed countries but yields opposite outcomes in developing nations. These findings show the necessity of considering various factors in the relationship between ICT and economic development, thereby highlighting the significance of elements such as countries' levels of development.

These studies have measured technological penetration through standard indicators of ICT that limit its scope to the existence and use of technological infrastructure, such as internet usage, mobile cellular subscriptions, and fixed-broadband subscriptions. However, as economies evolve and countries develop, there is a need for a broader perspective that includes infrastructure investments aimed at developing technological infrastructure, research and development (R&D) activities, integration of technologies like AI, IoT, and big data into existing technologies, and the development of human capital to enhance the effectiveness of technology use. From this perspective, there are studies that examine the relationship between development and technology in terms of "frontier technologies." For instance, Bakouan and Sawadogo (2023) analyzed the interaction between frontier technologies and BioTrade in relation to income inequality in 131 countries from 2010 to 2019 and found that these interactions had a negative effect on income inequality. On the other hand, Maswana (2024) in his study on the impact of AI technology dissemination on economic growth in Sub-Saharan African countries stated that frontier technology has a significant and positive effect on economic growth. Višić (2023) analyzed the effect of frontier technologies on productivity in 15 OECD countries from 2012 to 2019 using panel data analysis and concluded that an increase in frontier technologies negatively affected productivity.

The discussions above highlight the negative effects of frontier technologies on development factors such as productivity, income distribution, and economic growth. Frontier technology represents evolutionary technological tools that encompass processes such as the development of existing technological infrastructure and its integration with new technology fields like AI. Therefore, it would not be true to assert that countries that have fully adapted to this evolutionary process and that posses the institutional infrastructure are capable of leveraging the benefits of these technologies. The negative impact of these technologies on the mentioned factors can be attributed to the dynamic and changing structures of frontier technologies. On the other hand, a significant factor in the adaptation and use of rapidly evolving and changing frontier technologies is the disparity among countries. Analyses should take these differences among countries into account, and countries that are not successful in adapting these technologies should examine the institutional infrastructure and policies of successful nations as models. By adapting the strategies that have been effective in successful countries to their own national contexts, it is possible to mitigate the negative effects of technology use disparities between countries and enhance the positive impacts of these technologies.

Data and Research Methodology

This study investigates the causal relationship between frontier technologies and economic development in the new EU members for the 2008–2021 term by means of causality test. Frontier technologies (FRONTECH) are proxied by frontier technology readiness index of UNCTADSTAT (2024) and the index indicates a country's readiness for using, adopting and adapting frontier technologies and is calculated as a combination of ICT deployment, skills, R&D activity, industry activity and access to finance through principal component analysis (UNCTADSTAT, 2024). On the other hand, economic development (ECONDEV) is represented by GDP per capita based on constant 2015 US$ and taken from the World Bank database (World Bank, 2024). The causality analysis is conducted for the period of 2008–2021 because frontier technology readiness index is available for this period. The sample of the study includes 11 new EU members (Bulgaria, Croatia, Czechia, Estonia, Hungary, Latvia, Lithuania, Poland, Romania, Slovak Republic, and Slovenia).

The econometric analyses are performed through Stata 15.0 and Eviews 12.0 statistical programs. The causal relationship between frontier technologies and economic development is examined via Juodis, Karavias, and Sarafidis (JKS) (2021) causality test.

The summary statistics of frontier technology readiness index and real GDP per capita are shown in Table 1. The mean of frontier technology readiness index and real GDP per capita are US$0.6943 and US$14195.87, respectively. Frontier technology readiness index is quite stable among the new EU members, but real GDP per capita is highly volatile during the same period.

The JKS (2021) causality test is designed for application across both homogenous and heterogeneous panels. Additionally, this test incorporates the Half-Panel Jackknife (H.P.J.) technique as introduced by Dhaene and Jochmans (2015) to mitigate the Nickell bias associated with the pooled estimator. This causality test operates in the following way:

Table 1 Summary statistics of FRONTECH and ECONDEV

Summary statistics	FRONTECH	ECONDEV
Mean	0.6943	14195.87
Std. Dev	0.0532	4225.115
Maximum	0.814	24786.79
Minimum	0.5773	6294.433

Source Authors' calculations

$$y_{it} = \pi_{oi} + \sum_{k=1}^{k} \delta_{pi} y_{it-k} + \sum_{q=1}^{Q} \varphi_{qi} X_{it-k} + \epsilon_{it}$$

For each country $i = 1, \ldots, N$ and year $t = 1, \ldots, T$. The equation describes a model where X_{it} is a scalar variable, and δ_{pi} and φ_{qi} represent heterogeneous autoregressive effects and Granger causality effects, respectively. The JKS (2021) causality test supposes that y_{it-k} signifies an autoregressive distributed lag process that maintain the null hypothesis that $\varphi_{qi} = 0$ for all i and q, which facilitates the use of a pooled estimator. To address the bias inherent in pooled estimations, the test employs a Half-Panel Jackknife (H.P.J.) estimator. Furthermore, in presence of cross-sectional dependence within panel data, the variance of the H.P.J. estimator can be ascertained via bootstrapping techniques. The resulting bias-corrected estimations yield Wald statistics for testing Granger non-causality.

In the subsequent section, the analysis will a couple of tests for cross-sectional dependence and homogeneity. Following the determination of cross-sectional dependence, the appropriate category of unit root tests either first-generation or second-generation will be applied based on the identified characteristics of the data. Concluding this process, the model will be estimated, and the empirical findings will be presented.

Results and Discussions

In this part of the paper, cross-sectional dependence between frontiers technologies and economic development is checked through tests of Lagrange Multiplier (LM) test by Breusch and Pagan (1980), LM adjusted test by Pesaran et al. (2008), and Pesaran's (2004) LM cross-section dependence tests and findings are displayed in Table 2. The null hypothesis of cross-sectional independency is rejected regarding the probability values of cross-section dependence tests and the entity of cross-sectional dependency is discovered between two series. This is important because the existence of cross-sectional dependency implies that any shock in any of these countries would also affect another country in the panel. It is also important as it shows that countries are dependent each other in technology diffusion and economic development.

Table 2 Cross-sectional dependence tests' results

Summary statistics	Test statistics	P values
LM (Breusch & Pagan, 1980)	466.1	0.0000
LM adj. (Pesaran et al., 2008)	78.54	0.0000
LM CD (Pesaran, 2004)	21.2	0.0000

Source Authors' calculations

Table 3 Homogeneity tests' results

Summary statistics	Test statistics	P values
$\tilde{\Delta}$	2.519	0.012
$\tilde{\Delta}_{adj.}$	2.842	0.004

Source Authors' calculations

Table 4 Results of CIPS unit root test

Variables	Test statistics	
	Constant	Constant + Trend
ECONDEV	−0.937	−0.193
D(ECONDEV)	−5.840***	−5.099***
FRONTECH	−0.484	−0.093
D(FRONTECH)	−5.991***	−4.625***

*** Significant at 1% level
Source Authors' calculations

The homogeneity is explored via delta tilde tests of Pesaran and Yamagata (2008) and their results are represented in Table 3. The homogeneity is declined regarding the probability values of the tests and entity of heterogeneity is disclosed.

The stationarity of ECONDEV and FRONTECH is investigated by CIPS unit root tests of Pesaran (2007) due to the cross-sectional dependency between two series and its results are shown in Table 4. The test results identify that ECONDEV and FRONTECH are I(1).

The causal association between frontiers technologies and economic development is examined by means of JKS (2021) causality test and the findings are given in Table 5. The JKS causality test shows that there is a unidirectional causal relationship running from frontier technologies to economic development. In other words, frontiers technologies have a significant influence on economic development.

As causality test hypothesizes the non-existence of causal relationship between the variables at hand, it is important as it demonstrates any change in any of the variable affects the other one. In other words, the causality analysis is usually used to uncover whether any change in one of the variables led to any change in the other variable in question. Therefore, it allows us to analyze the existence of any causal link between the variables. According to causality findings, there is causality from frontier technologies to economic development, but the reverse is not true. Changes in frontier technologies explain variations in economic development. This situation indicates

Table 5 JKS (2021) causality test results

Null hypothesis	HP Wald test statistic	P value
FRONTECH ↛ECONDEV	79.1008	0.0000
ECONDEV ↛FRONTECH	1.2689	0.2600

Source Authors' calculations

that frontiers technologies that have multidimensional factors associated with technology infrastructure investment, enhancement of technological literacy, R&D, and the integration of new technologies like AI into existing technologies can have significant impacts on economic development. This outcome is not surprising. The mere presence of technology infrastructure and its use will contribute to economic development to some extent. Because it seems that changes in economic development does not granger cause frontier technologies in new EU member countries. However, as countries progress in development, they will require new and advanced technologies to meet economic and social demands. Frontier technologies serve precisely to address this shortage of new and modern technologies required for development. In this unilateral causal relationship from frontier technologies to economic development in new EU member countries, frontier technologies seem to be the key drivers of economic development.

This result appears to contradict findings in the literature that identify a negative relationship between frontier technologies and other factors of development, such as income distribution, economic growth, and productivity (Maswana, 2024; Sawadogo, 2023; Višić, 2023). This situation underlines the importance of enhancing the existing technological infrastructures of new EU member countries, as well as adapting and improving their capabilities to use advanced technologies to eliminate economic disparities with other EU member countries. In other words, for new EU member countries to reach the levels of the current EU member states, they should adapt comprehensive technological innovations that will positively impact their economic growth. However, new EU member countries, which have recently joined the EU and have already some institutional challenges, may face some difficulties in making progress in frontier technologies solely through deploying necessary economic resources because it is expected that these countries have deficiencies in the use and adaptation of such technologies. That is to say, the new EU member countries should first create an environment that they would more benefit from frontier technologies rather than prioritizing economic concerns.

Additionally, these results are consistent with the studies finding that technological dissemination has positive effects on economic development (Bayar et al., 2023; Nasab & Aghaei, 2009; Vu, 2011). Indeed, the dissemination of technology facilitates the delivery of education, health, and other public services in a more cost-effective and efficient way. Thus, it can be said that technology not only contributes to economic growth but positively impacts economic development by reducing factors that cause environmental degradation that make education accessible to many people and enhance the quality of health services. Therefore, it is possible to see the technological penetration as a factor that incorporates economic, social, and environmental dimensions.

Conclusions

ICT are a vital component of the process of information society. The information society is characterized by economic, social, and cultural complements aiming to enhance countries' welfare and support their development. From this perspective, technological penetration appears as an important part of the pathway from economic growth on a smaller scale to broader economic development. This implies that ICT play a foundational role in achieving long-term sustainable development by facilitating access to information, improving efficiency in various sectors, and enabling the creation of new services and industries.

The relationship between technological penetration and economic development has been previously studied in the literature. The development of technology infrastructure and the ability of many individuals to benefit from this infrastructure to keep pace with the information society contribute to increased productivity in production and labour, thereby enhancing the welfare of countries. On the other hand, individuals becoming more knowledgeable and adapting faster to technological advancements of the technological era is a critical factor in this process. However, social and economic development necessitate the presence of new technologies that have the potential in enhancing production and meeting new needs. In this regard, the mixture of existing ICT tools with new technologies like AI and the necessity of individuals' ability to adapt to these new technologies have highlighted the significance of "frontier technologies." These are technological tools that open the doors to the new era and play an important role in shaping the future of economic development and social progress.

This study, therefore, analyzes the causal relationship between frontier technologies and economic development, which is proxied by economic growth, in new EU member countries for the years 2008–2021 by deploying the Juodis et al. (2021) causality analysis. The selection of these years is due to data availability. The reason for focusing on new EU member countries is their recent inclusion in the EU framework and their potential to implement new technological reforms to support their economic and institutional infrastructures, thereby reducing the disparity with other countries in the EU cohesion process. According to the causality results, a causal relationship exists from frontier technologies to economic growth. This finding suggests that investments in frontier technologies in new EU member countries can have significant impacts on their economic development. The non-causality from economic growth to frontier technologies indicates that in the absence of critical factors such as R&D activities, technological literacy, and countries' capabilities to adapt new technology, economic factors alone may not explain the changes in new technologies. This emphasizes the importance of a comprehensive approach that includes enhancing R&D activities, improving technological literacy, and developing capacities to adapt to and integrate new technologies as pivotal for leveraging the potential of frontier technologies in driving economic development.

This study has several limitations. Firstly, only the period between 2008 and 2021 has been analyzed due to data availability. Secondly, focusing solely on new EU member countries means that the analysis results and inferences are limited to these countries. Future studies could conduct a more extensive analysis by considering other factors that may affect this relationship. Additionally, analyzing a similar relationship in existing EU countries and comparing it with new EU member countries could provide insights for designing policies related to the cohesion process. Such comparative analyses could offer policymakers guidance on how to leverage technological advancements effectively across the EU and help them create a more integrated and cohesive economic and social development strategy. For future research, we plan to explore through which channels frontier technologies may impact economic development on a country-by-country basis by using alternative and more detailed analyses. Our aim is to understand how the connection between economic development and frontier technologies operates and how it varies from one country to another.

References

Aghaei, M., & Rezagholizadeh, M. (2017). The impact of information and communication technology (ICT) on economic growth in the OIC Countries. *Economic and Environmental Studies, 17*(2(42)), 257–278.

Aksentijević, N. K., Ježić, Z., & Zaninović, P. A. (2021). The effects of information and communication technology (ICT) use on human development—a macroeconomic approach. *Economies, 9*(3), 1–12.

Alfaro Cortés, E., & Alfaro Navarro, J. L. (2011). Do ICT influence economic growth and human development in European Union countries? *International Advances in Economic Research, 17*, 28–44.

Arendt, Ł. (2015). The digital economy, ICT and economic growth in the CEE countries. *Olsztyn Economic Journal, 10*(3), 247–262.

Bayar, Y., Remeikienė, R., & Sart, G. (2023). ICT penetration and human development: Empirical evidence from the EU transition economies. *Journal of the Knowledge Economy*. https://doi.org/10.1007/s13132-023-00875-7.

Breusch, T. S., & Pagan, A. R. (1980). The Lagrange multiplier test and its applications to model specification in econometrics. *The Review of Economic Studies, 47*(1), 239–253.

Dhaene, G., & Jochmans, K. (2015). Split-panel jackknife estimation of fixed-effect models. *Review of Economic Studies, 82*(3), 991–1030.

Fanoro, M., Božanić, M., & Sinha, S. (2021). A review of 4IR/5IR enabling technologies and their linkage to manufacturing supply chain. *Technologies, 9*(4), 77.

Juodis, A., Karavias, Y., & Sarafdis, V. (2021). A homogeneous approach to testing for Granger non–causality in heterogeneous panels. *Empirical Economics, 60*(1), 93–112. https://doi.org/10.1007/s00181-020-01970-9.

Khan, F. N., Sana, A., & Arif, U. (2020). Information and communication technology (ICT) and environmental sustainability: A panel data analysis. *Environmental Science and Pollution Research, 27*, 36718–36731.

Lau, C. K., Mahalik, M. K., Pal, S., & Gozgor, G. (2023). The impact of technology frontier on the total factor productivity growth in African economies: The role of human capital. *Economic Research-Ekonomska Istraživanja*. https://doi.org/10.1080/1331677X.2022.2164324.

Li, B. (2021). Dynamic evaluation and system coordination degree of the integration of artificial intelligence and real economy. *Complexity, 2021*, 1–9. https://doi.org/10.1155/2021/5539793.

Lucas, R. E., Jr. (1988). On the mechanics of economic development. *Journal of Monetary Economics, 22*(1), 3–42.

Maswana, J. C. (2024). *Exploring the growth effects of Artificial Intelligence in developing countries in Africa using a semi-endogenous growth model.*

Morawczynski, O., & Ngwenyama, O. (2007). Unraveling the impact of investments in ICT, education and health on development: An analysis of archival data of five West African countries using regression splines. *The Electronic Journal of Information Systems in Developing Countries, 29*(1), 1–15.

Nasab, E. H., & Aghaei, M. (2009). The effect of ICT on economic growth: Further evidence. *International Bulletin of Business Administration, 5*, 46–56.

Nimrod, G. (2018). Technostress: Measuring a new threat to well-being in later life. *Aging & Mental Health, 22*(8), 1086–1093.

Papaioannou, S., & Dimelis, S. (2007). Information technology as a factor of economic development: Evidence from developed and developing countries. *Economics of Innovation and New Technology, 16*(3), 179–194. https://doi.org/10.1080/10438590600661889.

Pesaran, M. H. (2007). A simple panel unit root test in the presence of cross-section dependence. *Journal of Applied Econometrics, 22*(2), 265–312.

Pesaran, M. H., & Yamagata, T. (2008). Testing slope homogeneity in large panels. *Journal of Econometrics, 142*(1), 50–93.

Pesaran, M. H., Ullah, A., & Yamagata, T. (2008). A bias-adjusted LM test of error cross-section independence. *The Econometrics Journal, 11*(1), 105–127. https://doi.org/10.1111/j.1368-423X.2007.00227.

Pesaran, M. H. (2004). General diagnostic tests for cross section dependence in panels. CESifo working paper series no. 1229; IZA discussion paper no. 1240. SSRN: http://ssrn.com/abstract=572504.

Romer, P. M. (1990). Endogenous technological change. *Journal of Political Economy, 98*(5, Part 2), S71–S102.

Sawadogo, R. (2023). Biotrade and income inequality: Frontier technology readiness matter? SSRN: 4499272.

Solow, R. M. (1956). A contribution to the theory of economic growth. *The Quarterly Journal of Economics, 70*(1), 65–94.

Tcheng, H., Huet, J. M., Viennois, I., & Romdhane, M. (2007). Telecoms and development in Africa: The chicken or the egg. *Convergence Letter, 8*(16), 16.

UN. (2021). Technology and innovation report. Retrieved April 01, 2024, from https://unctad.org/page/technology-and-innovation-report-2021#:~:text=The%20Technology%20and%20Innovation%20Report,also%20could%20accentuate%20initial%20inequalities.

UNCTADSTAT. (2024). Data centre-technology and innovation. Retrieved April 01, 2024, from https://unctadstat.unctad.org/datacentre/.

Višić, J. (2023). Impact of ageing and technological progress on labour productivity. *UTMS Journal of Economics, 14*(2), 212–222.

Vu, K. M. (2011). ICT as a source of economic growth in the information age: Empirical evidence from the 1996–2005 period. *Telecommunications Policy, 35*(4), 357–372.

Wang, L., Huang, Y., Ghafoor, A., Hafeez, M., & Salahodjaev, R. (2023). Asymmetric macroeconomic determinants of renewable energy consumption: Do financial institutions and ICT trade matter? *Environmental Science and Pollution Research, 30*(4), 9841–9851.

Welfens, P. J. (2005). Rational regulatory policy for the digital economy: theory and EU policy options. Universitätsbibliothek Wuppertal, University Library, No. disbei136.

World Bank. (2024). GDP per capita (constant 2015 US$). Retrieved April 01, 2024, from https://data.worldbank.org/indicator/NY.GDP.PCAP.KD.

Xu, Z. (2022). The influence of robot-assisted industry using deep learning on the economic growth rate of manufacturing industry in the era of artificial intelligence. *Wireless Communications and Mobile Computing, 2022*, 1–12. https://doi.org/10.1155/2022/4594858.

The Impact of Environmental, Social, and Governance Factors on the Performance of Romanian Listed Companies

Cristina Ioana Balint

Abstract The primary objective of this paper is to examine the impact of environmental, social and governance (ESG) factors on the financial performance of Romanian companies listed on the Romanian Stock Exchange. Using content analysis, ESG performance data were collected from annual and sustainability reports of the sampled companies. The study uses multiple linear regression to analyse the relationship between the dependent variables: operational performance (Return on Assets), financial performance (Return on Equity), market performance (Market-to-Book Value) and market value (Price-Earnings Ratio)—and the independent variable, ESG score. Control variables include company size, financial leverage, company age, and foreign institutional investment. The results of the model indicate that the ESG score is not positively associated with the companies' performance but does have an impact on their market value. Among the control variables, size and leverage have the most influence, with both positive and negative effects. It is important to note that the findings are specific to the context of this study and the limited number of Romanian companies listed on the Bucharest Stock Exchange during the period 2022–2023. Given the short time frame to assess lag effects and the limited availability of ESG scores for companies, this research has limitations.

Keywords ESG · Company performance · Market value · Romanian companies

Introduction

With the introduction of the United Nations Principles for Responsible Investment (UNPRI), the concept of environmental, social and governance (ESG) has gained significant popularity and is now widely adopted within the investment sector. Globally, there has been a remarkable increase in the integration of ESG as a sustainable

C. I. Balint (✉)
Faculty of Business, Babeş-Bolyai University, Cluj-Napoca, Romania
e-mail: cristina.balint@ubbcluj.ro

A. L. Negruşa et al. (eds.), *Sustainable and Resilient Businesses in the Global Economy*, Springer Proceedings in Business and Economics,
https://doi.org/10.1007/978-3-031-75883-6_18

investment strategy, indicating that investors and asset managers are progressively incorporating relevant ESG considerations into their decision-making processes (Bourghelle et al., 2009; Friede et al., 2015; Jemel-Fornetty et al., 2011). Furthermore, stakeholders are increasingly recognising the need for corporate transparency with respect to both ESG and financial information (Siew et al., 2013a, 2013b). In response to the expectations of managers and investors around the world, companies are actively integrating ESG factors into their operations (Eccles & Serafeim, 2011; Lokuwaduge & Heenetigala, 2017).

Investors are more interested in companies operating in accordance with ESG principles because compliance with ESG principles is much more sustainable, requires longer-term development resources, optimises activities and improves financial performance (Egorova et al., 2022). As a result, the United Nations Principles on Responsible Investment promote the inclusion of ESG factors in investment decisions and encourage investors to take greater responsibility for investments. Additionally, several companies are ready to integrate ESG practices into different areas of their business (Landi et al., 2022).

Furthermore, scientists and researchers have been paying more and more attention to ESG in recent years.

This paper aims to provide new empirical data on the environmental, social and governance (ESG) system in Romania. It examines the impact of ESG on Romanian publicly traded companies and further explains the link between ESG and company performance. The study developed hypotheses to examine the potential relationships between several key variables and their impact on financial, operational, and market performance, as well as market valuation. These variables include: ESG ratings, company size, financial leverage, company age and foreign institutional investment.

By formulating hypotheses around these factors, the study aimed to systematically investigate their individual and collective influence on various aspects of company performance and market value.

Furthermore, the findings of this research will be of significant use to corporate managers, investors, and other stakeholders, as they shed light on various interconnected aspects of this subject matter.

The structure of this research begins with an introduction that lay the groundwork for the study. This is followed by a review of the literature and the formulation of hypotheses that integrate relevant theories and establish the proposed hypotheses. Next, the research methodology will be outlined, along with the data processing techniques used. Finally, the study will conclude with a discussion of the findings and their implications.

Literature Review

Environmental, Social, and Governance (ESG) criteria have become integral to corporate decision-making and investor perceptions worldwide. ESG considerations reflect a company's commitment to sustainable practices, ethical conduct, and governance standards. Integrating ESG into business strategies is now widely recognised as crucial for long-term value creation and risk management.

The literature on the influence of ESG on listed firms utilises several theoretical frameworks. Stakeholder theory posits that businesses should consider the interests of all stakeholders, including the community, shareholders, employees, and customers. This aligns with the principles of ESG. Agency theory emphasises the role of governance mechanisms in aligning the interests of managers and shareholders, making ESG practices a significant aspect of corporate governance.

A study by Bassen and Kovács (2020) demonstrated that investors can utilise environmental, social, and governance (ESG) metrics to gather nonfinancial data about companies and assess their performance. Recently, there has been a growing focus among shareholders and other stakeholders on the performance of ESG factors. This increased interest stems from the recognition that transparency is vital to the long-term sustainability and growth of organisations. Chen and Xie (2022) also noted that investors tend to favour companies with strong ESG performance, as these firms often yield better investment returns. Consequently, leveraging ESG data to formulate strategies and objectives is an effective approach to foster organizational growth.

Financial Performance

The relationship between the financial performance of publicly traded companies and their ESG performance has been the subject of numerous studies. Research by Friede et al. (2015) and Khan et al. (2016) indicates that companies with higher ESG scores generally demonstrate superior financial performance, which includes increased profitability and reduced capital costs. These findings suggest that strong ESG practices can lead to improved operational efficiency.

The ESG score evaluates efficiency and performance based on publicly available data. A higher ESG rating correlates with more sustainable company performance (Melinda & Wardhani, 2020).

Wang et al. (2023) conducted a recent review of the literature on the impact of ESG performance on the market value and performance of listed companies. Their study found that ESG performance significantly improved the market value and book value. The research highlights the importance of ESG performance in improving operational efficiency, increasing profits, and reducing specific risks for companies, ultimately improving overall value.

Although the empirical literature on ESG and financial performance is extensive, it remains inconclusive. Studies have reported positive, negative, and neutral relationships between ESG and corporate financial performance. Many studies indicate positive correlations (Pham et al., 2022; Kim & Li, 2021), while others show negative relationships (Lee et al., 2009; Velte, 2017) or neutral relationships (Atan et al., 2018; Junius et al., 2020).

Operational Performance

ESG factors significantly influence operational performance through multiple channels. Companies that prioritise environmental sustainability often achieve cost reductions by optimising resource use and adhering to regulatory requirements. Implementing socially responsible practices can improve employee satisfaction and retention, increase productivity, and reduce turnover expenses. Additionally, strong governance structures promote transparency and accountability, which in turn build investor trust and help mitigate legal and reputational risks (Eccles & Serafeim, 2011).

The review conducted by the Stern Center for Sustainable Business and Rockefeller Asset Management at NYU, which analysed over 1,000 studies published between 2015 and 2020, revealed that 58% of corporate studies focusing on operational metrics indicated a positive correlation between ESG factors and financial performance. On the contrary, only 21% of these studies identified a negative relationship.

The study published by Frontiers in Environmental Science found that ESG performance significantly and positively influences return on assets (ROA), which is a crucial indicator of operational efficiency. Their regression analysis indicated that the beneficial effect of ESG on financial performance remains robust, even when addressing potential endogeneity issues through lag effects.

Market Valuation

ESG considerations influence market perceptions and investor behaviour. Studies by Eccles et al. (2014) and Kotsantonis and Serafeim (2017) indicate that companies with strong ESG performance are perceived favourably by investors, leading to higher market valuations and reduced cost of capital. ESG disclosures, including sustainability reports and ratings, provide stakeholders with valuable insight into a company's risk management practices and growth potential, influencing stock prices and market-to-book value ratios.

Sector-Specific Studies

Sector-specific analyses reveal differences in the impact of ESG factors in various industries. For example, energy companies are scrutinized for their carbon emissions and environmental stewardship, while financial institutions are evaluated based on their governance practices and ethical standards. Additionally, technology firms are increasingly assessed on their data privacy and cybersecurity measures, reflecting the evolving ESG priorities within specific sectors (Weber, 2014).

In the financial sector, ESG factors play a crucial role in influencing investment decisions. Institutions assess a company's ESG performance when making choices about investing, lending, and underwriting (Loew et al., 2024).

Regarding the real estate sector, Morri et al. (2024) used the regression model to examine the relationship between sustainability, particularly environmental factors, and financial performance. Their results indicated that governance aspects are linked to negative financial outcomes.

Regional Differences

The impact of ESG integration differs between regions due to variations in regulatory frameworks, cultural norms, and market dynamics. In Europe, companies are often influenced by stringent ESG regulations and active investor participation, leading them to prioritise sustainability disclosures and stakeholder participation (Dyck et al., 2020). On the contrary, emerging markets may face challenges in implementing ESG practices due to regulatory gaps and resource limitations, which can adversely affect their competitive positioning and attractiveness to investors.

Research conducted by Said and ElBannan (2024) in emerging countries indicated a positive association between firms' stock performance and their levels of environmental and social performance in the region. However, their findings did not provide empirical evidence to establish a connection between ESG performance and the perception of the stock market, measured by the price-to-sales ratio.

Investor Behaviour and ESG Integration

Institutional investors, such as asset managers and pension funds, play an essential role in driving the integration of ESG considerations within publicly traded companies. These investors are increasingly incorporating ESG factors into their portfolio allocation decisions, prioritising sustainable investment strategies and actively engaging with companies as shareholders (Raithatha & Bapat, 2012). Through shareholder resolutions and proxy voting initiatives, institutional investors promote corporate accountability and transparency, thereby influencing the ESG practices and disclosures of listed companies.

Investor preferences are essential in shaping sustainable investment strategies. Research has explored the motivations and preferences of investors regarding ESG factors. For example, a survey conducted by Barber et al. (2019) revealed that a significant majority of individual investors showed interest in sustainable investments, motivated by both financial returns and ethical considerations. The study emphasised the necessity for more accessible and customised information on ESG factors to support informed investment choices.

Challenges and Limitations

Despite increasing acknowledgment of the importance of ESG, challenges remain in consistently measuring and reporting ESG metrics. Methodological issues, such as data reliability and comparability of information between companies and sectors, impede accurate assessments of ESG performance (Khan, 2019).

ESG data are often reported by companies themselves, which leads to problems in data quality, reliability, and comparability between companies. Furthermore, the absence of standardised ESG reporting frameworks and the divergence in regulations worldwide complicate investor decision-making and corporate benchmarking efforts. The lack of consistency in ESG reporting standards across jurisdictions makes it difficult for investors to compare and evaluate the ESG performance of companies, hindering their ability to make informed investment decisions. Similarly, the lack of a unified regulatory approach to ESG reporting creates challenges for companies in benchmarking their ESG performance against their peers, both within and across national borders.

Future Directions

Future research on the impact of ESG on publicly traded companies should focus on developing methodological approaches to measure ESG performance, enhancing comparisons between sectors and regions, and exploring the long-term effects of ESG integration on corporate resilience and innovation. Furthermore, increased collaboration among stakeholders, including governments, businesses, and civil society, is essential to promote sustainable development goals and effectively address emerging challenges and financial risks of ESG.

Research Methodology

This paper will use quantitative methodologies, specifically multiple regression analysis, to examine the secondary data collected. The purpose of this analysis is to evaluate the hypothesis presented in the study, which investigates the relationship between environmental, social and governance (ESG) factors and performance.

Hypotheses Development

- **ESG and company performance**

The primary objective of this research is to examine whether there is a positive association between ESG performance and firm performance, hypothesising that higher ESG scores result in better overall company performance. This hypothesis is further divided into three sub hypotheses, each corresponding to different performance metrics: operational, financial, and market performance.

Operational performance, measured by return on assets (ROA), is hypothesised to be linked to ESG performance, which is influenced by the transparency of sustainability reporting. Enhanced disclosure in sustainability reports is considered integral to a company's reputation and goodwill. Consequently, improved reporting is expected to increase the company's ESG score, potentially enhancing its operational performance (Buallay, 2018). Accordingly, the first hypothesis posits:

H1a: There is no significant association between ESG and operational performance

Regarding financial performance, Buallay (2018) suggests that participating in sustainable development activities can increase demand for products and services, thus promoting company growth and reducing business risks. Margolis et al. (2009), as cited in Buallay (2018), also provide evidence of a positive correlation between ESG performance and financial performance. Thus, the second hypothesis is as follows:

H1b: There is no significant association between ESG and financial performance

The studies conducted by Barth et al. (2017), as referenced in Buallay (2018), indicate a positive correlation between market performance and ESG performance. This relationship is believed to be the result of increased cash flow generated by sustainable development projects highlighted in sustainability or CSR reports, which in turn enhances market value. Similar findings have been reported by Tarigan and Semuel (2014), Wagner (2010), and Velte (2017). Consequently, the third hypothesis states:

H1c: There is no significant association between ESG and the market performance

- **ESG and market value**

Most of the earlier research on the relationship between ESG performance and overall performance has focused on this relationship. However, market value is another important aspect that requires further investigation. Research by Banerjee et al. (2009), who examined how corporate governance influenced the Price-Earnings Ratio (PER), a measure of market value, supports this view. Their findings suggest a positive association between these characteristics, as organisations with stronger corporate governance typically have a higher PER.

The Price-Earnings Ratio (PER) is closely related to a company's reputation (Little & Little, 2000). This is supported by the claims made by Clarkson et al. (2008), as cited in Velte (2017), which indicate that a company's reputation can be enhanced by better ESG scores. Consequently, these factors are positively associated with each other. Therefore, the fourth hypothesis can be expressed as follows:

H2: There is no significant association between ESG and the market value

Sampling

The sample for this study consists of companies listed on the Bucharest Stock Exchange (BVB). These companies were selected based on data available in the ESG section of the BVB Research Hub, which aims to present relevant ESG parameters derived from the ESG analysis reports of BVB-listed issuers. The ESG scores for 2021 are published voluntarily, with the consent of the issuers. These analysis reports are independently generated by Sustainalytics, a leading global provider of ESG ratings and analysis, in collaboration with the Bucharest Stock Exchange since 2020. Scores are calculated using publicly accessible reports and information.

This study utilizes four dependent variables, four control variables, and one independent variable (see Table 1).

In this study, the independent variable is the ESG rating, which is measured using the ESG score. The primary objective is to investigate whether ESG has a positive influence on company performance and market value. Company performance is assessed using three indicators: operational performance, measured by return on assets (ROA), financial performance, measured by return on equity (ROE), and market performance, measured by market-to-book value (MBV). The study also examines the impact of ESG on market value, which is measured using the price-to-earnings ratio (PER).

Regarding the control variables, by incorporating them, the study aims to isolate the effects of the main independent variable (ESG performance) on the dependent variables (operational, financial and market performance, as well as market value) while accounting for the potential confounding influences of firm size (measured by total assets), age (since establishment), leverage (measured by total debt-to-total asset) and foreign institutional investment (% of FIIs in shareholding) on disclosure practices.

These control variables were selected because they influence the performance of companies. The total assets serve as a proxy for firm size, which can significantly impact financial performance metrics. Larger firms often have different operational

Table 1 List of variables and calculation method

Variable	Proxy	Formula
Independent variable		
ESG rating	ESG	Economic + Social + Governence
Dependent variables		
Operational performance	Return on Assets (ROA)	Net income/total assets
Financial performance	Return on Equity (ROE)	Net income/shareholders equity
Market performance	Market to Book Value (MBV)	Market value/asset value
Market value	Price-Earnings Ratio (PER)	Price per share/earnings per share
Control variables		
Company size	Total Assets (SIZE)	Log of total assets
Financial leverage	Debt to Assets Ratio	Total Debt/total assets
Company age	Company's Age (LOGAGE)	Company's age since establishment (in years)
Foreign institutional investment	Foreign Institutional Investment stake (FII)	% of FIIs in shareholding

Source Author's own processing

capacities, market power, and access to resources compared to smaller firms, which can influence their overall performance outcomes.

As for the age, older companies typically have more experience, which can lead to better management practices, established processes, and a deeper understanding of their markets. Companies that have been in operation for a longer period of time often build a stronger reputation and trust with customers, suppliers, and investors. As companies age, they tend to develop more formalised organisational structures and systems, which can lead to increased efficiency and effectiveness in operations.

With regard to financial leverage, highly leveraged firms may have higher interest expenses, which can reduce net income and returns on equity and assets. The degree of financial leverage reflects the risk profile of a company. Firms with higher leverage have greater financial risk and may be more vulnerable to economic downturns.

The last control variable, namely foreign institutional investment, can play an important role in monitoring and improving corporate governance practices, which can ultimately impact firm performance. FIIs often have access to more information and resources compared to domestic investors. Their presence can reduce information asymmetry and improve transparency, which can positively influence firm performance. Studies have shown that FIIs can encourage firms to increase their voluntary disclosure practices, which can improve investor confidence and firm valuation.

Data Analysis

To explore the relationship between company market value and their operational, financial, and market performance in relation to the independent variable (ESG), a multiple regression analysis was performed on these factors.

The assumptions of multicollinearity and autocorrelation must be met by the regression model for it to be considered valid. When there is a high degree of correlation between the independent variables, multicollinearity is a serious problem. For every regression equation, the variance inflation factor (VIF) and the collinearity tolerance were calculated to assess multicollinearity. Since the Variance Inflation Factor (VIF) for each variable was well below the threshold of 10 (which indicates significant collinearity), according to Gujarati (2004), the results indicated that multicollinearity was not a concern.

The Durbin-Watson statistic was used to evaluate autocorrelation. There is no autocorrelation in the residuals when the Durbin-Watson value is 2 or almost 2 (Brooks, 2019). For every equation in the sample, the Durbin-Watson values were close to 2. Thus, for all the regression equations, neither multicollinearity nor autocorrelation posed a substantial problem (Table 2).

The regression equations are as follows:

$$OA = \alpha_1 + \beta_1 ESG + \beta_2 SIZE + \beta_3 LEVERAGE + \beta_3 AGE + \beta_4 FII + \varepsilon_i \quad (1a)$$

$$ROE = \alpha_1 + \beta_1 ESG + \beta_2 SIZE + \beta_3 LEVERAGE + \beta_3 AGE + \beta_4 FII + \varepsilon_i \quad (1b)$$

$$MBV = \alpha_1 + \beta_1 ESG + \beta_2 SIZE + \beta_3 LEVERAGE + \beta_3 AGE + \beta_4 FII + \varepsilon_i \quad (1c)$$

Table 2 Description of variables of regression model

Variable	Description
ROA	Return on assets for the period 2022–2023
ROE	Return on equity for the period 2022–2023
MBV	Market-to-book value for the period 2022–2023
PER	Price Earnings Ratio for the period 2022–2023
ESG	Overall ESG score for the period 2022–2023
SIZE	Logarithmic values of total assets for the period 2022–2023
LEVERAGE	Debt-to-Assets ratio for the period 2022–2023
AGE	Logarithmic values of number of years since establishment
FII	FII stake for the period 2022–2023
α_1	Intercept term
β_n	Coefficient
ε_i	Error term

Source Author's own processing

$$PER = \alpha_1 + \beta_1 ESG + \beta_2 SIZE + \beta_3 LEVERAGE + \beta_3 AGE + \beta_4 FII + \varepsilon_i \quad (2)$$

Results and Discussions

Descriptive Statistics

The descriptive statistics (Table 3) include measures such as the mean, minimum, maximum and standard deviation of the dependent, independent, and control variables selected. The average ESG score is 60.2273 out of 100, indicating moderate sustainability practices, but with significant room for improvement. This moderate average indicates that the commitment of the sample companies to sustainable development is not particularly strong as only a limited number of them have published sustainability reports. Even so, companies exhibit a wide range of ESG scores, from 20 to 90, suggesting a diverse set of priorities and commitments to ESG.

In terms of operational and financial performance, the average ROA is 7.91%, indicating a reasonable level of asset utilisation. However, the range of ROA values, from 1.54% (Banca Transilvania, 2023) to 23.44% (Nuclearelectrica, 2022), points to significant differences in operational efficiency among companies. Similarly, the average ROE of 14.34% suggests moderate profitability, but the wide range from 2.63% (Rompetrol Well Services, 2022) to 27.91% (Romgaz, 2022) highlights the variability in financial performance.

The market valuation metrics, MBV and PER, show that on average, the companies are valued at 1.54 times their book value and 18.16 times their earnings. However, the broad ranges (MBV: 0.59–3.91; PER: 2.75–141) suggest that the market perceives

Table 3 Summary statistics of variables

	Mean	Minimum	Maximum	Standard deviation
ESG	60.2273	20.0000	90.0000	21.6398
ROA	7.9109	1.5394	23.4417	6.8397
ROE	14.3432	2.6254	27.9074	8.3543
MBV	1.5386	0.5900	3.9100	0.8107
PER	18.1604	2.7500	141.0000	31.0006
SIZE	9.7827	8.0908	11.2089	0.9475
LEVERAGE	41.5395	10.6613	94.0593	26.7677
AGE	1.4114	1.1761	1.5051	0.0933
FII	18.1733	0.0000	83.6908	30.2881

Source Author's own processing

companies very differently, likely based on factors such as growth potential, risk, and investor sentiment.

The average company size, measured by the logarithm of total assets, is 9.78, with a range of 8.09–11.21. This indicates a mix of small, medium, and large companies in the sample. The average leverage ratio is 41.54%, suggesting a moderate reliance on debt financing, but the range (10.66–94.06%) highlights the varying financial risk profiles of the companies.

The average company age is 1.41 (in logarithmic years), indicating a relatively mature sample, with the youngest company being 1.18 years old and the oldest being 1.51 years old. This narrow age range suggests that the companies are in similar stages of development.

Finally, the average foreign institutional investment (FII) is 18.17%, with a range from 0 to 83.69%. This wide variation in foreign ownership suggests differing levels of international investor interest and confidence in companies.

In general, descriptive statistics reveal a diverse sample of companies with varying levels of ESG performance, operational efficiency, market valuation, and financial characteristics. This diversity provides an opportunity to explore the relationships between these factors and gain insight into the drivers of company performance.

Multiple Regression Analysis

To examine the relationship between company performance, market value and ESG scores from 2022 to 2023, four regression equations were formulated and analysed using a multiple regression model. Table 4 presents the correlation matrix for all dependent variables (representing company performance), independent variables, and control variables.

The results do not show a significant correlation between ROA and ESG scores (0.056, $p = 0.806$), indicating that ESG performance does not appear to affect operational efficiency as measured by ROA in this sample. There is a significant positive correlation (0.425, $p = 0.049$) between ESG scores and ROE, suggesting that companies with higher ESG scores tend to have better returns on equity. This relationship indicates that strong ESG performance may contribute to enhanced profitability for shareholders. MBV has no significant correlation with ESG scores (-0.132, $p = 0.559$), suggesting that market valuations relative to book values are not influenced by ESG performance in this data set. Size shows a moderate positive correlation with ROE (0.648, $p = 0.001$), indicating that larger companies tend to have better returns on equity. Leverage has a significant negative correlation with ROA (−0.632, $p = 0.002$), indicating that higher debt levels are associated with lower operational efficiency. Age shows a positive correlation with ESG scores (0.412, $p = 0.057$), approaching significance. This suggests that older companies may have more established ESG practices. FII has a significant positive correlation with ESG scores (0.499, $p = 0.018$), indicating that companies with higher ESG scores attract more

Table 4 Correlation matrix for company performances and independent variables

	ESG	ROA	ROE	MBV	SIZE	LEVERAGE	AGE	FII
ESG	*1.000*							
ROA	*0.056* *(0.806)*	1.000						
ROE	*0.425* *(0.049)*	0.588 (0.004)	1.000					
MBV	*−0.132* *(0.559)*	−0.111 (0.621)	−0.062 (0.783)	1.000				
SIZE	*0.351* *(0.109)*	0.053 (0.814)	0.648 (0.001)	−0.327 (0.137)	1.000			
LEVERAGE	*0.267* *(0.229)*	−0.632 (0.002)	0.219 (0.327)	0.105 (0.643)	0.613 (0.002)	1.000		
AGE	*0.412* *(0.057)*	−0.180 (0.423)	0.102 (0.651)	0.243 (0.277)	−0.143 (0.527)	0.211 (0.347)	1.000	
FII	*0.499* *(0.018)*	−0.147 (0.513)	−0.052 (0.817)	−0.082 (0.716)	−0.094 (0.677)	0.076 (0.737)	0.412 (0.057)	1.000

Source Author's own processing

foreign institutional investment. This suggests that investors may favour companies with strong ESG performance.

A multiple regression analysis was performed to investigate the relationship between independent variables and company performance. The results of this analysis are presented in Fig. 1, which displays the coefficients for each independent and control variable in Eq. 1a, where ROA serves as the dependent variable.

The ESG score has a low, positive, and insignificant effect on ROA, while total assets and debt-to-assets are the most significant factors influencing operational

Model	R	R Square	Adjusted R Square	Std. Error of the Estimate	Durbin-Watson
1a	.801[a]	.742	.661	3.9818243	2.009

Model		Unstandardized Coefficients B	Unstandardized Coefficients Std. Error	Standardized Coefficients Beta	t	Sig.	Collinearity Statistics Tolerance	Collinearity Statistics VIF
1a	(Constant)	-50.631	23.545		-2.150	.047		
	ESG	.026	.057	.082	.458	.653	.500	2.001
	SIZE	5.281	1.430	.732	3.692	.002	.411	2.433
	LEVERAGE	-.289	.045	-1.130	-6.394	.000	.516	1.936
	AGE	12.567	11.916	.172	1.055	.307	.610	1.639
	FII	-.024	.036	-.105	-.660	.518	.643	1.555

a. Dependent Variable: ROA

Fig. 1 Multiple regression analysis for Eq. 1a. *Source* Author's own processing

Model	R	R Square	Adjusted R Square	Std. Error of the Estimate	Durbin-Watson
1b	.766[a]	.587	.458	6.1526696	1.683

Model		Unstandardized Coefficients B	Unstandardized Coefficients Std. Error	Standardized Coefficients Beta	t	Sig.	Collinearity Statistics Tolerance	Collinearity Statistics VIF
1b	(Constant)	-95.976	36.382		-2.638	.018		
	ESG	.072	.088	.187	.823	.423	.500	2.001
	SIZE	7.598	2.210	.862	3.438	.003	.411	2.433
	LEVERAGE	-.128	.070	-.410	-1.834	.085	.516	1.936
	AGE	26.743	18.413	.299	1.452	.166	.610	1.639
	FII	-.043	.055	-.156	-.781	.446	.643	1.555

a. Dependent Variable: ROE

Fig. 2 Multiple regression analysis for Eq. 1b. *Source* Author's own processing

performance. Specifically, a p-value of less than 0.05 indicates a positive and significant impact for SIZE, while a p-value of less than 0.05 reveals a low and negative effect for LEVERAGE.

The correlation coefficient (R) of 0.861 indicates a significant relationship between the predictor variables. The coefficient of determination, R^2, represents the proportion of outcome variability that can be explained by the predictors. For Eq. 1a, Fig. 1 shows an R2 value of 0.742, which means that independent variables account for 74.2% of the variation in ROA.

To assess collinearity, tolerance and variance inflation factor (VIF) were used. All VIF values are well below the threshold of 10 (Gujarati, 2004), suggesting that there are no multicollinearity issues. The Durbin-Watson statistic was used to test for autocorrelation in the residuals. The 2.009 result indicates that there is no autocorrelation in the residuals (Brooks, 2019).

To further explore the relationships indicated by Eqs. 1b (shown in Fig. 2) and 1c (shown in Fig. 3), multiple regression analysis was also conducted.

Figure 2 shows that the ESG score has a low, positive, and insignificant effect on financial performance (ROE), similar to the impact observed in Eq. 1a. With a p-value less than 0.05, the SIZE control variable, determined by total assets, emerges as the most significant factor positively influencing ROE. On the contrary, the other control variables do not significantly impact financial performance, with the exception of LEVERAGE, which demonstrates a low and negative effect, but for a significance level of 10%.

There is a strong correlation among the predictor variables, as indicated by the correlation coefficient (R) of 0.766. For Eq. 1b, the coefficient of determination (R^2) is 0.587, which means that the independent variables explain 58.7% of the variation in ROE. Tolerance and variance inflation factor (VIF) were used to assess collinearity, and all VIF values are well below the threshold of 10, indicating that there are no issues. A Durbin-Watson statistic score of 1.683 suggests that the residuals do not exhibit autocorrelation.

Model	R	R Square	Adjusted R Square	Std. Error of the Estimate	Durbin-Watson
1c	.623a	.389	.197	.7262426	1.868

Model		Unstandardized Coefficients B	Unstandardized Coefficients Std. Error	Standardized Coefficients Beta	t	Sig.	Collinearity Statistics Tolerance	Collinearity Statistics VIF
1c	(Constant)	6.989	4.294		1.627	.123		
	ESG	.015	.010	.404	1.461	.163	.500	2.001
	SIZE	-.714	.261	-.835	-2.738	.015	.411	2.433
	LEVERAGE	.016	.008	.537	1.974	.066	.516	1.936
	AGE	.105	2.173	.012	.048	.962	.610	1.639
	FII	-.011	.007	-.408	-1.673	.114	.643	1.555

a. Dependent Variable: MBV

Fig. 3 Multiple regression analysis for Eq. 1c. *Source* Author's own processing

The ESG score continues to have a minimal effect on market performance (MBV) in Eq. 1c. The control variables that influence MBV are SIZE and LEVERAGE, consistent with the other equations. SIZE shows an indirect and significant impact on MBV, with a p-value of less than 0.05, while LEVERAGE exhibits a very low positive effect at a significance level of 10%.

The correlation coefficient (R) is 0.623, indicating a strong relationship between the predictor variables. For Eq. 1c, the coefficient of determination (R^2) is 0.389, representing the lowest explained variation among all equations tested, which means that 38.9% of the variation in MBV is explained by independent variables. Collinearity was evaluated using tolerance and the variance inflation factor (VIF), which did not indicate any issues, as all VIF values are below the threshold of 10. A Durbin-Watson statistic value of 1.868 suggests that the residuals do not exhibit autocorrelation.

A multiple regression analysis was performed to examine the impact of the ESG score on the market value of the companies analysed. The findings presented in Table 5 indicate a significant negative correlation between ESG and the Price-Earnings Ratio (PER). Additionally, none of the control variables demonstrate significant correlations with the company's market value.

In Eq. 2, as shown in Fig. 4, the ESG score exhibits a significant negative effect on market value (PER) at a level of significance of 5%. On the contrary, none of the control variables (SIZE, LEVERAGE, AGE, FII) show a significant impact on PER.

There is a moderate correlation among the predictor variables, as indicated by the correlation coefficient (R) of 0.498. The coefficient of determination (R^2) for Eq. 2 is 0.248, which is lower than that of Eq. 1c, indicating that the independent variables explain 24.8% of the variation in PER. Collinearity was assessed using tolerance and the variance inflation factor (VIF), with all VIF values falling below the threshold of 10, suggesting that there are no issues. A Durbin-Watson statistic score of 2.107 indicates that the residuals do not show autocorrelation.

Table 5 Correlation matrix for the market value and independent variables

	ESG	PER	SIZE	LEVERAGE	AGE	FII
ESG	1.000					
PER	*−0.315 (0.039)*	1.000				
SIZE	− 0.109 (0.631)	−0.211 (0.345)	1.000			
LEVERAGE	−0.460 (0.031)	−0.026 (0.907)	0.613 (0.002)	1.000		
AGE	−0.352 (0.108)	−0.023 (0.920)	−0.143 (0.527)	0.211 (0.347)	1.000	
FII	−0.430 (0.046)	0.008 (0.971)	−0.094 (0.677)	0.076 (0.737)	0.412 (0.057)	1.000

Source Author's own processing

Model	R	R Square	Adjusted R Square	Std. Error of the Estimate	Durbin-Watson
2	.498[a]	.248	.013	30.7910835	2.107

Model		Unstandardized Coefficients B	Unstandardized Coefficients Std. Error	Standardized Coefficients Beta	t	Sig.	Collinearity Statistics Tolerance	Collinearity Statistics VIF
2	(Constant)	75.456	182.075		.414	.684		
	ESG	-.772	.439	-.539	-1.758	.048	.500	2.001
	SIZE	-4.411	11.061	-.135	-.399	.695	.411	2.433
	LEVERAGE	.264	.349	.228	.755	.461	.516	1.936
	AGE	12.113	92.147	.036	.131	.897	.610	1.639
	FII	.238	.277	.232	.859	.403	.643	1.555

a. Dependent Variable: PER

Fig. 4 Multiple regression analysis for Eq. 2. *Source* Author's own processing

Conclusions

ESG reporting serves as a means of communication with stakeholders, including shareholders, investors, employees, and clients, facilitating transparency regarding a company's performance. The findings of this study will significantly enhance the existing knowledge about the influence of ESG on business success within the empirical literature. Additionally, this research contributes to understanding of whether ESG factors play a role in improving a company's operational and financial performance. The study utilised both performance indicators and a market valuation metric to assess a firm's financial success and to draw conclusions about its future prospects.

This study seeks to demonstrate the positive influence of the ESG score on company performance and market value. The independent variable is the ESG score (ESG), while the dependent variables are evaluated through operational performance

Table 6 Summary of hypotheses

Null hypotheses	Result of Eq. 1a	Result of Eq. 1b	Result of Eq. 1c	Result of Eq. 2
There is no significant association between ESG and companies' performance	Accepted	Accepted	Accepted	
There is no significant association between ESG and companies' market value				Rejected

Companies' performance is measured as ROA in Eq. 1a, ROE in Eq. 1b and MBV in Eq. 1c
Source Author's own processing

(ROA), financial performance (ROE), market performance (MBV), and market value (PER). The results of the regression tests, summarised in Table 6, indicate that three of the hypotheses were supported, as no significant effects were found on the company's performance. However, the fourth hypothesis (Eq. 2) was not supported, as it indicated an effect on the company's market value.

The results indicate that while the ESG score affects the company's market value (PER), it does not significantly influence overall company performance, which is primarily driven by the control variable SIZE (measured by total assets). Although there is a positive association between ESG and company performance, this relationship is not statistically significant.

These findings suggest that awareness of ESG scores or sustainable development initiatives in the studied companies has not significantly impacted the performance of the companies, positively or negatively.

A limitation of this study is its focus on analysing the lag effect over a one-year period (t-1). The regression analysis for t-1 yielded results consistent with the initial period, implying that the effects of ESG performance can extend beyond a single year. Additionally, the study did not examine each individual ESG component separately; instead, it focused on the overall ESG score.

Contributions and Implications of the Study

The findings of this study can provide valuable information to various stakeholders, such as regulators and the government. This research can serve as a reference for the government to establish benchmarks for categorising companies based on their ESG performance, thus improving public awareness and transparency.

Given Romania's historical context of low voluntary involvement in corporate social responsibility and corporate governance reporting, this study may also prove beneficial for regulatory agencies and organisations in identifying company-specific attributes that contribute to improved ESG disclosure. The growing concerns surrounding ESG factors have led to a recent surge in ESG reporting. In addition to assisting the government in evaluating ESG initiatives through corporate social responsibility reporting, this study sheds light on the scope of ESG reporting practices in Romanian companies. Government agencies and regulators are among the groups of stakeholders who may find value in the research's conclusions.

Furthermore, this research is expected to encourage companies, as economic entities, to engage in sustainability reporting. For investors, the implications of this study suggest the need for companies to disclose and incorporate ESG performance indicators as integral components of their overall performance assessment.

By highlighting the relationship between ESG ratings and performance, company managers can gain insight into which areas require improvement to enhance overall performance. The findings of this study also have significant implications for shareholders, investors, and other stakeholders on the Romanian market.

For shareholders and managers, it is important to note that ESG initiatives typically align with long-term shareholder interests. Although investments in ESG do not yield immediate returns, benefits begin to materialise once a certain level of ESG performance is achieved. Therefore, until this benchmark is reached, it is advisable for company managers to focus on strategic allocation of capital investments and cost control to mitigate any short-term negative impacts of ESG initiatives on overall performance.

For investors, the results of this study serve as a reminder that strong ESG performance does not automatically equate to strong company performance. Investors are recommended to consider additional factors when evaluating the performance of a company before making investment decisions.

This study also contributes to the existing literature by identifying key determinants of ESG performance among Romanian listed companies. In conclusion, the research aims to improve understanding of ESG outcomes within the Romanian context, illustrating how ESG factors influence various financial and nonfinancial attributes of companies.

Recommendations for Future Research

Considering the findings and limitations outlined in this study, several observations and recommendations are proposed for future research efforts.

First, the sample size is limited; therefore, future studies could benefit from increasing the sample size or incorporating listed companies from various countries. This is important because the size of the company influences performance, and ESG requirements can differ significantly between countries.

Furthermore, future research could explore moderating variables to assess their impact on the relationship between environmental, social and governance (ESG) factors and corporate performance. In addition, researchers might investigate disclosure practices through surveys and interviews. Conducting a longitudinal study on ESG performance and examining its determinants could also provide valuable information.

References

Atan, R., Alam, M. M., Said, J., & Zamri, M. (2018). The impacts of environmental, social, and governance factors on firm performance: Panel study of Malaysian companies. *Management of Environmental Quality, 29*(2), 182–194. https://doi.org/10.1108/meq-03-2017-0033.

Banerjee, A. S., Gokarn, M., Pattanayak, & Sinha, S. K. (2009). *Corporate governance and market value: Preliminary evidence from Indian companies*. Standard & Poor.

Barber, B. M., Morse, A., & Yasuda, A. (December 12, 2019). Impact Investing. Available at SSRN: https://ssrn.com/abstract=2705556 or http://dx.doi.org/10.2139/ssrn.2705556

Barth, M., Cahan, S., Chen, L., & Venter, E. (2017). *The economic consequences associated with integrated report quality: Capital market and real effects*. Accounting, Organizations and Society, Forthcoming. https://ssrn.com/abstract=2699409 or https://doi.org/10.2139/ssrn.2699409.

Bassen, A., & Kovács, A. M., et al. (2020). Environmental, social and governance key performance indicators from a capital market perspective. In T. Beschorner (Ed.), *Wirtschafts- und Unternehmensethik* (pp. 809–820). Springer Fachmedien.

Bourghelle, D., Jemel, H., & Louche, C. (2009). *The integration of ESG information into investment processes: Toward an emerging collective belief?* Vlerick Leuven Gent Management School.

Brooks, C. (2019). *Introductory econometrics for finance*. Cambridge University Press.

Buallay, A. (2018). Is sustainability reporting (ESG) associated with performance? Evidence from the European banking sector. *Management of Environmental Quality: An International Journal, 30*(1), 98–115. https://doi.org/10.1108/MEQ-12-2017-0149.

Chen, Z., & Xie, G. (2022). ESG disclosure and financial performance: Moderating role of ESG investors. *International Review of Financial Analysis, 83*, 102291.

Clarkson, P. M., Li, Y., Richardson, G. D., & Vasvari, F. P. (2008). *Revisiting the relation between environmental performance and environmental disclosure: An empirical analysis. Accounting, Organizations and Society, 33*(4–5), 303–327. ISSN 0361-3682, https://doi.org/10.1016/j.aos.2007.05.003.

Dyck, A., Lins, K., Roth, L., & Wagner, H. F. (2019). Do institutional investors drive corporate social responsibility? *International evidence, Journal of Financial Economics, 131*(3), 693–714. ISSN 0304-405X, https://doi.org/10.1016/j.jfineco.2018.08.013.

Eccles, R., Ioannou, I., & Serafeim, G. (2014). The impact of corporate sustainability on organizational processes and performance. *Management Science, 60*, 2835–2857. https://doi.org/10.1287/mnsc.2014.1984.

Eccles, R., & Serafeim, G. (2011). Accelerating the Adoption of Integrated Reporting. In F. de Leo, & M. Vollbracht (Eds.), *CSR INDEX*. InnoVatio Publishing Ltd. https://ssrn.com/abstract=1910965.

Egorova, A. A., Grishunin, S. V., & Karminsky, A. M. (2022). The impact of ESG factors on the performance of information technology companies. *Procedia Computer Science, 199*, 339–345.

Friede, G., Busch, T., & Bassen, A. (2015). ESG and financial performance: Aggregated evidence from more than 2000 empirical studies. *Journal of Sustainable Finance & Investment 5*(4), 210–233. https://doi.org/10.3389/fenvs.2023.1256052. (Front. Environ. Sci., 2023. Sec. Environmental Economics and Management, 11)

Gujarati, D. N. (2004). *Basic econometrics*. McGraw-Hill Companies.

Jemel-Fornetty, H., Louche, C., & Bourghelle, D. (2011). Changing the dominant convention: The role of emerging initiatives in mainstreaming ESG. In W. Sun, C. Louche, & R. Perez (Eds.), *Finance and sustainability: Towards a new paradigm* (pp. 85–117). Emerald Group Publishing.

Junius, D., Adisurjo, A., Rijanto, Y. A., & Adelina, Y. E. (2020). The impact of ESG performanceto firm performance and market value. *Jurnal Aplikasi Akuntansi, 5*(1), 21–41. https://doi.org/10.29303/jaa.v5i1.84.

Khan, M. (2019). Corporate governance, ESG, and stock returns around the world. *Financial Analysts Journal, 75*(2019), 103–123. https://doi.org/10.1080/0015198X.2019.1654299.

Khan, M., Serafeim, G., & Yoon, A. (2016). Corporate sustainability: First evidence on materiality. *The Accounting Review, 91*(6), 1697–1724. https://doi.org/10.2308/accr-51383.
Kim, S., & Li, Z., (frank). (2021). Understanding the impact of ESG practices in corporate finance. *Sustainability, 13*(7), 3746. https://doi.org/10.3390/su13073746.
Kotsantonis, S., & Serafeim, G. (2019). Four Things no one will tell you about ESG data. *Journal of Applied Corporate Finance 31*(2), 50–58. https://ssrn.com/abstract=3420297.
Landi, G. C., Iandolo, F., Renzi, A., & Rey, A. (2022). Embedding sustainability in risk management: The impact of environmental, social, and governance ratings on corporate financial risk. *Corporate Social Responsibility and Environmental Management, 29*(4), 1096–1107. https://doi.org/10.1002/csr.2256.
Lee, D. D., Faff, R. W., & Langfield-Smith, K. (2009). Revisiting the vexing question: Does superior corporate social performance lead to improved financial performance? *Australian Journal of Management, 34*(1), 21–49. https://doi.org/10.1177/031289620903400103.
Little, P. L., & Little, B. L. (2000). Do perceptions of corporate social responsibility contribute to explaining differences in corporate price earnings ratios? A research note. *Corporate Reputation Review, 3*(2), 137–142. https://doi.org/10.1057/palgrave.crr.1540108.
Loew, E., Endres, L., & Xu, Y. (2024). *How ESG performance impacts a company's profitability and financial performance.* European Banking Institute Working paper series no. 162. SSRN: https://ssrn.com/abstract=4700173.
Lokuwaduge, C. S. D. S., & Heenetigala, K. (2017). Integrating environmental, social and governance (ESG) disclosure for a sustainable development: An Australian study. *Business Strategy and the Environment, 26*(4), 438–450.
Margolis, J. D., Elfenbein, H. A., & Walsh, J. P. (2009). *Does it Pay to Be Good...And Does it Matter? A meta-analysis of the relationship between corporate social and financial performance.* https://ssrn.com/abstract=1866371 or https://doi.org/10.2139/ssrn.1866371.
Melinda, A., & Wardhani, R., 2020. The effect of environmental, social, governance, and controversies on firms' value: Evidence from Asia. In *Advanced issues in the economics of emerging markets* (pp. 147–173). Emerald Publishing Limited.
Morri, G., Yang, F., & Colantoni, F. (2024). Green investments, green returns: exploring the link between ESG factors and financial performance in real estate. *Journal of Property Investment & Finance,* ahead-of-print No. ahead-of-print. https://doi.org/10.1108/JPIF-09-2023-0084.
Pham, T. N., Tran, P. P., Le, M.-H., Vo, H. N., Pham, C. D., & Nguyen, H.-D. (2022). The effects of ESG combined score on business performance of enterprises in the transportation industry. *Sustainability, 14*(14), 8354. https://doi.org/10.3390/su14148354.
Raithatha, M., & Bapat, V. (2012). Corporate governance compliance practices of Indian companies. *Corporate Governance, 3*(8), 19–26.
Said, M. T., & ElBannan, M. A. (2024). Do ESG ratings and COVID-19 severity score predict stock behavior and market perception? Evidence from emerging markets. *Review of Accounting and Finance, 23*(2), 222–255. https://doi.org/10.1108/RAF-03-2023-0083.
Siew, R. Y., Balatbat, M. C., & Carmichael, D. G. (2013a). A review of building/infrastructure sustainability reporting tools (SRTs). *Smart and Sustainable Built Environment, 2,* 106–139.
Siew, R. Y., Balatbat, M. C., & Carmichael, D. G. (2013). *The relationship between sustainability practices and financial performance of construction companies.* Smart and Sustainable Built Environment.
Tarigan, J., & Semuel, H. (2014). Pengungkapan sustainability report Dan Kinerja Keuangan. *Jurnal Akuntansi Dan Keuangan, 16*(2), 88–101. https://doi.org/10.9744/jak.16.2.88-101.
Velte, P. (2017). Does ESG performance have an impact on financial performance? Evidence from Germany. *Journal of Global Responsibility, 8*(2), 169–178. https://doi.org/10.1108/jgr-11-2016-0029.
Wagner, M. (2010). The role of corporate sustainability performance for economic performance: A firm-level analysis of moderation effects. *Ecological Economics, 69*(7), 1553–1560. https://doi.org/10.1016/j.ecolecon.2010.02.017.

Wang, Z., Sun, Z., & Teo, B.S.-X. (2023). Literature review on the impact of corporate ESG performance on corporate value. *Journal of Innovation and Development, 2*(2), 1–4.

Weber, O. (2014). The financial sector's impact on sustainable development. *Journal of Sustainable Finance & Investment, 4*(1), 1–8.

The Regulation of Telemedicine in Romania. Issues of Legal Limitations in Digital Medical Development

Ionuț Vida-Simiti

Abstract In Romania, telemedicine was introduced by the amendments to Law no. 95/2006 by Government Emergency Ordinance no. 8/2018 and Government Emergency Ordinance no. 196/2020 with the implementing rules of Government Decision no. 1.133/2022. The present study analyses these regulations in the context of Romanian medical law as a whole, from which three significant problems emerge. First, the provision of medical assistance by telemedicine can only be done by health units accredited by the National Authority for Quality Management in Health Care through doctors specialized in the medical fields expressly and limitatively listed in Annex 1 of the GD. no. 1.133/2022. They may provide medical services by telemedicine only in compliance with Article 32 of the Code of Medical Ethics, which implies remote contact between doctor and patient assisted by another doctor, who in turn must be a specialist in a field listed in Annex 1 of GD. no. 1.133/2022 and employed at a health unit accredited for telemedicine services. Secondly, the fact that Article 4(2) of Annex 2 to Government Decision No 1.133/2022 instils the possibility that proof of obtaining the patient's consent for the provision of the medical service itself may be provided by the health establishment by any means whatsoever constitutes a breach of Article 660 of Law No 95/2006 and Article 13 of Law No 46/2003, which require that the patient's consent must be expressed in writing only. Thirdly, a significant aspect is the establishment of a specific form of liability for medical malpractice, since Government Decision No 1.133/2022 establishes the liability of the health unit for the medical service provided by telemedicine and, in this regard, expressly confers on health units the prerogative to assess the effectiveness of telemedicine medical services, depending on the diagnosis and pathological evolution of the patient, and, if the assessment is unfavorable, the patient will be indicated alternative ways of providing the medical service.

Keywords Malpractice · Telemedicine · Patient consent · Confidentiality

---

I. Vida-Simiti (✉)
University of Medicine and Pharmacy "Iuliu Hațieganu" Cluj-Napoca, Cluj-Napoca, Romania
e-mail: Ionut.Vida-Simiti@umfcluj.ro

A. L. Negrușa et al. (eds.), *Sustainable and Resilient Businesses in the Global Economy*, Springer Proceedings in Business and Economics,
https://doi.org/10.1007/978-3-031-75883-6_19

Introduction

In recent decades, as technological development has facilitated the increasingly complex provision of remote healthcare services, concerns have arisen about the regulation of this area which was initially accepted only as an exception to allow access to healthcare for people in remote areas. Practical solutions vary from country to country in terms of licensing healthcare providers, ensuring confidentiality of data and quality of medical care, and liability for malpractice (Nittari et al., 2020).

Literature Review

The Romanian legal framework enabling the provision of healthcare services through telemedicine is represented by very recent legislation Government Emergency Ordinance No 8/2018 and Government Emergency Ordinance No 196/2020 with implementing rules in Government Decision No 1.133/2022. Consequently, research into these legal provisions is still at an early stage. The delay with which the Romanian legislator has paid attention to this field and the consequences of a quasi non-existent regulation were pointed out in the study by Mr. Cosmin Şovar (2020) published before the adoption of Government Emergency Ordinance No 196/2020, which regulates telemedicine in Romania. The paper presents the European Commission's Communication 689 of 4 November 2008 sent to the European Parliament, the Council, the European Economic and Social Committee, and the Committee of the Regions with reference to telemedicine (which subsequently justified the issuing of Government Emergency Ordinance No 196/2020), to which the Romanian legislator responded only with a very vague provision included in some methodological rules that would have allowed the transmission of certain parameters via the Single National Emergency Call System 112. Subsequent to the adoption of Government Emergency Ordinance No 196/2020 and Government Decision No 1.133/2022, without any reference to these normative acts, "brief considerations on telemedicine—a new challenge in the field of medical civil liability" are made in the work of Ms. Alina Doina Tănase, in which the need for a broader study is stressed, which goes beyond the framework of this work (Tănase, 2024). Another pioneering study is the dissertation of the Master of Medical Malpractice of the University of Medicine and Pharmacy "Iuliu Haţieganu" in Cluj-Napoca, by Mrs Adina-Raluca Zugravu (without year).

Research Methodology

In Romania, telemedicine has been introduced by the amendments to Law no. 95/2006 by Government Emergency Ordinance no. 8/2018 and Government Emergency Ordinance no. 196/2020 with the implementing rules of Government Decision no. 1.133/2022. The present study analyses these regulations in the Romanian medical law as a whole by following the three elements of the medical legal relationship: the parties (obviously, especially the healthcare provider), the object (which medical services can be provided through telemedicine and in what way) and the content of the relationship (the specific rights and obligations of the patient who is the beneficiary of telemedicine healthcare). The aim of the analysis is to identify situations of liability for malpractice when the provision of the telemedicine medical service is not carried out by an authorised provider or the patient's rights are violated.

Results and Discussions

Based on Article 16(2^2) of the Law no. 95/2006 on health care reform, the health care information and information system is composed of the health care reporting system and databases managed by public authorities and institutions, as well as service providers within the health care system. In order to implement and use the components of the health information and information system in an integrated and interoperable manner, the legislator has tasked the Ministry of Health with establishing the system's policy and strategies. Also, through Government Emergency Ordinance No 8/2018 on the regulation of certain measures in the field of health, the legislator intervened directly in the structure of the system by introducing, as of 2018, the rural telemedicine information system and the defence telemedicine information system, projects of public utility of national interest, ensuring their implementation and interoperability with the other health information and informatics systems.

The Romanian Government considered that maintaining the existing situation at that time would perpetuate the risks identified (inaccurate and time-consuming diagnosis, less effective interventions, and the possibility of complications), which could be reduced by rural telemedicine and defence telemedicine systems, respectively. However, the lack of this legal framework prevents the operationalisation and implementation of both systems. Consequently, the need to introduce telemedicine in Romania through a Government Emergency Ordinance was motivated by the need for efficient management of human and financial resources in the field of public health and in order to capture closely and promptly the real health needs of the population in the communities, it is necessary to urgently establish the legal framework for the rural telemedicine system and the defence telemedicine information system.

According to Article 16(2^3) of Law No 95/2006, the rural telemedicine information system involves the provision of remote medical services, using information technology and modern communication systems, between the offices of rural family doctors and specialist doctors in county hospitals or health institutions. Its implementation should have been carried out by the Ministry of Health in stages. The first stage of the pilot program would have consisted in the operationalization of the telemedicine system, under the conditions to be set out in a Government Decision, with funding from the budget of the Ministry of Health, in the first semester of 2018. And the second operational stage of implementation of the telemedicine system under the conditions set out in the Government Decision would have continued in the second semester of the same year.

Thus, the first regulation of telemedicine in Romania did not allow its use by all health care providers, but limited the exchange of information only between family medicine offices in rural areas and specialists in county hospitals or health institutions. The family doctor's office is a health care provider with a separate regulation provided for in Articles 75–89 of Law 95/2006 and with a special character in relation to the common law provisions of Ordinance 124/1998 *on the organisation and functioning of medical offices*. The family doctor working in their office may only provide the services mentioned in Articles 76–84 of Law No 95/2006, and exceeding the competence (except in emergency situations) is an act of malpractice. Consequently, the scope of telemedicine is reduced to primary health care.

Under Article 16(2^4) of Law No 95/2006, the Defence Telemedicine Information System involves the provision of remote medical services using information technology and modern communication systems between operational medical structures, medical structures in military units and health units in the Ministry of National Defence's own health network, which was to implement it from the second half of 2018 under the terms of a Government decision.

Based on Article 118(2) of the Romanian Constitution: "*The structure of the national defence system, the preparation of the population, the economy and the territory for defence, as well as the status of the military personnel, shall be established by an organic law.*". These constitutional provisions have been concretized by Law No. 80/1995 *on the status of military personnel*. According to Article 9 of Law No. 80/1995: "*Serving military personnel have the right to: (…) b) equipment, food, medical care, medicines, service housing, free of charge, paid leave and medical exemptions, under the conditions established by Government decision*". In order for Romania to finance the fulfillment of the obligation to fulfill the entitlement of military personnel (and their families), Ordinance No 56/1998 (approved by Law No 458/ 2001 [10]) was issued under Article 1(1) of the Regulation. (1): "*The Health Insurance House for Defence, Public Order, National Security, and the Judiciary is hereby established as a public, non-profit institution with legal personality, which carries out social health insurance activities for the staff of ministries and institutions with their own health networks in the fields of defence, public order, national security, and the judiciary*". The Ministry of National Defence's own health network (in which telemedicine was to be implemented) is described in Article 5 GO no. 56/1998: "*(1) Health service providers in the field of defence, public order, national security, and judicial authority are: hospitals and polyclinics, medical centers for diagnosis and treatment, county and municipal medical centers, medical centers for rehabilitation and rest, medical cabins, military and civilian spa complexes for treatment and rest, their own ambulance services and other health structures established according to legal provisions. (1) shall have sole competence in making decisions on the medical and military situation of persons with military obligations*". Precisely because of the specific nature of military hospitals, according to Article 185(7^2) of Law no. 95/ 2006: "*By exception to the provisions of para. (7), in the clinical wards, clinical laboratories and clinical medical services of hospitals belonging to the health network of the Ministry of National Defence, the position of head of ward, head of laboratory and head of medical service shall be held by a serving military officer (...)*".

In GEO no. 8/2018, the need for the urgent implementation of the telemedicine system in the health network of the Ministry of National Defence is justified by the NATO commitments undertaken in the field of health care by the Romanian Army, the risks to which Romanian and foreign military personnel are exposed through participation in complex military exercises carried out in polygons/training bases, as well as missions carried out by military personnel in theatres of operations, in remote areas or in areas that are difficult to access; the effect of the regulation is to achieve the medical act through the rapid diagnosis of military personnel and the rapid application of medical protocols.

From this, it can be deduced that telemedicine in the health network of the Ministry of National Defence is limited only to the provision of medical services to soldiers during complex military exercises carried out in polygons or training bases, as well as missions carried out by soldiers in theatres of operations, in remote or hard-to-reach areas, but not to the other categories of patients in the hospitals of the Ministry of National Defence's own network.

Thus, under Article 23 of Law No 80/1995: '*Members of the families of serving military personnel are entitled free of charge to: (a) medical care and medicines within the health network of the Ministry of National Defence or other health networks, with the expenses paid by this Ministry*'. The provisions of the latter article are also referred to in Article 26: '*Military officers, military masters and non-commissioned officers in reserve and retired, military pensioners, shall be entitled to free medical care and medicines under the conditions of Article 23(2)* of the Constitution. *1 letter (a) (...). The same rights are also enjoyed by officers, military masters and non-commissioned officers in reserve and retired, former military invalidity pensioners due to accidents in service or illnesses contracted during and as a result of the performance of military duties, who subsequently opted for a state social security pension. Family members of retired and reserve military pensioners are entitled to free health care under Article 23(2) of the Social Security Act. 1 letter (a)*'. Moreover, according to Article 5(3) of GO No 56/1998: "*The health care providers referred to in paragraph 1 of* Article 5 of GO No 56/1998*shall be entitled* to provide the following services. *(1) shall conclude, in accordance with the law, contracts also with the Health Insurance House of the Defence, Public Order, National Security and Judicial Authority, with the county and municipal health insurance houses respectively, as well as with other natural or legal persons requesting medical services.*" However, military hospital contracts will be distinct from the framework contracts adopted by Government Decision according to Article 224(1) of Law no. 95/2006, because under Article 5^1(1) GO no. 56/1998: "*The National Health Insurance House* *and the Health Insurance House of Defence, Public Order, National Security, and Judicial Authority, together with the ministries and institutions in the field of defence, public order, national security, and judicial authority, will draw up, by joint order, their own rules for the application of the annual framework contract, adapted to the specifics of the organization of health care in the health networks of these ministries and institutions*".

The regulation of telemedicine was extended from primary health care in rural areas to military health care in operational areas by Government Emergency Ordinance No. 196/2020 *amending and supplementing Law No. 95/2006 on health care reform*. The Government justified the adoption of this emergency ordinance on two grounds. The first argument was that maintaining the legislative situation at that time would have led to the perpetuation of the risks of the spread of infections caused by the SARS-CoV-2 virus, and that the urgent adoption of the general legal framework enabling the provision of health care through telemedicine by all health units met the real health needs of the population in the communities, through the efficient management of human and financial resources in the health sector.

Based on the European Commission's communications to the European Parliament and the European Economic and Social Committee, in which telemedicine is described as the provision of healthcare services based on the use of information and communication technology, in situations where the healthcare professional or two healthcare professionals and the patient are in different locations, Article[301] (2) of the Treaty on European Union provides that telemedicine is to be used for the provision of healthcare services in situations where the healthcare professional or two healthcare professionals and the patient are in different locations. (2) of Law no. 95/2006, telemedicine in Romania was legally defined as: "*the totality of medical services provided remotely, without the simultaneous physical presence of the medical staff and the patient, for the purpose of establishing a diagnosis, indicating the treatment, monitoring certain conditions or indicating methods of disease prevention, in a secure manner, by means of information technology and electronic means of communication*".

Telemedicine services are provided in Article 30^1 (6) and defined in Article 30^2-30^7 of Law no. 95/2006: 1. teleconsultation; 2. tele-expertise; 3. teleassistance; 4. teleradiology; 5. telepathology; 6. telemonitoring. Teleconsultation is medical assistance carried out in relation to the patient, in remote contact with the doctor by any means of communication, in order to establish the diagnosis, treatment and/or measures necessary for the prevention of diseases and their complications, all in the interest of promoting the patient's health. Teleexpertise is an exchange of medical opinions between several doctors carried out by any means of remote communication, with the aim of confirming a diagnosis and/or therapeutic course of action, based on data in the patient's medical records, and which does not involve the physical presence of the patient. Telecare is the remote assistance provided by a doctor either to a colleague in the performance of a medical or surgical procedure or to a healthcare professional providing an emergency medical service. Teleradiology (remote radiology and medical imaging) involves the electronic transmission of radiological images using digital image capture technologies for interpretation by radiology-imaging specialists and diagnosis. Telepathology is the remote transmission of microscopic data and images for interpretation by pathologists and diagnosis. Telemonitoring enables a medical specialist to remotely monitor and interpret medical data about the patient transmitted by the patient via remote electronic communication means, necessary for medical follow-up, and adjustment of therapeutic, lifestyle, and hygienic-dietary behavior related to the care of this patient, as appropriate. The recording and transmission of medical data may be automated by means of approved medical devices that measure and transmit medical information on the patient's condition or performed by the patient, physician, or other medical personnel.

Unlike telemedicine services covered by GEO 8/2018, which were to be covered by the Ministry of Health and the Ministry of National Defence respectively, telemedicine services covered by GEO 196/2020 are to be paid for by health units. They may be public or private and are not subject to any contractual relationship with the health insurance company. However, in the case of public health units, the expenses must be included in their own income and expenditure budget and are borne from the income generated in accordance with the provisions of Articles 140 and 190 of Law No 95/2006. Thus, public hospitals are public institutions financed entirely from their own income and operate on the principle of financial autonomy, which implies approval by the management of the unit with the agreement of the hierarchical superior authorizing officer of appropriations of an own budget drawn up on the basis of the evaluation of own income for the budget year and the distribution of expenses on the basis of the substantiated proposals of the departments and compartments of the hospital structure. Since Government Decision No 521/2023 *approving the service packages and the framework contract governing the conditions for the provision of health care, medicines and medical devices under the social health insurance system* does not provide for the settlement of telemedicine expenses, public hospitals will have to cover these expenses from other revenues provided for in Article 140 of Law No 95/2006: contracts concluded with private insurers; contracts concluded with local public administration authorities; the cost of services provided to patients in the framework of services not contracted with third party payers and borne by them; research and teaching contracts; donations, sponsorships, etc.

However, in order to be able to conclude contracts for telemedicine services, healthcare providers had to wait for the methodological rules, which, under Article II of GEO 196/2020, were to be drawn up by the Ministry of Health within 45 days from the date of entry into force of the emergency ordinance, and subsequently approved by a Government decision. It was to provide for: "*(a) medical specialties and the list of services covered by telemedicine services; (b) conditions for the organization and operation of telemedicine; (c) responsibility for establishing the quality of the telemedicine service and verifying compliance with it, under the terms of Law no. 185/2017 on quality assurance in the health system, as subsequently amended and supplemented; d) the method of providing telemedicine services.*"

Government Decision No 1.133/2022 *approving the methodological rules for implementing the provisions of Government Emergency Ordinance No 196/2020 amending and supplementing Law No 95/2006 on health reform* was adopted only after 2 years on 14 September 2022. Only after its entry into force could telemedicine medical services be legally provided in the specialties expressly and restrictively listed in Annex 1: (1) Pathological anatomy; (2) Allergology and clinical immunology; (3)Infectious diseases; (4) Cardiology; (5) Paediatric cardiology; (6) Cardiovascular surgery; (7) General surgery; (8) Oral and maxillofacial surgery; (9) Paediatric surgery; (10) Aesthetic plastic surgery and reconstructive microsurgery; (11) Thoracic surgery; (12) Vascular surgery; (13) Dermatovenerology; (14) Diabetes, nutrition and metabolic diseases; (15) Endocrinology; (16) Gastroenterology; (17) Paediatric gastroenterology; (18) Medical genetics; (19) Geriatrics and gerontology; (20) Haematology; (21) Paediatric haematology; (22) Family medicine; (23) Internal medicine; (24) Laboratory medicine; (25) Occupational medicine; (26) Nuclear medicine; (27) Sports medicine; (28) Emergency medicine; (29) Nephrology; (30) Paediatric Nephrology; (31) Neonatology; (32) Neurology; (33) Neurosurgery; (34) Paediatric Neurology; (35) Obstetrics and Gynaecology; (36) Ophthalmology; (37) Medical Oncology; (38) Paediatric Oncology; (39) Paediatric Orthopaedics; (40) Orthopaedics and Traumatology; (41) Otorhinolaryngology; (42) Pediatrics; (43) Pulmonology; (44) Pediatric Pulmonology; (45) Psychiatry; (46) Pediatric Psychiatry; (47) Medical Imaging Radiology; (48) Radiotherapy; (49) Medical Rehabilitation; (50) Rheumatology; (51) Urology.

The conditions for the organization and operation of telemedicine and the modalities for the provision of telemedicine services listed in Annex 1 are set out in Annex 2 of GD. no. 1.133/2022. Thus, telemedicine medical assistance can be provided either between medical staff and patient present simultaneously, but at a distance, by immediate transmission of information, in real time, through a means of remote communication; or between medical professionals by immediate transmission of information for subsequent analysis in order to interpret the transmitted data, establish a diagnosis and/or treatment. In the first situation, telemedicine services will be programmed in a similar way to face-to-face medical services so that during the provision of remote medical services, medical staff have access to the patient's record, i.e., information relevant to the patient's health. It is not possible to provide medical care through telemedicine services if the medical staff and the patient do not speak the same language at a level sufficient to ensure good communication. In the second scenario of collaboration between healthcare professionals, telemedicine allows two or more doctors to form remote multidisciplinary groups to analyze, interpret, and diagnose patients.

The information is transmitted in the form of data, digital images, audio, videos, recordings or web applications; and consists of clinical and health data. Pursuant to Article 10 of Annex 2 of GD. no. 1. 133/2022, the doctor shall make a written record in the consultation register and/or the patient's file, as the case may be: "*(a) the report on the performance of remote medical consultations/services, indicating the means of communication used; (b) the date and time interval in which they were carried out; (c) the medical prescription and documents issued as a result of the telemedicine service; (d) the identity of the medical staff participating in the telemedicine service; (e) technical incidents occurring during the telemedicine service, if any*". After the telemedicine medical service has been terminated, facilities may issue to the patient or the patient's proxy, documents either by electronic means, as agreed between the healthcare facility and the patient, or in physical format when electronic transmission is not possible or sufficient. These documents include electronic prescriptions, simple prescriptions, referrals for specialist medical tests, referral or admission tickets, treatment schedules, vaccination schedules, sick leave certificates, medical certificates for pregnancy and childcare, medical certificates for examinations and various competitions, medical certificates for schools and kindergartens, medical certificates for the health care system, medical certificates for the health care system and for the health care provider.

A first issue raised by the present re-regulation of telemedicine in Romania is the issue of which healthcare providers can provide healthcare services via telemedicine. Thus, under Article 2(2) of GD. no. 1.133/2022, healthcare providers must be specifically accredited for this mode of healthcare delivery. To this end, the National Authority for Quality Management in Health Care must develop standards for medical services provided via telemedicine, the methodology for evaluation, and accreditation of services provided via telemedicine. These standards for the provision of services through telemedicine must include mandatory conditions imposed by art. 2 of Annex 2 of GD. no. 1. 133/2022 (a) provision of infrastructure, personnel using information and communication technology, appropriate procedures and mechanisms for the secure use of audio or video platforms, cable, optic fiber, radio or satellite networks; (b) training of health professionals in the use of communication and data transmission systems and platforms; (c) archiving and securing all information and documents from the patient or documents resulting from the medical service provided through telemedicine, in an electronic patient health record, respecting their retention periods according to the provisions of the National Archives Law No 16/1996, republished; (d) informing the patient of the doctor's right to decide on the appropriateness of providing/continuing the medical service via telemedicine, i.e., the need to perform the medical service in a doctor's office, face-to-face. The quality of telemedicine health services is verified during an accreditation cycle by the health service quality control structures of the Ministry of Health and the public health directorates.

Within an accredited health facility, not all salaried physicians will be able to provide health care via telemedicine, but only specialists in the fields expressly and restrictively listed in Annex 1 of GD no. 1.133/2022. For other medical specialties, in the case of specific specialties, specific rules for the provision of services by telemedicine may be issued and approved by the order of the Minister of Health.

A doctor specializing in a field listed in Annex 1 to Government Decision No 1.133/2022 and employed at a health establishment accredited for telemedicine services must comply with the Code of Medical Ethics, which in Article 32 of the *Remote Medical Act* states that: "*Remote medical investigation or intervention, in whatever form and by whatever means, is permitted only if the patient is directly assisted by his or her physician and the purpose of the investigation and procedures to which the patient is subjected is to assist the physician in determining the diagnosis, establishing the treatment or taking any other medical measure necessary to complete the medical act or medical intervention in the case of operations. The exception is in emergency situations*". It follows from this that telemedicine involves remote contact between doctor and patient assisted by another doctor, who in turn must be a specialist in a field listed in Annex 1 of GD. no. 1.133/2022 and employed at a health establishment accredited for telemedicine services.

This situation does not pose any problem for tele-expertise, teleassistance, teleradiology and telepathology services, as they are linked to each other by means of a connection between specialists. There is also no problem in the provision of health care by teleconsultation, because the legal interpretation is that Article 32 of the Code of Medical Ethics explains the application of Article 30^2 of Law No 95/2006. Thus, Article 302 of Law no. 95/2006 establishes the possibility of a remote discussion between the doctor and the patient, and Article 32 of the Code of Medical Ethics details that in this discussion of the specialist with the patient, the latter is "*assisted directly by his doctor*". However, the problem arises in the case of telemonitoring, which is permitted by Article^{307} of Law No. 95/2006 precisely "*for medical follow-up, adjustment of therapeutic conduct, lifestyle and hygienic-dietetic regime relating to the care of this patient, as appropriate*".

A second problem is the patient's rights. According to Article 30^9 of Law no. 95/2006, medical units providing services by telemedicine must respect patients' rights provided for by Law no. 46/2003 on *patients' rights* and in addition: (a) establish the identity and professional qualifications of the doctors providing these services; (b) inform the patient about the medical services available, the purpose and implications of the medical act, and the means of carrying it out, including the technical means used for data transmission; (c) respect the patient's right to express his/her consent in a free and informed manner; (d) to respect medical confidentiality; (e) to ensure the technical conditions for the transmission of data and the conditions for the processing of such data by medical staff in such a way as to guarantee their confidentiality; (f) to secure and enter in the patient's electronic health record the documents resulting from these services; (g) to ensure continuity of care.

Among the patient's rights, Article 30^9 of Law no. 95/2006 insists on the right to confidentiality by insisting on securing information related to the patient's state of health throughout the procedures of collection, processing, use, and storage of personal data and to respect the fundamental rights and freedoms relating to the protection of privacy and personal data, medical data, in accordance with the legal provisions in force. In this regard, Article 6 Annex 2 of GD. no. 1.133/2022 data processing must comply with Regulation (EU) 2016/679 of the European Parliament and of the Council of 27 April 2016 on the protection of individuals with regard to the processing of personal data and on the free movement of such data and repealing Directive 95/46/EC (General Data Protection Regulation), as well as Law No 190/2018 on measures implementing Regulation (EU) 2016/679 of the European Parliament and of the Council of 27 April 2016 on the protection of individuals with regard to the processing of personal data and on the free movement of such data and repealing Directive 95/46/EC (General Data Protection Regulation), as amended.

The second patient right that has special provisions in the rules governing telemedicine is the right to consent. Thus, according to Article 30^9 (2) letter c. of Law no. 95/2006 detailed by art 4 Annex 2 of GD. no. 1.133/2022 except in cases of medical emergency, the express consent of the patient is required for accessing the platform or for the provision of the medical service by telemedicine, and the patient may refuse the medical service provided by telemedicine. For this consent to be legally valid, it must be free, i.e., not vitiated by error or fraud. Accordingly, Article[309] para. (1) and (2) lit. a. and b. of Law no. 95/2006 explained by Article 3 and 5 Annex 2 of GD. no. 1. 133/2022 establishes the obligation of prior information to the patient on how the medical service through telemedicine will be carried out by means of instructions for using the platform and information such as: "*(a) the names and qualifications of the professionals providing the medical services; (b) the doctor's contact details; (c) the charges for the medical services provided by telemedicine, as well as the possible methods of payment by patients; (d) details of the technical support service and how to connect; (e) the limits of the telemedicine services and the possibility for the doctor and patient to transform the service at any time into a traditional face-to-face service; (f) the conditions for processing personal data.*"

But the problem raised by Article 4 (2) Annex 2 of GD. no. 1.133/2022 arises: "*Proof of fulfillment of the obligation under para. (1) is incumbent on the units referred to in Article 30 para. (1) of Law no. 95/2006, republished, as subsequently amended and supplemented, and shall be made by any means of evidence demonstrating that express consent has been obtained.*" This is because GD. no. 1.133/2022 can only elaborate or explain provisions of Law no. 95/2006. But Article 660 of Law no. 95/2006 and Article 13 of Law no. 46/2003 make it obligatory to express the patient's consent only in written form (Vida-Simiti, 2017). It follows that the provisions of Article 4(2) Annex 2 of GD No 1.133/2022 are unlawful.

A third significant aspect is the establishment of a specific form of liability for medical malpractice. Thus, under Article 2 (1) GD. no. 1.133/2022: "*Ensuring the quality of medical services provided by telemedicine is the responsibility of the health units providing the medical service in question.*" For this reason, according to Article 7 Annex 2 of GD. no. 1.133/2022, health units must evaluate the efficiency of telemedicine medical services, according to the diagnosis and pathological evolution of the patient, and, if the evaluation is unfavorable, the patient will be indicated alternatives for the medical service.

Also, according to art. 655 Law 95/2006, health establishments are liable for damages caused by known defects of medical devices and appliances used improperly, without being repaired, the use of sanitary materials, medical devices, assistive technologies and devices, medicinal and sanitary substances, after the expiry of the warranty period or their validity period, as the case may be; acceptance of medical equipment and devices, assistive technologies and devices, health materials, medicinal and sanitary substances from suppliers without the insurance required by law, as well as subcontracting medical or non-medical services from suppliers without medical liability insurance; and under Article 655 Law 95/2006, the same health units are liable together with manufacturers of medical equipment and devices, assistive technologies and devices, medicinal substances and sanitary materials are liable according to civil law for damages caused to patients in the activity of prevention, diagnosis, and treatment, directly or indirectly caused by hidden defects of medical equipment and devices, assistive technologies and devices, medicinal substances and sanitary materials, during the warranty/validity period, according to the legislation in force (Vida-Simiti, 2013). For these reasons Article 5 in Annex 2 of GD. no. 1.133/2022 states that: "*The audio or video platforms, cable, optic fiber, radio or satellite networks used by health units for the provision of the medical service by telemedicine must ensure the possibility of verification, periodically or whenever necessary, by ICT personnel, of the accuracy and security of data transmission*".

Conclusions

The analysis of the current legal framework regulating telemedicine in Romania reveals three significant problems.

Firstly, the provision of medical assistance by telemedicine can only be done by health units accredited by the National Authority for Quality Management in Health Care through doctors specialized in the medical fields expressly and limitatively listed in Annex 1 of GD. no. 1.133/2022. They may provide medical services by telemedicine only in compliance with Article 32 of the Code of Medical Ethics, which implies remote contact between doctor and patient assisted by another doctor, who in turn must be a specialist in a field listed in Annex 1 of GD. no. 1.133/2022 and employed at a health unit accredited for telemedicine services.

Secondly, the fact that Article 4(2) of Annex 2 to Government Decision No 1.133/2022 instils the possibility that proof of obtaining the patient's consent for the provision of the medical service itself may be provided by the health establishment by any means whatsoever constitutes a breach of Article 660 of Law No 95/2006 and Article 13 of Law No 46/2003, which require that the patient's consent must be expressed in writing only.

Thirdly, a significant aspect is the establishment of a specific liability for medical malpractice because GD no. 1.133/2022 establishes the liability of the health unit for the medical service provided by telemedicine and, in this regard, expressly confers on health units the prerogative to evaluate the efficiency of telemedicine medical services, depending on the diagnosis and pathological evolution of the patient, and if the evaluation is unfavorable, the patient will be indicated alternatives for performing the medical service.

References

Code of Medical Ethics of the Romanian College of Physicians published in the Official Gazette of Romania no. 298 of 07.05.2012.

Constitution of Romania, published in the Official Gazette of Romania no. 767 of 31 October 2023.

Emergency Ordinance no. 196/2020 *for amending and supplementing Law no. 95/2006 on health care reform* published in the Official Gazette of Romania no.1108 of 19.11.2020.

Government Decision No 1.133/2022 *approving the methodological rules for implementing the provisions of Government Emergency Ordinance No 196/2020 amending and supplementing Law No 95/2006 on health reform* published in the Official Gazette of Romania No 909 of 15.09.222.

Government Decision No 521/2023 *approving the service packages and the Framework Contract governing the conditions for the provision of health care, medicines and medical devices under the social health insurance system* published in the Official Gazette of Romania No 466 of 26.05.2023.

Government Emergency Ordinance no. 8/2018 *on the regulation of certain measures in the field of health* published in the Official Journal of Romania no. 190 of 01.03.2018.

Law No 190/2018 *on measures implementing Regulation (EU) 2016/679 of the European Parliament and of the Council of 27 April 2016 on the protection of individuals with regard to the processing of personal data and on the free movement of such data and repealing Directive 95/46/EC (General Data Protection Regulation)*, as amended. Published in the Official Journal of Romania No 652 of 28.08.2015 in the Official Journal of Romania No 651 of 26.07.218

Law No 458/2001 *approving Government Ordinance No 56/1998 on the establishment, organization, and operation of the Defence, Public Order, National Security, and Judicial Authority Health Insurance Fund*. Published in the Official Gazette of Romania No 418 of 27.07.2001.

Law No 80/1995 *on the status of military personnel*, published in the Official Gazette of Romania No 155 of 20 July 1995.

Law no. 46/2003 *on patients' rights* published in the Official Gazette of Romania no. 652 of 28.08.2015 in the Official Gazette of Romania no. 51 of 29.01.2003.

Law no. 95/2006 *on health care reform*. Published in the Official Journal of Romania no. 652 of 28.08.2015.

Nittari, G., Khuman, R., Balani, S., Pallotta, G., Battneni, G., Sirignano, A., et al. (2020). Telemedicine practice: Review of the current ethical and legal challenges. *Telemedicine and e-Health, 26*(12), 1427–1437.

Ordinance No 124/1998 *on the organisation and functioning of medical practices* published in the Official Gazette of Romania No 568 of 01.07.2002.

Regulation (EU) 2016/679 of the European Parliament and of the Council of 27 April 2016 *on the protection of individuals with regard to the processing of personal data and on the free movement of such data and repealing Directive 95/46/EC (General Data Protection Regulation)* published in Official Journal I. 119 of 04.05.2016.

Şovar, C. (2020). Telemedicine in Romania: A myth. When does it become reality? *Curierul Juridic, 8*, 468–471.

Tănase, A. D. (2024). *Civil liability for medical malpractice* (pp. 304–305). Universul Juridic.

Vida-Simiti, I. (2013). *Civil liability of the doctor* (pp. 210–213). Bucharest: Hamangiu.

Vida-Simiti, I. (2017). Capacity to exercise consent to medical acts. *Pandectele Române, 5*, 81–93.

Zugravu, A. R. (without date). Clinica virtuală- beneficii, limitări e pericole (pp. 21–44). Cluj-Napoca, University of Medicine and Pharmacy "Iuliu Haţieganu".

The Interplay Between Information and Communications Technology Solutions Development and Millennial Visitors' Behavior in Refining Smart Tourism Destinations

Otilia Cramariuc and Carmen Eugenia Nastase

Abstract The concept of smart tourism destinations is related not only to the development of new technologies and the provision of various solutions in this regard but also to the correct positioning of the needs and capabilities of tourists themselves in using the respective technology. Thus, the studies carried out on the Millennial generation (the digital generation) revealed several points that should be considered when configuring the technological solutions for a smart tourist destination. This paper is a review of specific literature and aims to present the correlation between the development of various intelligent solutions for transforming tourist destinations into smart tourist destinations (STD) and travelers' generational traits. Furthermore, the paper will delineate the potential limitations a strict technological interpretation of the proposed smart destination concept might have on its intended beneficiaries. Moreover, several studies have shown that Millennials, a generation viewed as "receivers of new technologies" (Benckendorff & Xiang in Tourism information technology, 2019), can be described as a generation that intensively uses ICT for travel planning (Xiang Zheng in Journal of Retailing and Consumer Services 22:244–249, 2015). Therefore, Millennials represent the most important target market for the travel sector. Thus, the design of smart tourist destinations should take into account the main characteristics of such digital natives, including their idiosyncrasies, overall philosophy, preferences, level of knowledge and biases, especially their fear of losing control (Williams & Allan in Innovation and smart destinations: Critical insights, 2020). By developing a high level of knowledge of the human dimension involved in choosing and using new technologies related to smart destinations, tourism agencies and operators can refine existing technologies, increase the attractiveness of tourist destinations, or offer personalized services. Therefore, emerging technologies will be tailored to meet the specific needs of tourists, while intelligent destination strategies will establish a framework that positions technology as a crucial instrument for enhancing tourist satisfaction and increasing management efficiency.

O. Cramariuc (✉) · C. E. Nastase
Ştefan Cel Mare University of Suceava, Suceava, Romania
e-mail: otilia.cramariuc@usm.ro

A. L. Negruşa et al. (eds.), *Sustainable and Resilient Businesses in the Global Economy*, Springer Proceedings in Business and Economics,
https://doi.org/10.1007/978-3-031-75883-6_20

Keywords Smart tourism destinations (STD) · Millennials · Intelligent ICT solutions · Management implications · Target market

Introduction

The digital revolution has led to the development of smart cities which involves conceptualizing the framework for smart tourism destinations (STD). Such destinations represent complex systems and the ongoing digital revolution enables better collaborations between tourism companies/travel agencies and tourists, who exchange and share information and knowledge (Dobrica, 2019). Managing a destination has always been considered a complex task, and the accelerated development process of new technologies gives entrepreneurs more tools for properly managing a tourist destination (Buhalis Dimitrios, 2014).

Answering what makes a city a smart destination or smart city for tourism involves a nuanced, yet straightforward perspective. From the standpoint of an independent or smart tourist, having a smartphone and Internet access might be enough for a comprehensive travel experience. This premise, coupled with the responsibility to provide affordable, high-speed Internet, marks the initial indication of a tourist destination evolving into a smart destination. While this concept appears straightforward, numerous challenges often go unnoticed or are overlooked by Destination Management Organizations (DMOs), museums, hotels, and tourism companies, which can exacerbate these issues (CenterSmartTourism, n.d.).

The concept of smart tourism destinations is related not only to the development of new technologies and the provision of various solutions in this regard, but also to the correct positioning of the needs and capabilities of tourists themselves in using the respective technology. Thus, the studies carried out on the millennial generation (the digital generation) revealed several points that should be considered when configuring the technological solutions for a smart tourist destination.

Literature Review

The Client journey concept

In classic marketing and Internet marketing, there is a simple concept of the Customer Journey—the customer's path from knowing to buying and getting the product. A smart tourist's journey to a particular destination starts with **Step 1—choosing that destination**. At this stage, according to a report by Expedia Media Solutions, even if tourists have decided on a destination, they visit an average of 38 different websites (Abrihan, 2017). Therefore, the websites and social networks of destination management organizations, visitor centers and local travel agencies are very important.

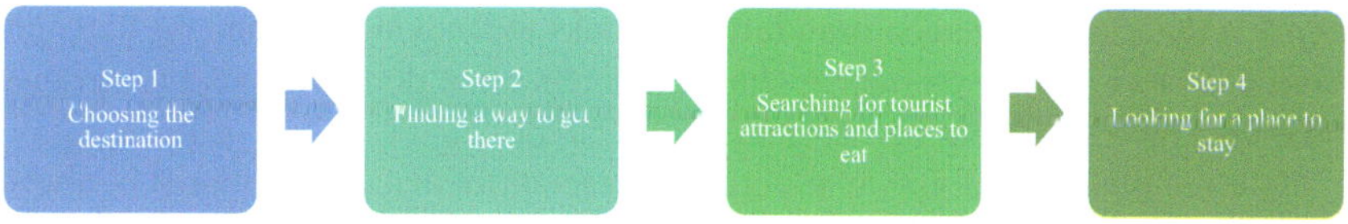

Fig. 1 The Client journey steps (Tin, Dmitriy, https://centersmarttourism.world/news/expert-opinions/expert-articles-about-smart-tourism/our-definition-of-smart-city-and-smart-destination/)

However, these websites do not always have all the necessary information. Moreover, such information can be obsolete, it loads slowly, and the hosting websites have a whole list of technical and content problems, still unsolved. For example, websites require regular checks to provide up-to-date, quality information (Fig. 1).

Step 2 is looking for a way to get to your chosen location. Although online services for global transportation, especially airlines, are generally satisfactory, local rail and bus services often lag behind. Tourists frequently face difficulties purchasing tickets online, accessing timetables, or finding Internet connectivity on public transport. Moreover, ordering a taxi online can be challenging, particularly when language barriers exist. In such scenarios, it's crucial for destination management organizations to collaborate with transportation companies to address these issues.

Step 3 is to search for tourist attractions in the chosen destination (sights, tours and museums you can visit), as well as places to dine. Tourists often seek out reviews from other travelers on TripAdvisor and various social media platforms, encountering challenges when many companies fail to effectively manage their online presence. This includes not responding to reviews, failing to post photos, or not answering questions. Additionally, a notable issue is the lack of sufficiently trained staff, particularly among business owners in areas with tourism potential. To mitigate these problems, destination management organizations can play a pivotal role by organizing training for local tourism businesses and assisting them in the development of Internet-based marketing strategies.

Museums hold a unique position as they serve not only as centers of entertainment but also as carriers of cultural and historical heritage. However, tourists often encounter several challenges, including the lack of information available in multiple languages, insufficient content, and poorly managed social media presence. These issues could be effectively addressed by introducing virtual tours and mobile audio guides. Such solutions enable tourists to explore museums virtually in their preferred language before their visit and utilize their smartphones as audio guides during their museum visits. While some tourists might look up information on Tourist Information Points' websites, these websites frequently resort to producing and distributing paper catalogues, brochures, and maps. This approach, however, is environmentally unfriendly and inefficient. Thus, a shift towards more digital solutions, like downloadable electronic guides for smartphones or audio guides for cities and their vicinities, would be more eco-friendly, cost-effective, and space-saving than traditional paper-based materials.

Step 4 is the search for accommodation. Thanks to Booking.com and the like, this is often an acceptable experience, although hotels often do not respond to reviews and do not indicate additional services.

In the realm of smart tourism, it's crucial to recognize that a smart tourist is, above all, a *responsible* tourist. They often share their travel experiences on social media platforms, and a negative review from a dissatisfied visitor can significantly deter others from visiting a certain destination. Consequently, Destination Management Organizations (DMOs) and the travel industry at large must vigilantly manage their online reputation. This involves promptly addressing any negative feedback with detailed, well-reasoned responses and taking concrete steps to rectify the underlying issues. While future technological advancements will inevitably shape the tourism sector, the current priority is to acknowledge and address the industry's challenges. By leveraging smart technologies, there's an opportunity to rectify these issues, transforming more destinations into Smart Destinations. The goal is to ensure that every aspect of the Customer Journey contributes to a memorable and positive experience for every tourist.

Smart solutions for transforming tourist destinations into SMART DESTINATIONS

Many locations claim to be smart, but 32% of destination management organizations (DMOs) do not have a guide app. We are talking about a tool (guide app) in a world where 90% of people do all their holiday research online (Felipe, 2021). While many destinations recognize the importance of smart tourism, there is still a long way to go before it becomes standard worldwide.

Transforming the status of a city/town from a tourist destination to a smart destination is a process that requires investment and coordinated work between the private and public sectors. Digitizing the entire tourism offer is perhaps the first step for a destination to become a Smart Destination. This process involves not one but a set of solutions. Eight of the most popular (Felipe, 2021) are:

- Travel Marketplace;
- Self-guided itineraries;
- Virtual Reality;
- Augmented Reality;
- Artificial Intelligence;
- Chatbots;
- Internet of Things;
- Big Data.

Travel Marketplace serves as an integrated platform that consolidates all the points of interest, tours, attractions, activities, and experiences available at a destination. Its core aim is to facilitate a direct connection between visitors and service providers. Designed with user-friendliness in mind, the platform streamlines the booking process for various tourist services and attractions. This approach not only simplifies the reservation experience but also promotes lesser-known locales, aiding in the equitable distribution of tourists and financial benefits across the destination. However, the development of such an innovative solution from the ground up can be expensive and time-intensive when undertaken solely by an individual destination management organization (Felipe, 2021).

Self-guided Itineraries, typically available as mobile apps, offer travelers a way to explore a location or tourist attraction at their own pace, much like a traditional tour but with a highly personalized experience. Users of these tours have the flexibility to select from various languages and departure times and can tailor the itinerary to suit their specific interests. This level of customization caters to travelers seeking a personalized adventure without the high costs often associated with bespoke tours. Destinations with their own tour apps can collaborate with tour operators to develop these self-guided itineraries, providing a valuable opportunity for local businesses to increase their revenue streams (Felipe, 2021).

Virtual Reality (VR) offers a unique way to enhance the traveler's experience both before and after they arrive at their destination. Unlike Augmented Reality (AR), VR requires the use of a special headset to immerse the user in a completely digital environment, opening up a realm of possibilities for Destination Management Organizations (DMOs). Essentially, VR introduces the "try before you buy" concept to the travel industry, serving as an excellent marketing tool for DMOs. By utilizing interactive videos and 360° images of key attractions, DMOs can provide travelers with a virtual preview of the destination. This approach helps to diminish the uncertainty that often accompanies the purchase of a travel experience and streamlines the decision-making process. Furthermore, smart destinations can leverage VR not only for marketing purposes but also to facilitate a seamless booking experience. While virtually navigating a destination, tourists can make reservations for hotels, restaurants, and other services, thereby enhancing the convenience and appeal of planning a trip (Felipe, 2021).

Augmented Reality (AR) employs technology that alters an individual's perception of their physical environment when viewed through a specific device, like a smartphone, by overlaying digital information onto the real world. This capability holds significant promise for smart destinations looking to enrich the travel experience. AR is especially well-suited to self-guided tours, turning a simple city stroll into an engaging interactive adventure with the addition of virtual elements, such as historical figures and objects. Companies like Questo are at the forefront of developing these innovative solutions. Tourist offices can leverage AR to disseminate valuable information across points of interest, for instance, displaying restaurant reviews or details about historical landmarks directly in the visitor's field of view, thereby enhancing the exploration and discovery process for travelers (Felipe, 2021).

Artificial Intelligence (AI) represents the pinnacle of complexity among the technologies discussed, referring to computer systems and programs capable of learning patterns, making decisions, and interpreting data sets autonomously, without human intervention. A hallmark of AI is its rapid and efficient problem-solving capabilities, boasting a significantly lower error rate than that of human counterparts. The progression of AI technology in recent years has seen it become both more sophisticated and reliable, positioning it as a formidable tool for enhancing both administrative and customer service operations within the travel industry. AI's pattern recognition abilities enable travel services to analyze tourist behaviors, thus proactively facilitating the delivery of tailored experiences. Moreover, AI finds a critical application in tourism for managing visitor flows by allowing authorities to preemptively redistribute tourists across locations to avoid overcrowding, thereby improving the visitor experience and preserving the integrity of attractions (Felipe, 2021).

Chatbots powered by AI are automated messaging applications designed to communicate with users autonomously, negating the need for human oversight. Urban centers equipped with dedicated portals or apps can integrate chatbots to elevate customer service and information accessibility for tourists. These bots can be programmed to respond to common inquiries about the destination, including details about landmarks and attractions. Furthermore, they assist travelers in planning their journeys, facilitating the booking of accommodations, flights, and various activities, thereby streamlining the travel planning process (Felipe, 2021).

The **Internet of Things (IoT)** represents a significant advancement in technology, especially with the advent of 5G networks. IoT encompasses devices connected to the Internet that can transmit and receive data to automatically execute tasks. High-speed internet enables tourists to maintain contact with their families, friends, and social media networks, sharing their travel experiences instantly. This connectivity not only enriches their journey but also serves as an effective marketing tool for destinations through user-generated content. The application of IoT in tourism can range from smart energy meters to autonomous vehicles, elevating the travel experience to unprecedented levels. For instance, envision tourists navigating through the historic heart of a city. As they near a renowned landmark, their smartphone receives a notification detailing the landmark's visiting hours and current visitor count. This feature not only streamlines the tourist's itinerary but also aids in evenly distributing visitors across various attractions, optimizing the overall experience (Felipe, 2021).

Big Data isn't a tool exclusive to tech giants like Facebook and Google; it's increasingly accessible to smaller travel entities and destinations as well. This technology enables the collection of vast amounts of data, from hotel bookings and tour reservations to video views and website visits. By mixing both internal and external data sources, stakeholders can forecast travel trends and behaviors, allowing for more strategic actions. Destination Management Organizations (DMOs) can leverage Big Data to effectively target and attract the ideal visitors at the optimal times and through the most suitable channels. For instance, a promotion could be launched targeting a specific market showing increased interest in the destination. Furthermore, public authorities can utilize Big Data to gain insights into tourist behavior upon arrival, enabling them to enhance services and overall visitor experiences (Felipe, 2021).

Multi-generational travel trends

The analysis of technology tools to provide the framework for smart tourism destinations should first and foremost take into account how they can be tailored to the needs and technological knowledge of tourists. In this sense a cross-generational analysis (in terms of the dominant characteristics of the different categories) could provide valuable information.

A study conducted by Expedia Media Solutions (Inteligence, 2017) has uncovered insights into the motivations and attitudes of travelers from the UK, France, and Germany across various generations, including Baby Boomers, Generation X, millennials, and Generation Z (Table 1).

Despite stable global smartphone penetration rates in recent years, there's a notable increase in mobile searches and bookings. Data comparison between 2022 and 2019 shows a 15% increase in the share of bookings made via mobile devices, including phones and tablets, with a significant rise attributed to mobile phone bookings (Group, Travel Insights Report, 2023).

With the growing prevalence of mobile search and booking, a noticeable shift is observed from mobile web to mobile app searches. In 2022, the search share via mobile apps rose by 30% compared to 2019, emphasizing the significance of

Table 1 Multi-generational travel trends

Indicator	Boomers	Generation X	Generation Y/ Millennials	Gen Z
Born	Between 1943 and 1960	Early 1960s to late 1970s	Early 1980s to early 1990s	1995–2010
Main characteristic	Millennials tend to be less budget-conscious and are more inclined to spend on upscale hotels than other generations	Family-oriented travelers, affinity for outdoor activities	Frequent travelers who seek deals, Millennials have diverse interests ranging from fine dining and wine tasting to outdoor adventures	Open minded, mobile savvy and budget conscious travelers looking to cross off things on their bucket list
When planning a trip, they value/ Are motivated/ influenced by	Informative content and helpful reviews (influenced by experts) less likely to be influenced by deals in ads	Read reviews of destinations before making travel decisions, relying heavily on online travel agents, travel review websites, and search engines	Blogs or articles about travel destinations	Pictures posted by their friends on social media
Average trips per year	1	2.5	4.3	2.5
Preferred				
For planning a trip	*Computer*	*OTA* *Travel review websites* *Search engines*	*Search engines* *OTA* *Travel review websites*	*Smartphone*
For booking a trip	*Tablet*	*OTA*	*OTA*	*Smartphone*

Source (Xiang Zheng, 2015)

engaging potential travelers across various devices, channels, and platforms (Group, Travel Insights Report, 2023).

It can be observed that of all categories such as Boomers and Generation X travel the least over a year and make the least use of technology tools associated with smart destinations, while Generation Z is still partially financially dependent and connected with Generation Y through jointly chosen destinations. Also, Generation Y has the highest average number of trips per year. This leads us to take a closer look at the travel habits and use of smart technologies associated with Generation Y or Millennials, as they now account for most of the world's travel market.

Digital natives, particularly Millennials and Generation Z individuals who grew up with the Internet and social media, tend to perceive new tools as highly useful and easy to learn compared to older generations like Boomers or seniors. This group is also more inclined to show interest in emerging health technologies, such as personalized genetic testing and robotic surgery, seeing them as intriguing advancements rather than intimidating innovations.

Tourism practices of Generation Y or Millennials

In the following, we will focus on the analysis of Generation Y because, on the one hand, they have active tourism behavior and are representative in terms of share in the world tourism market, and, on the other hand, they have the knowledge and articulate use of technology related to smart destinations. This group is commonly identified as millennials or the digital/net/Web generation (Benckendorff & Xiang, 2019). Their innate comfort with technology has earned them the nickname "digital natives." Traits such as frequent use of online intermediaries, search engines, and a keen interest in user-generated content on social media distinguish Generation Y. These characteristics might make them the "ideal tourist" for a Smart Destination (SD). Yet, these broad traits haven't been directly compared to the specific expectations that SDs hold for tourists (Femenia-Serra Francisco, 2018).

The concept of generational theory, introduced by sociologist Karl Mannheim, emphasizes the importance of cohort analysis in studying societies and their diverse generational components (Karl, 1972). A generation consists of individuals born and raised within a specific period, sharing a common socio-cultural, political, and economic background. These shared experiences influence their values, behaviors, and attitudes towards various aspects of life, including work, family, consumption, leisure, and travel.

As societal generations evolve, so do their expectations and demands for services. Presently, Generation Y, or millennials, who constitute a significant portion of the global workforce, are becoming increasingly influential in the travel industry. The "EY—Global Hospitality Insights Top Thoughts" study highlights that millennials account for half of the global workforce and are expected to dominate tourism spending in the coming years, representing over a third of the world's travel market. Extensive research on millennials reveals their distinct characteristic of valuing connectivity with others. Their interests lie primarily in friendships, entertainment, and digital culture. Millennials are known for their confidence, laid-back attitude,

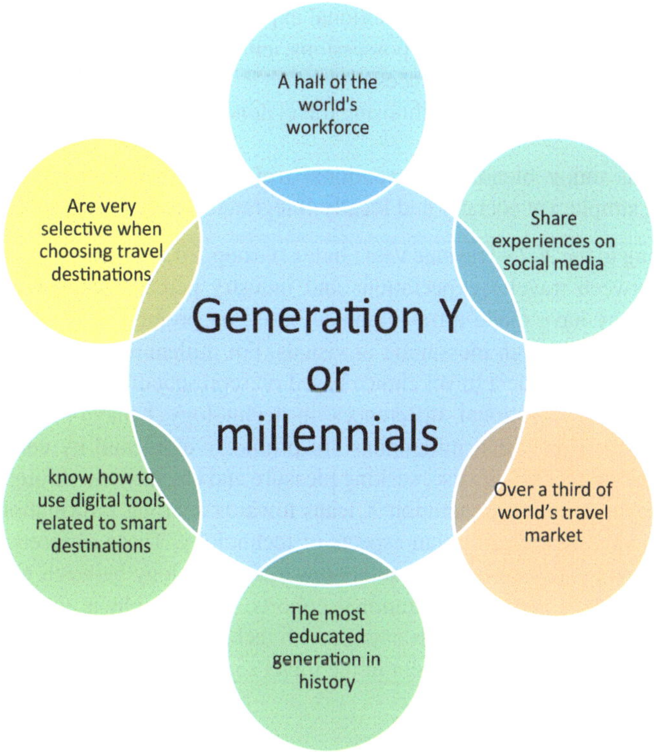

Fig. 2 Main characteristics of Generation Y or Millennials (Inteligence, 2017; Angheluţă, 2017)

relatively conservative viewpoints, and notably, being the most educated generation to date (Fig. 2).

According to IPK International, one-third of millennials' holidays are city trips (AdventureTravelNews, 2018), highlighting their preference for urban exploration and experiences. Additionally, millennials are responsible for half of all luxury holidays worldwide, demonstrating their significant economic impact on the travel industry (CBI, https://www.cbi.eu/market-information/tourism/luxury-tourism-0/market-potential, 2023a). Given that many travel influencers are millennials, GlobalData (GlobalData, n.d.) emphasizes the importance of monitoring and engaging with their social media content. Their influence on travel trends and decisions is substantial, making it crucial for the travel sector to connect with these influencers across various platforms. During the pandemic, YouTube and Twitter were the primary channels for travel promotion, but platforms like TikTok are emerging as new avenues for marketing.

Several characteristics of Generation Y related to tourism (CBI, https://www.cbi.eu/market-information/tourism/ generation-y-tourism/market-potential, 2023b) include:

- their need for connectivity and emotional experiences.
- preference for experiences over possessions, authenticity, fulfillment, and sustainability in personalized experiences.
- millennials are meticulous in choosing travel done through online research before booking.
- millennials enjoy blending conventional trips with their enhanced lifestyles, favoring simple, yet socially and locally-integrated travel experiences.

According to the 2023 Traveler Value Index (Group, 2023 Traveler Value Index—The gap between traveler expectations and industry perceptions) nearly 78% of people say they have made a travel choice based on promotions or ads they feel represented them through messaging or visuals. For millennials, that figure is even higher—84% have made a travel choice based on representation in advertising.

A study on generational differences in technology behavior (Calvo-Porral Cristina, 2019) reveals that millennials are primarily motivated by entertainment when it comes to technology use, seeking pleasure and engagement through hedonic motivations. In contrast, Generation X leans towards a strong information-seeking motivation, valuing the utilitarian aspects of technology. These differences underscore the varying attitudes and behaviors towards technology between generations, influenced by their distinct generational cohorts. This insight is vital for understanding how different age groups engage with technology and the implications for marketing and product development in the travel sector.

Research Methodology

This paper is a review of specific literature regarding smart solutions for transforming tourist destinations into smart destinations (SD). We have studied more than 50 articles related either to ICT solutions used in smart destinations or to Millennial behavior concerning technologies developed in the tourism sector, or articles that present these aspects together. Finally we've selected the most relevant articles, reports, studies, and web resources according to the references presented at the end of this paper. The specific purpose of this article is to establish a causal relationship between the development of ICT solutions for smart tourism destinations and the behavioral patterns of the tourist generations, with a focus on Generation Y or Millennials.

We chose generation Y because they are known as "digital natives" for their natural acceptance of technology as part of their lives (Benckendorff & Xiang, 2019), use many online intermediaries, search engines and pay particular attention to user-generated content on social media, aspects that qualify them as the "perfect tourists" for smart destinations (Femenia-Serra Francisco, 2018). If we add to this characterization the fact that they are very selective in their choice of travel destinations using numerous online information sources and various ICT products and represent over a third of the world's travel market and half of luxury holidays worldwide (CBI, https://www.cbi.eu/market-information/tourism/luxury-tourism-0/market-pot

ential, 2023a), then the specific interest for Millennials concerning the development of ICT solutions for smart destinations is justified.

In arguing for the relationship that exists between the development of ICT solutions related to smart tourism destinations and the purchasing behavior of tourism packages and exploration of holiday destinations we have used the following structure to develop this idea:

- presenting the Client journey concept in exploring a tourism destination.
- presenting eight of the smart solutions used for transforming tourist destinations into smart tourism destinations.
- multi-generational travel trends.
- tourism practices of Generation Y, behavioral aspects and motivation.

Starting from this structure, an attempt was made to find a way to link the development of smart tourism solutions with the specific trends and characteristics of Millennials in order to make the best use of technology for a more articulate and personalized response towards the dominant demand of the tourism market.

Results and Discussions

Benefits of investing in smart tourism and ways to effectively engage with consumer-centric technology

Younger generations, particularly Generation Y, are notably more engaged and proactive in their travel planning processes. They utilize a wide array of information and communication devices along with various online channels. They seek out information from diverse sources, including television, documentaries, videos, and social media platforms. Online reservations are more common among them, and they tend to use online travel agencies (OTAs) more frequently. They consider a broader range of potential destinations, actively search for activities, shopping venues, and dining options, and show interest in a wide variety of tourist experiences such as festivals, events, and spectator sports. Additionally, they engage with videos, live chats, and podcasts and are more open and responsive to online advertising (Xiang Zheng, 2015).

Investing in smart tourism offers cities a range of benefits, including:

- Enhancing the competitiveness of the destination by leveraging technology to attract more visitors;
- Increasing travel accessibility through user-friendly digital solutions, making it easier for tourists to explore;
- Enabling smarter resource management, allowing destinations to optimize their use of energy, water, and other resources efficiently;
- Fostering the growth of the local tourism ecosystem by integrating various stakeholders and services;

- Assisting destinations in strategizing and implementing targeted actions to boost tourism effectively;
- Elevating the quality of visitor experiences while simultaneously improving the quality of life for local residents by reducing congestion and preserving natural and cultural assets.

To effectively engage with consumer-centric technology, the industry should (EY, 2018):

1. Prioritize clarity in data utilization. Surveys conducted in the United States reveal that the willingness of individuals to share travel related information significantly hinges on the transparency surrounding the usage of their data as outlined in consent documents
2. Opt for universally appealing incentives. Vouchers and complimentary services in exchange for their reviews and social media shares. Studies show a high level of interest across various age demographics in receiving personalized travel insights and results, with at least 85% of participants across all age groups indicating this as the most compelling reason for data sharing

- Adopt user-friendly design principles. Overcoming generational skepticism towards technology ("This isn't for me, I'm too old for this") is essential for widespread acceptance. Technologies that operate seamlessly, supported by an interface that's straightforward to navigate and offers robust functionality across the entire technological proficiency range, can attract both older adults and younger generations alike

Managerial implication

To cater effectively to Millennial travelers, tourism operators can adopt the following strategies:

- Prioritize transparency and social responsibility to meet Millennial travelers' expectations. Demonstrating your commitment to ethical practices and sustainability can significantly appeal to this demographic;
- Enhance traveler confidence by openly communicating about the safety and security measures in place. Millennials value their safety and want reassurance that their chosen destinations and accommodations are secure;
- Embrace technology. Ensure you have a robust online presence, facilitating guests to use their mobile devices seamlessly during their travels. Enhance your online visibility by collaborating with professionals for reviews, such as reputable bloggers, social media influencers, or journalists from well-known publications like National Geographic, Business Insider, the Travel Channel, and TripAdvisor. Offer incentives like vouchers and complimentary services in exchange for their reviews and social media shares, which can be mutually beneficial. Use platforms like NoxInfluencer to keep track of influencer statistics and their impact;
- Target specific groups more effectively, including solo travelers, large families, friend groups, singles, or those with common interests like yoga, boot camps, culinary arts, entrepreneurship, mountain biking, birdwatching, or scientific research.

Tailoring the services to these niches can attract more dedicated and interested visitors.

- Customize services and emphasize the consumer experience throughout all stages, including the orientation and purchasing phases. Offering personalized options and attentive service can greatly enhance the appeal to Millennial travelers, who value unique and tailored experiences.

Hotels could enhance their appeal to guests by offering:

- Accommodations with stylish, innovative, and smart designs, similar to the boutique options found on platforms like Airbnb. Ensure strong Wi-Fi connectivity and incorporate smart technologies wherever feasible;
- Opportunities for blending business with leisure, catering to the growing trend of 'bleisure' travel, with the aid of platforms like WeWork, facilitating a seamless transition between work and relaxation;
- Invitations to online influencers for complimentary stays, leveraging their experience to attract followers through engaging promotions;
- Flexibility in their offerings to cater to Millennials' diverse interests in activities and experiences, which often span multiple days and seek to avoid overcrowded spots. Further insights can be gained from studies on free independent traveler (FIT) tourism;
- Initiatives that underline environmental stewardship and contribute to improving local quality of life, aligning with travelers' increasing interest in sustainable and responsible tourism.

Smart destinations and the companies involved in the development of related technologies should develop marketing strategies and segment their offer according to the specific characteristics and different motivations of different generations of tourists. Technology-based businesses could use the user cohort as a variable for segmenting their customers to develop specific marketing strategies (Calvo-Porral Cristina, 2019). In setting up smart tourist destinations, the reluctance of the tourist, technology literacy, and fear of losing control should also be taken into account. This aspect is not usually considered, although it has an influence on the ICT configuration of tourist destinations (Williams & Allan, 2020) (Fig. 3).

In the quest to create smart, tailored travel experiences across different generations, Destination Management Organizations (DMOs) encounter obstacles in gathering data. Notably, Millennials show reluctance in sharing personal details. Therefore, to obtain such information and craft tourism offerings that align with the distinct requirements of travelers, DMOs must focus on establishing trust with their clientele. This trust will, in turn, motivate clients to actively engage in the collaborative creation of these services. Practically speaking, the research indicates that Destination Management Organizations (DMOs) ought to prioritize technologies that, while not always the most conspicuous, provide tangible advantages to tourists. The adoption of such technologies should be informed by a thorough cost–benefit analysis. It is advised to utilize proven and popular technologies, including an up-to-date official website, reliable public Wi-Fi, an innovative smart tourism office, smart cards,

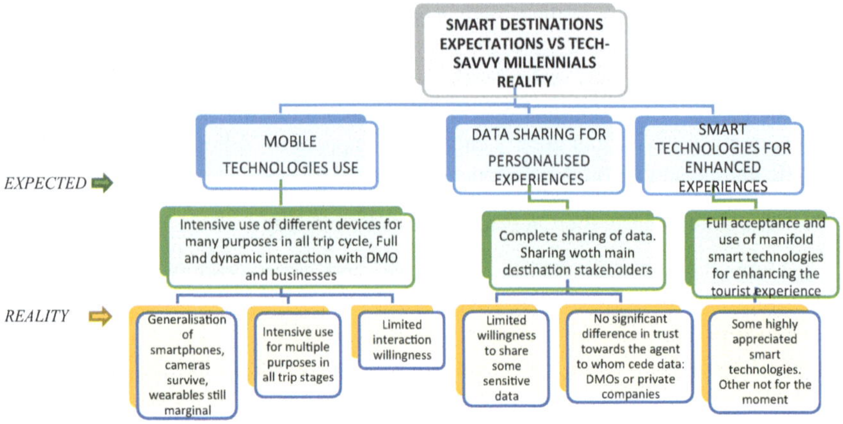

Fig. 3 Smart Destinations expectations vs tech-savvy millennials reality. *Source* (Femenia-Serra Francisco, 2018, p. 77)

effective smartphone applications, and regularly updated social media profiles for the destination (Femenia-Serra Francisco, 2018).

The success of businesses in the travel industry hinges on understanding and leveraging consumer behavior trends. The use of the Internet for trip planning is now ubiquitous across all customer segments, with "eTourism" becoming standard practice. However, this widespread adoption presents a challenge for marketers needing to differentiate their offerings in a saturated market. Particularly noteworthy is the behavior of Generation Y travelers, who actively engage in trip planning through diverse digital channels and are more open to online advertising. They utilize a wide range of sources, including social media, to inform their travel decisions, and show a preference for experiences over possessions, seeking out activities and destinations that offer personal enrichment.

The evolution of technology, especially social media and mobile devices, has introduced new behaviors among travelers, who now seek comprehensive and unbiased information from various channels to make informed travel decisions. This shift emphasizes the importance of "eWOM" (electronic word-of-mouth) in shaping consumer attitudes and choices. Moreover, the convenience of mobile Internet access has changed trip planning behavior, with travelers making decisions on-the-go rather than in advance.

Interestingly, while the Internet's role in booking primary travel services like flights and accommodations has plateaued, its value in planning the tourist experience, such as purchasing museum tickets or making dining reservations, is on the rise. This trend indicates a shift in the Internet's perceived economic value towards facilitating a broader range of travel experiences (Xiang Zheng, 2015).

Conclusions

The concept of smart tourism destinations is related not only to the development of new technologies and the provision of various solutions in this regard but also to the correct positioning of the needs and capabilities of tourists in using the respective technology. Due to the fact that generation Y or Millennials represent over a third of world's travel market with strong knowledge in using digital tools related to smart destinations, they are a category that can and must be taken into account when designing new technological solutions for smart tourist destinations. The common generational traits of Generation Y influence their values, behaviors, and attitudes towards various aspects of life, including leisure and travel. Their influence on travel trends and decisions is substantial, making it crucial for the travel sector to connect with these influences across various platforms.

Millennial travelers make significant use of mobile technologies across all phases of travel, although they demonstrate a limited inclination to use these technologies for engaging dynamically with various stakeholders. Their approach to sharing data presents a contradiction: they are open to sharing general information but are cautious about sharing more personal data that Smart Destinations (SDs) aim to utilize for enhancing the customization of their experiences. Their views on smart technologies as facilitators of their travel experiences exhibit a dual nature. While they adopt certain smart technologies, other more pioneering technologies vital for Smart Destinations' development are not embraced uniformly.

To establish a causal relationship between the development of ICT solutions and Millennial tourist behavior, it is essential to consider several key factors:

- Temporal Relationship: longitudinal studies can track changes in Millennial travel behavior over time in response to the implementation of new ICT solutions. By observing how travel patterns evolve before and after the introduction of specific technologies, researchers can identify causal links
- Controlled Experiments: conducting controlled experiments, such as A/B testing different ICT solutions within a destination, can provide direct evidence of their impact on tourist behavior. For instance, offering a group of tourists an enhanced mobile app with personalized recommendations and comparing their behavior to a control group using a standard app can reveal the effectiveness of such solutions.
- Path Analysis: using statistical techniques like path analysis and structural equation modeling (SEM) can help quantify the strength and direction of the relationship between ICT development and travel behavior. These methods allow for the control of confounding variables and provide a clearer understanding of the causal mechanisms at play.

In conclusion the development of ICT solutions for smart tourism destinations has a profound impact on the behavioral patterns of Millennial tourists. Enhanced connectivity, personalized experiences, immersive technologies, and social media integration are key factors that drive this relationship. By leveraging these technologies, destinations can better cater to the preferences of Millennials, thereby boosting tourism engagement and satisfaction. Establishing a causal relationship through rigorous research methods further validates the significance of ICT in shaping modern travel behavior, guiding future developments in the tourism industry.

The tourism industry should prioritize clarity in data utilization and enhance travelers' confidence related to safety and security, opt for appealing incentives and adopt user-friendly design principles. By focusing on the generational particularities, smart destinations should target specific groups more effectively and customize their services depending on the type of customer. By offering personalized options and attentive service, smart destinations and tourism operators can enhance the appeal of their services towards Millennial travelers who value unique and customized experiences.

References

Abrihan, R. (2017). https://www.startupcafe.ro/afaceri/turism-site-clienti.htm.

AdventureTravelNews. (2018). https://www.adventuretravelnews.com/itb-berlin-and-ipk-international-luxury-trips-are-increasing-among-millennials.

Angheluţă, B. (2014). *How generation Y is changing the rules of modern tourism.* Business Magazin. Retrieved from https://www.businessmagazin.ro/analize/turism/cum-schimba-generatia-y-regulile-din-turismul-modern-12020322/?p=2.

Benckendorff, P. J., & Xiang, Z. (2019). *Tourism information technology.*

Buhalis Dimitrios, A. A. (2014). *Smart Tourism Destinations.* https://doi.org/10.1007/978-3-319-03973-2_40.

Calvo-Porral Cristina, P.-S.R. (2019). Generational differences in technology behavior: Comparing Millennials and Generation X. *Kybernetes, 49*, 2755–2772.

CBI. (2023a). https://www.cbi.eu/market-information/tourism/generation-y-tourism/market-potential.

CBI. (2023b). https://www.cbi.eu/market-information/tourism/luxury-tourism-0/market-potential.

CenterSmartTourism. (n.d.). https://centersmarttourism.world/news/expert-opinions/expert-articles-about-smart-tourism/our-definition-of-smart-city-and-smart-destination/.

Dobrica, J. Z. (2019). From the traditional understanding of tourism destination to the smart tourism destination. *Current Issues in Tourism, 22*(3). https://doi.org/10.1080/13683500.2017.1313203.

EY. (2018). *Global hospitality insights top thoughts.*

Felipe, F. (2021). https://orioly.com/8-solutions-that-are-turning-cities-into-smart-destinations/.

Femenia-Serra Francisco, P.-R.F.-B. (2018). Smart destinations and tech-savvy millennials tourists: Hype versus reality. *Tourism Review, 74*, 77.

GlobalData. (n.d.). https://www.globaldata.com/media/travel-tourism/millennial-influencers-will-be-indispensable-in-travel-sector-recovery/suggests.

Group, E. (2023). *Travel insights report.* file:///E:/CONFERINTA%20UBB%20TURISM/Q1-2023-Traveler-Insights-Report.pdf.

Group, E. (n.d.). *2023 Traveler value index-the gap between traveler expectations and industry perceptions.*

Inteligence. (2017). https://globetrender.com/2017/05/19/travel-trends-across-generations/.

Karl, M. (1972). *The problem of generations* (pp. 276–322). Routledge.

Tin, D. (n.d.). *Our definition of smart city and smart destination.* Retrieved from https://centersmarttourism.world/news/expert-opinions/expert-articles-about-smart-tourism/our-definition-of-smart-city-and-smart-destination/.

Williams, M., & Allan, R. I. (2020). Innovation and smart destinations: Critical insights. *Annals of Tourism Research*. https://doi.org/10.1016/j.annals.2020.102930.

Xiang Zheng, M. P. (2015). Information technology and consumer behavior in travel and tourism: Insights from travel planning using the internet. *Journal of Retailing and Consumer Services, 22*, 244–249.

The manufacturer's authorised representative in the EU is Springer Nature Customer Service Centre GmbH, Europaplatz 3, 69115 Heidelberg, Germany. If you have any concerns regarding our products, please contact ProductSafety@springernature.com

Printed and bound by CPI Group (UK) Ltd, Croydon, CR0 4YY
01/10/2026
02225960-0013